Design and Emotion
The Experience of Everyday Things

Design and Emotion

The Experience of Everyday Things

Edited by

Deana McDonagh
Paul Hekkert
Jeroen van Erp
Diane Gyi

LONDON AND NEW YORK

First published 2004 by Taylor & Francis

2 Park Square, Milton Park, Abingdon, Oxfordshire OX14 4RN
52 Vanderbilt Avenue, New York, NY 10017

Routledge is an imprint of the Taylor & Francis Group, an informa business

First issued in paperback 2019

Publisher's note

This book has been produced from camera-ready copy supplied by the authors

British Library Cataloguing in Publication Data
A catalogue record for this book is available from the British Library

Library of Congress Cataloging in Publication Data
A catalog record for this book has been requested

ISBN 978-0-415-30363-7 (hbk)
ISBN 978-0-367-39490-5 (pbk)

CONTENTS

GENERATIVE TOOLS

EVALUATIVE TOOLS

AFFECTIVE USABILITY

ATTACHMENT

PRODUCT CHARACTER

DESIGN AND EMOTION: THEORETICAL AND ETHICAL ISSUES

PREFACE

Design and emotion has gained significant interest within design practice and design research over the last ten years. We can no longer ignore the important role that emotions play in the generation, development, production, purchase, and final use of products that we surround ourselves with.

In 1999, the Department of Industrial Design of Delft University of Technology organised a two-day conference on 'design and emotion'. Invited speakers from all over the world shared their views and data on a range of topics, such as design for the senses, engaging experiences, irrational aspects of technology, mixed emotions, and Kansei engineering. As broad as these topics may seem, they are all related to our affective interaction with and responses to human artefacts. At the seminar it was widely recognised that this new perspective in research and design is fruitful and deserves special attention. In order to stimulate more work done in this field, the birth of a society was launched on the final day of the conference: the Design and Emotion society.

With a strong base in the Dutch design world, the society started a range of initiatives to *promote, stimulate, and contribute to experience/emotion driven product design.* The society aims to be a platform for researchers, designers, and other practitioners in the field of product design, to share their common interest in the relationship between design and emotion and to exchange insights and experiences. Among the society's initiatives were a series of emotion-driven design workshops, the construction of a website (www.designandemotion.org), and the co-organisation of a 2nd conference on design and emotion ('Emotionale'), hosted by the Department of Design at the University of Applied Sciences Potsdam, June 2000. This book represents the proceedings from the 3rd international conference held at Loughborough University, UK.

At the two previous events, only researchers and designers, known for their knowledge and experience in design and emotion related issues, were invited to present their work. Unlike those events, this conference was open to papers from everyone feeling the ability and urge to share their ideas or work in this field. However, efforts were made to avoid making it a traditional - and often soporific - academic conference. In line with the society's objectives, *practitioners* (designers, marketers and product developers from design studios and industry) were specifically invited to present their work. Many of them did, and so did academics from various backgrounds, turning this volume into a colourful collection of ideas, approaches, methods, and tools representing the state of the art in design and research on 'design and emotion'. To guide the reader through this field of views, the papers have been presented under the following themes/topics:

Experience driven design

In practice, designers face the challenge of evoking emotions and experiences in the prospective users of their products. How do they deal with these incalculable phenomena? Do they follow particular strategies or is it merely intuition that guides their design work? Lee Crossley, for example, describes a case study in which explorative techniques – like storytelling, taking pictures - were applied to understand the grooming experience of male teenagers. This approach must lead to the design of a 'cool' experience, which has very little to do with designing 'cool' products.

Generative tools

Experience or emotion driven design requires new tools and methods for designers. What kind of *information* or *inspiration* do designers need and how do they get it? In their paper, Pieter Jan Stappers and Elizabeth Sanders report an experiment in which they compare various techniques to inspire users to express their past, present, or future home experience. They conclude that symbol + word collages on paper, collages on a computer, or mindmaps each capture unique insights into the lives of people.

Evaluative tools

Experience or emotion driven design also requires new tools for evaluating designs. How do we know whether interaction with our artefact will result in the experience intended? Jennifer Downs and Jayne Wallace, for example, apply Kelly's Repertory Grid Technique to evaluate people's responses to their jewellery designs. The results provided them with a clearer and broader view of their own practice.

Emotive effects of visual properties

Visual aesthetics is traditionally the field in product design that deals with the emotional dimensions of pleasantness and unpleasantness. Words, colour, shape, and motion each have their own effect on our visual experience and these properties can interact and counteract to raise all kinds of feelings. Brian Stone *et al.*, for instance, demonstrate that typographic messages evoke stronger emotional responses when they are shown in motion than in a stationary manner.

Emotive effects of the other senses

A product can simply be beautiful to look at, but can also have an elegant sound, a pleasing touch, or an exiting smell. What is more, sensorial information can guide users or mislead them and can in these various ways contribute to an *experience*. What makes a tactile sensation delightful or a blurring sound confusing? Nicolas Bouché stresses the importance of manipulating the sensorial qualities of sporting goods to raise their sensuality and achieve customer satisfaction.

From design to emotion

Stating that products evoke emotions is one thing, finding out how they do so is a more complicated undertaking. Products not only evoke emotions, they can also be extremely powerful in communicating them. How are experiences communicated and can we design products that *adapt* to our emotional well-being? This latter issue is addressed in a paper by De Angeli and Johnson who present a framework for designing emotionally intelligent interactive systems.

Affective usability

Researchers in the field of human factors have started to realise that a usable or comfortable design does not univocally predict a particular user experience. What factors contribute to pleasurable usability? Can a pleasant experience even enhance usability? Porter *et al.*, for example, present the results of a questionnaire study to identify product characteristics that contribute to the various types of user pleasure.

Attachment

Sometimes, a product is so dear to us that we feel intense grief when loosing it or personally attacked when someone speaks belittlingly about it. How is it possible that we get so attached to material possessions? Is it in the designer's hand to establish such attachment? Battarbee and Mattelmäki collected personal stories from young Finnish adults and show that a strong bond is established when products are either seen as meaningful tools, evoke meaningful associations, or are perceived as animate or living objects.

Product character

"Gosh, it looks so cute!" "This car is far too masculine for me" We easily attribute personality characteristics to products and these tell us a lot about our feelings towards them. On what grounds do we consider products as having a personality? Can designers deliberately assign such characteristics to a non-living object? This latter question was addressed in a paper by Govers *et al.* who demonstrate that designers can deliberately attach personality characteristics, such as 'happy' or 'tough', to a familiar consumer product.

Design and emotion: Theoretical and ethical issues

Is all this talking about pleasurable or cool emotions, respect for user feelings, and an emphasis on soft or intangible aspects of design, *sincere* and genuine, or do we simply want to *sell more*? Is it really a new design paradigm or an old approach dressed as something new? Patlar and Kurtgozu address such issues at length in their historical overview of related claims on authentic user experiences by design(ers).

Emotion in design

Most work in the field of design and emotion is concerned with the emotional experience of the user or customer. But what about the designers themselves? What do they feel about this design focus, and how do they deal with their own emotions in the act of creation? Denton *et al.* report an evaluation among design students who carried out a design exercise in which emotional aspects addressed in particular. As so many at the conference, these students acknowledged the importance of these aspects for the field of design.

Keynotes

Next to the paper presentations, four distinguished keynote speakers presented their ideas and work in the field. This volume opens with their contributions. Mike Robinson, Director of Design at Centro Stile Fiat, extensively reflected on our century-old battle with automobiles, a battle that has 'moved us' in all sorts of directions. Pieter Desmet, fresh PhD laureate from Industrial Design Delft, explained us that our emotional responses to products are less intangible than they seem at first sight. The lawful processes underlying them offer interesting opportunities for designers. In his theoretical account of the two main paradigms in emotion psychology, Gerry Cupchik, professor of Psychology at the University of Toronto, destroyed the mind-body dualism and entered at length into the design implications of an experiential view. Jane Fulton Suri, Head of the human factors design and research discipline at IDEO, finally discussed how design practice has evolved from designing things to designing experiences and how this has affected the design process.

Workshops

On the final afternoon of the conference, three workshops were organised in which experience driven design was put into practice. This volume concludes with an account of these events, organised by design offices Fabrique and KVD (The Netherlands), the Design Transformation Group (UK), and Marieke Sonneveld (ID-Studiolab, Delft University of Technology).

We strongly believe that this volume will become a must-read for anyone seriously involved in product design. As was anticipated years ago in the editorial of the first proceedings, studying the relationship between design and emotion and employing experience-driven design has turned out to be much more than whim of fashion. The particular attention for emotional and experiential issues in design has become a strong and persistent research and design perspective, one that may profoundly affect design research and education, and, above all, the very nature of product design itself.

Deana McDonagh Paul Hekkert Jeroen van Erp Diane Gyi

KEYNOTE STORIES

The design of emotion

Gerald C Cupchik
University of Toronto, Canada

TWO STREAMS OF EMOTION THEORIES: ACTION AND EXPERIENCE

Consciousness serves as a sentient boundary between stimulation from the external physical or social worlds and the internal bodily world. Emotions are a part of consciousness and reflect the complex interaction of mind and body. Psychologists have always seemed to disagree about the nature of emotion, in other words, about how the mind and body interact. While a unified theory of emotion remains elusive, the main theories can be divided into complementary "action" and "experience" oriented groups. Three *action oriented* approaches to emotion, associated with centralism, behaviourism, and cognitivism, focus on the adaptive and purposive mind. In contrast, *experience oriented* theories related to peripheralism, psychodynamics, and phenomenology/existentialism, encompass bodily reactions to social meanings.

Action theories

The centralist view, first developed in the 19th century (see Gardiner *et al.*, 1970/1937), held that higher order centres in the brain were responsible for controlling emotion. This was followed by the behaviouralist approach which is expressed in the simple formula: Emotion = Intensity + Direction (Duffy, 1941, 1962) or Emotion = Arousal + Cognition (Schachter, 1966). From this perspective, there is nothing unique about emotional states, which are reduced to levels of activation or arousal. Moderate levels of bodily arousal facilitate the realisation of goals, though excessive excitation can interfere with performance. The cognitive viewpoint (Oatley, 1992) emphasises plans and expectations, and the way that bodily arousal modulates concentration (Pribram, 1984). Generally speaking, action oriented theories focus on the "new" brain's control over the "old" reptilian brain, serving as a gatekeeper that facilitates or inhibits signals from the viscera, thereby turning the subtle qualities of emotional experience into background noise. In sum, the content of the mind takes priority over a body, which is treated as the mere vehicle of goal attainment. Traditional images of the philosopher king and religious emphases on controlling bodily urges represent ethical variations on this theme of mind over body.

Experience theories

The peripheralist view was formalised by William James (1884) who emphasised the direct effects that feedback from the viscera and facial expressions have on colouring subjective emotional experiences. According to the psychodynamic viewpoint (Keen, 1977), emotional experiences of happiness, sadness, fear, and so on, are tied to distinctive and personally meaningful characters and social episodes. Subsequent emotional experiences are evoked by encounters with people and events that symbolically approximate these formative episodes. Phenomenology (Strauss, 1958) examines the structure of subjectivity, the ways that significant emotional episodes shape basic qualities of experience; space, time, materiality, causality, and pure sensation. Existentialism (Binswanger, 1958) assumes that these emotional experiences are an essential aspect of being human and are tied to the very meanings that govern our lives. Together, these theories confer legitimacy on subjectivity as an organismic response to meaningful life events.

THE COMPLEMENTARITY OF COGNITIVE AND EMOTIONAL PROCESSES

In this section we consider the ways that mind and body interact as implied by the Action and Experience perspectives on emotion. The simplest contrast is between top-down and bottom-up processes relating the mind and body in a figure/ground manner. Top-down processes are typical of the Action mode whereby the mind is the central *figure* dominating the body as *ground*. This is what happens when ideas and plans direct bodily activity; cognitions govern feelings. One can argue that in the Action mode, *feelings are the shadow of cognition*. When the pattern of ideas is coherent, then there is a feeling of calm or pleasure. When the ideas do not fit together harmoniously, there is the experience of tension. In a sense, feelings are almost mechanically (i.e., isomorphically) related to changes in the content of cognition. Given that feelings reflect the state of cognition, so to speak, artists and designers can use their feelings as an index of the state of their projects. An artist friend described to me how, when he first establishes the composition of a still-life painting, he looks for and *feels* the largest contrast in texture or tones. As the work unfolds, his changing feelings of tension, which are tied to the qualities of texture or tone, serve as a guide. Thus, his *feelings are a shadow of cognition* in that they are the bodily mirror of perceptual and conceptual qualities of the developing work. This same analysis can be extended to industrial design objects which the user must learn to manipulate, and to the bodily states of pleasure, tension, and calm which accompany the learning curve.

Bottom-up processes are more characteristic of the Experience mode in which the body is focal and the mind serves as a background context. This reflects the body's spontaneous emotional response, as if conditioned, to meaningful social events. From this perspective, *cognitions serve as a context for emotions*. In the bottom-up model, feedback from bodily states and muscular memories lend coherence to the overall experience, just as form or styles provide an overall structure for an artwork. In Proust's (1922/1960) book, *Swann's Way*, he described how the taste and smell of a madeleine pastry flooded him with childhood memories. This is what is called reintegration, a memory process wherein a single quality brings back the whole experience. The point here is that the critical cue is a sensation, odour or taste, that stimulates an entire package of cognitions from early life. Thus, an artists encounter with a scene, or a designer/users encounter with a product can bring back whole memories of lives lived. It is this greater depth to which the object/event is processed that distinguishes between a transitory feeling and a more profound emotional experience.

In sum, the interaction of mind and body can be characterised in complementary processes depending on whether the figure/ground relation is mind over body or body over mind. When the mind is dominant, then the body functions in terms of feelings, whereas when the body is dominant, it awakens the mind's eye with memories and symbolically meaningful experiences.

Designing for action

Design has an obvious instrumental side in that tools or images are used to manipulate the environment or convey information, respectively. This is very much a top-down process in which a series of acts enable a person to utilise an instrument that is related to adaptation. Such a planful process involves specifying the sequence of cognitions or actions needed to utilise the tool and complete the task. The body then executes these plans through the performance of stimulus - response sequences supported by concentrated attention and an appropriate level of excitation or arousal. When stimulus-response sequences are too difficult to execute serially, attention is diffuse, and arousal is either excessive or inadequate, the action mode breaks down and the design process fails. Design for the *action mode* will be successful to the extent that complementary relationships exist between (1) action plans associated with the tool (its function) and (2) immediacy of on-line adjustments. Thus, the designer and user are responding (i.e., attending) to the same mental algorithm without the physical manipulation of the instrument interfering with the process. For example, novice drivers over steer the vehicle until a correlation is clearly developed (i.e., overlearned) between rotation of the steering wheel and displacement of the car around a curve. Nervousness (i.e., excessive arousal) in the novice driver draws attention to the *manipulandum* (i.e., the steering wheel) and interferes with the driving process. In a sense, the instrument must disappear into the algorithm.

Design also has an aesthetic side to the extent that the form of tools or images embody sensory qualities that shape experience and are related to each other in a coherent manner. This is a bottom-up process in which sensation is valued in and of itself (i.e., intrinsically) and not because it conveys information related to the practical (i.e., extrinsic) use of the instrument. This sensory experience can also have a connotative aspect in that qualities expressed in the design (e.g., smoothness, roughness, and so on) are meaningfully (i.e., metaphorically) related to the denoted purpose of the object. Thus, a racing car can have the quality implied in the sleekness of its shape. While this point is almost a cliché, the general idea should be underscored. Function meets form to the extent that the denotative properties of one (i.e., its plan or algorithm) are implied by the connotative (i.e., expressive) properties of the other. When this complementarity is spontaneously experienced it fosters an overall experience of coherence in the instrument, tool, or image. The author would even propose a principle based on processes related to "aesthetic distance" (Bullough, 1912; Cupchik, 2002) that success of a design object will be heightened when its sensory impact is intense or engaging without affecting the user's ability to manipulate the instrument.

Designing for experience

The discussion thus far has focused on the interaction of designers and users with tools, instruments, and images that foster adaptation. But addressing emotion requires that we place design objects, designers, and users in social contexts. While objects in isolation can be treated in a *monovalent* fashion as serving a particular function in accordance with specific algorithms, design objects in a social context are *polyvalent* (Schmidt, 1982) and potentially embody multiple layers of meaning (Kreitler and Kreitler, 1972). Thus, a design object can possess various meanings for a designer having nothing to do with its function. The meanings can reflect political and professional circumstances of designers at various phases of their careers. Ideally one hopes for harmony between the designer's intrinsic motivation to create useful and valuable objects, and the extrinsic pressures of the professional work place. But professional pressures and fragmentation of the design process in working groups can lead to skewed performance and an emphasis on certain features of the design object to the exclusion of others. This reduces the potential for coherence in the design object. Just as excessive arousal can interfere with the use of tools in accordance with their planned function, fear and anger, which are consequent to disturbance in the work place, can have a deleterious effect on a designers creative functioning.

Can we actually design for emotion when it is understood to be a persons experience in a social context? And whose emotion will it be anyway? Analysis of popular and mass culture suggests that corporate decisions can lead a designer to induce emotions in the user in accordance with myths and image of a particular age. Is it the job of a psychologist to instruct people in how to do this or to raise peoples consciousness about the dangers of social manipulation through supposed image-making (Cupchik and Leonard, 2001)? I would prefer to imagine a great variability in the emotional responses to users depending on their life circumstances and not to think that the goal of designers is to manipulate them into experiencing particular emotions so that they become attached to products. It was one thing for electric lights and plate glass windows to enable the emerging urban bourgeoisie to view commodities from a close distance. But when coveting comes under the purview of mass culture and people's self-concepts become determined by large corporations, then individual development stops.

REFERENCES

Binswanger, L., 1958, The existential analysis school of thought. In *Existence*, edited by May, R., Angel, E., and Ellenberger, H.F., (New York: Simon and Schuster), pp. 191-213.

Bullough, E., 1912, `Psychical distance' as a factor in art and as an aesthetic principle, *British Journal of Psychology*, **5**, pp. 87-98.

Cupchik, G.C., The evolution of psychical distance as an aesthetic concept. *Culture and Psychology*, **8**(2), in press.

Cupchik, G.C. and Leonard, G, 2001. High and popular culture from the viewpoints of psychology and cultural studies. In *The psychology and sociology of literature: In honour of Elrud Ibsch*, edited by Schram, D. and Steen, G., (Amsterdam: John Benjamins), pp. 421-441.

Duffy, E., 1941, An explanation of "emotional" phenomena without the use of the concept "emotion." *Journal of General Psychology*, **25**, pp. 283-293.

Duffy, E., 1962, *Activation and behavior*, (New York: John Wiley).

Gardiner, H.M., Metcalf, R.C. and Beebe-Center, J.G. 1970, *Feeling and emotion: A history of theories*, (Westport, CT: Greenwood Press). (Originally published in 1937).

James, W., 1884, What is an emotion? *Mind*, **9**, pp. 188-205.

Keen, E. 1977. Emotion in personality theory. In *Emotion*, edited by Candland, D.K., Fell, J.P., Keen, E., Leshner, A.I., Plutchik, R., and Tarpy, R.M., (Monterey, CA: Brooks/Cole), pp. 213-252.

Kreitler, H., & Kreitler, S., 1972, *The psychology of the arts*, (Durham, NC: Duke University Press).

Oatley, K., 1992, *Best laid schemes: The psychology of emotions*, (New York: Cambridge).

Pribram, K.H., and McGuinness, D., 1975, Arousal, activation, and effort in the control of attention. *Psychological Review*, **82**, pp. 116-149.

Proust, M., 1960, *Swann's way*, (London: Part. Chatto & Windus). (Originally published in 1922) (trans., S. Moncrieff).

Schachter, S., 1966, The interaction of cognitive and physiological determinants of emotional state. In *Anxiety and Behavior*, edited by Spielberger, C.D., (New York: Academic Press), pp. 193-224.

Schmidt, S.J., (1982), *Foundations for the empirical study of literature*, (Hamburg: Helmut Buske Verlag.) (translated by R. de Beaugrande).

Strauss, E.W. 1958. Aesthesiology and hallucinations. In *Existence*, edited by May, R., Angel, E., and Ellenberger, H.F., (New York: Simon and Schuster), pp. 139-169.

From disgust to desire: how products elicit emotions

Pieter M A Desmet
Industrial Design Delft, Delft University of Technology, The Netherlands

INTRODUCTION

Our emotions enrich virtually all of our waking moments with either a pleasant or an unpleasant quality. Given the fact that a substantial portion of these day-to-day emotions is elicited by 'cultural products,' such as art, clothing, and consumer products (Oatley and Duncan, 1992), designers may find it important to include emotions in the intentions of their design efforts. In addition, emotional responses can incite customers to select a particular artefact from a row of similar products, and may therefore have a considerable influence on our purchase decisions. As a consequence, more and more producers are currently challenge designers to manipulate the emotional impact of their designs.

In design practice however, emotions elicited by product appearance are often considered to be intangible and therefore impossible to manipulate. This persistent preconception is partly caused by some typical characteristics of these 'product emotions.' Firstly, products can elicit all kinds of emotions. Moreover, the concept of product emotions is broad and indefinite. Emotions (and other types of affective states) are elicited not only by the product's aesthetics, but also by other aspects, such as the product's function, brand, behaviour, and associated meanings. Secondly, emotions are personal, that is, individuals experience different emotions towards the same product. Thirdly, products elicit 'mixed emotions.' Rather than eliciting one single emotion, products can elicit various emotions simultaneously.

However, designers *can* influence the emotions elicited by their designs because these emotions may not be as intangible as they seem. This claim is based on theories of emotion that maintain that although emotions are idiosyncratic, the conditions that underlie and elicit them are universal. Those theories indicate that each distinct emotion is elicited by an unique 'pattern of eliciting conditions' (Lazarus, 1991). The author has described such patterns for 14 particular 'product relevant' emotions, i.e., emotions that are often elicited by product appearance (Desmet 2002). On the basis of the variables that make up these patterns, product relevant emotions can be classified in one of five emotion types. This paper presents a synopsis of this classification, which can support designers to get a grip on the relationship between product design and emotional responses.

ELICITING CONDITIONS

The cognitive, functionalist position on emotion posits that emotions serve an adaptive function. In this view, emotions are considered the mechanisms that signal when events are favourable or harmful to ones concerns. This implies that every emotion hides a concern, a more or less stable preference for certain states of the world (Frijda, 1986). A product will only elicit an emotion if it either matches or mismatches a concern. Why did I feel attracted to an umbrella? Because I have a concern of staying dry. And why was I frustrated when my computer repeatedly crashed? Because I have a concern of efficiency. The process of signalling the personal relevancy of an event is commonly conceptualised as 'a process of appraisal.' An appraisal is a "direct, non-reflective, non-intellectual automatic judgement of the meaning of a situation" (Arnold, 1960, p. 170), in which our concerns serve as points of reference.

Many appraisal models advanced to date include small sets of appraisal types to differentiate between emotions. Each appraisal type (and related concern type) addresses a distinct evaluative issue. These various underlying appraisal and concern types are the key variables that have been used to classify emotions in one of the following five classes of product emotions: instrumental, aesthetic, social, surprise, and interest emotions.

Instrumental product emotions

Products are not objects of coincidence; they are designed, bought, and used with a purpose. We never buy a product without having some motive to invest our resources. Products can be regarded instrumental because we belief they can help us accomplish our goals. The concern type 'goal' refers to states of affairs that we want to obtain, i.e., how he would like things to be (Ortony *et al.*, 1988). Humans have numerous goals, which range between abstract (e.g., I want to be happy), and concrete (e.g., I want to have lunch). Our goals are the points of reference in the appraisal of motive compliance. A product that facilitates goal achievement will be appraised as motive compliant, and elicit emotions like *satisfaction*. Similarly, products that obstruct goal achievement will be appraised as motive incompliant, and elicit emotions like *disappointment*.

Also products that threaten to obstruct or promise to facilitate goal achievement elicit instrumental emotions. Each time we see a product, we anticipate its future use or possession. We predict the experiences of using the product and the consequences of owning it. These anticipations are based on knowledge about the type of product or the product brand, and on information conveyed by the product itself (e.g., appearance, price, packaging). When shopping for new shoes, one might, for example, anticipate that wearing a particular pair of elegant shoes will have the consequence of 'being attractive.' If this person has the goal to be attractive, he or she will appraise this particular pair of shoes as motive compliant and, for instance, experience *desire*. If the same person has the goal of 'comfortable walking,' he or she might appraise the anticipated discomfort as motive incompliant and experience *dissatisfaction*.

Aesthetic product emotions

As products are physical objects, they look, feel, smell, taste, and sound in a particular way. Each of these perceivable characteristics can both delight and offend our senses. Like all objects, products, or aspects of products, can be appraised 'as such' in terms of their appealingness. The concerns that are the points of reference in the appraisal of appealingness, are attitudes. Our attitudes are our dispositional likings (or dislikings) for certain objects or attributes of objects (Ortony *et al.* 1988). Like goals, we have many attitudes, of which some are innate (e.g., the innate liking for sweet foods), and others are learned (e.g., the acquired taste for oysters or wine). We have attitudes with respect to aspects or features of products, such as product colour or material. We also have attitudes with respect to product style. For instance, some people have developed an attitude for style of Japanese interior design, whereas others have a taste for Italian design.

A product that corresponds with (one of) our attitudes is appraised as appealing and will elicit emotions like *attraction*. A product that conflicts with (one of) our attitudes is appraised as unappealing and will elicit emotions like *disgust*. In some cases, the appealingness is based on characteristics of the product itself, such as shape, size, or particular details. As a result, a dispositional liking for a certain model will be generalizable to other products. Sometimes, however, the dispositional (dis) liking is restricted to only one specific product. In those cases, the liking results from previous usage or ownership of that particular exemplar. One can have a dispositional liking for a ring because it was a gift for someone special or for a particular backpack because one travelled with it to many different countries. In these cases, the attitudes are embedded with personal meaning and not applicable to other exemplars of the product type.

Social product emotions

Next to goals and attitudes, standards are a third type of human concerns relevant to product emotions. Our standards are how we believe 'things should be,' and how 'people should act' (Ortony *et al.*, 1988). For example, many of us have the standard that we should respect our parents, and eat fruit and vegetables. Most standards are socially learned, and represent the beliefs in terms of which moral and other kinds of judgmental evaluations are made. Products are embedded in our social environment; they are designed by people, used by people, and owned by people. Some even look like people (e.g. the Alessi eggcup "cico"). Because we cannot separate our view on products from our judgments of the people we associate them with, we apply our social standards and norms, and appraise products in terms of 'legitimacy.' Products that are appraised as legitimate elicit emotions like *admiration*, whereas those that are appraised as illegitimate elicit emotions like *indignation*.

The objects of social emotions are essentially agents. This agent can be either the product itself that is construed as an agent, or an associated agent, such as the designer or a typical user. Firstly, products are the result of a design process and the designer or company is the construed agent. While looking at a product, one can for example, praise its originality or blame the designer for a lack of product quality and experience *contempt*. Secondly, products are also often associated with particular users or user groups. Most of us have no difficulty in envisioning typical users of, for example, German cars, or skateboards. In those cases the typical user group or institution that is associated with the product is the object of appraisal. We can blame the user of a big car for not caring about environmental issues, or admire the owner of a digital agenda for their presumed time-efficiency. Thirdly, we also tend to apply our social standards to products themselves. Although products are not people, they can be treated as agents with respect to the presumed impact they generally (can) have on people or society. A person can, for instance, experience *indignation* towards mobile telephones because they blame these products for the disturbance they cause in public spaces such as train compartments.

Surprise product emotions

Any product (feature) that is appraised as 'novel', i.e., sudden and unexpected, will elicit a surprise response. Surprise emotions differ from the previous three emotion types because they are not related to a particular concern type. Instead, *pleasant surprise* is elicited by a sudden and unexpected match with any concern (i.e., a goal, attitude, or standard), and *unpleasant surprise* is elicited by a sudden and unexpected concern mismatch.

Products that are totally new to us can surprise us. A person can, for instance, be pleasantly surprised when first encountering a wireless computer mouse (that unexpectedly matches the concern of comfort). Besides totally new products, also product aspects or details can elicit surprise. A printer, for instance, can be surprisingly fast, or a door handle surprisingly soft. In the latter case, we expect door handles to be rigid, and are pleasantly surprised because this particular handle disconfirms that expectation. Once we have become familiar with the novel aspect of the product, it will no longer elicit surprise. Therefore, these are often *one-time-only* emotions.

Interest product emotions

The fifth product emotion type comprises emotions like *fascination*, *boredom*, and *inspiration*. These emotions are all elicited by an appraisal of *challenge combined with promise* (Tan, 2000) and all involve an aspect of (a lack of) stimulation.

Those are products that make us laugh, stimulate us, or motivate us to some creative action or thought. A well-established psychological principle is that people are 'intrinsically' motivated to seek and maintain an optimal level of arousal. A shift away from this optimal level is unpleasant. Since low arousal levels seem to be disliked, we appear to have a 'stimulus hunger.' Products that are appraised as not holding a challenge and a promise will elicit emotions like boredom (either because they do not provide us with any bodily sensation or leave nothing to explore). Products that are appraised as stimulating because they bring about some question or because they require further exploration will elicit emotions like fascination and inspiration. Interest emotions are similar to aesthetic emotions because in both cases the object of emotion is the product 'as such.'

DISCUSSION

The five types of product emotions are not claimed to cover all possible emotional responses towards products. Nevertheless, they do illustrate that products have many different layers of emotional meaning, and that some of these emotional meanings can be predicted. Designers that are aware of the patterns that underlie emotional responses (and the concerns and appraisals that make up these patterns) can therefore influence the emotions elicited by their designs. A designer can use these patterns for developing their personal sensitivity for recognising the eliciting conditions of product emotions. To this end, two tools have been developed (Desmet, 2002). The [product & emotion] navigator is an inspirational tool that aims to illustrate the patterns of eliciting conditions that underlie 14 distinct product relevant emotions, and the product emotion measurement instrument (PrEmo) is a non-verbal self-report tool that can be used to measure these 14 emotions.

Note that it is not assumed that to serve humans' well-being, designers should create products that elicit *only* pleasant emotions. Instead, it may be interesting to design products that elicit 'paradoxical emotions,' that is, positive and negative emotions simultaneously. Frijda (1996) stated that in experiencing art, it are precisely these paradoxical emotions that we seek. It may be interesting for designers to investigate the possibilities of designing paradoxical emotions because this may result in products that are unique, innovative, rich, and more challenging or appealing than those that elicit only pleasant emotions.

ACKNOWLEDGEMENTS

This research was funded by Mitsubishi Motor R&D, Europe GmbH, Trebur, Germany. Paul Hekkert and Jan Jacobs are acknowledged for their contribution to this research.

REFERENCES

Arnold, M.B., 1960, *Emotion and personality*, (New York: Colombia University Press).

Desmet, P.M.A., 2002, *Designing emotion*, Unpublished doctoral dissertation, Delft University of Technology.

Frijda, N.H., 1986, *The emotions*, (Cambridge: Cambridge University Press).

Lazarus, R.S., 1991, *Emotion and adaptation*, (Oxford: Oxford University Press).

Oatley, K. and Duncan, E., 1992, Structured diaries for emotions in daily life. In *International review of studies in emotion*, Vol. 2, edited by Strongman, K.T., (Chichester: Wiley), pp. 250-293.

Ortony, A., Clore, G.L. and Collins, A., 1988, *The cognitive structure of emotions*, (Cambridge: Cambridge University Press).

Tan, E.S.H., 2000, Emotion, art, and the humanities. In *Handbook of emotions (2nd ed.)*, edited by Lewis, M. and Haviland-Jones, J.M., (New York: The Guilford Press), pp. 116-134.

Design expression and human experience: evolving design practice

Jane Fulton Suri
IDEO, USA

NEW KINDS OF INFLUENCE

A few years ago it was simple. Designers designed things: objects like lamps, chairs, computer mice, cars, buildings, signage, page and screen layouts. We always knew that the things we designed affected people's experience. But still, it was enough to design the thing. The work of designers was to bring skills, creativity and insight to "designing things right."

Beyond design of things

As established products have become more similar in technology, functionality, price and quality, companies have turned to design to differentiate their offerings through human-centred innovation and to create stronger emotional connections with their customers. More companies have followed the example of Apple, Braun and Philips, recognising design as a strategic function in their business — not one subservient to marketing, manufacturing or engineering. As designers, we are now challenged to help companies explore and visualise directions for their future offerings — "the right things to design."

There is another expansion of design influence. Increasingly we find ourselves designing to support complex and dynamic interactions integrating hardware and software, spaces and services. A design project today is likely to involve connected products such as mobile digital devices, or systems of linked interactions, such as those comprising a train journey or an Internet banking transaction.

This expansion of opportunity for design is due partly to advances in technology. It is also due to a maturing confidence in the human-centred design profession that challenges the wisdom of focussing on the individual artefacts themselves when people's interactions can be better supported by thinking about design opportunities more holistically. It is also partly a result of new business strategies in which companies seek competitive advantage through more integrated offerings, with differentiation through all points of customer contact that express their brand. In any case, design is asked to influence, not just the look and feel of individual things, but the quality of experience that people have as they live their lives through time and space, encountering the designed world.

Focus on experience

As a result, phrases like *"design of experience,"* *"user experience"* and *"customer experience"* are frequently used in the design and business communities. Pine and Gilmore (1998) represent the design of experience as a new kind of economic offering, distinct from the design of products and services. They identify Niketown and Disney World as examples of staged branded environments achieved by integrating multiple designed elements including architecture, media and personnel. Even in these highly controlled situations it is too much to talk of *"designing experience."* Experience is personal and subjective; designers can neither predict nor control the experiences that people will have (Sutton, 1992). But designers can, and do, work with *concern* for the quality of people's experiences relating to individual products and to systems of things. In this way, designers have the ability to influence positively both the quality of these interactions and, perhaps more deeply than traditional advertising, affect people's perceptions of the company offering them.

New demands

With these opportunities for greater influence come new demands. Here are four.

First, to design with people's experience in mind, we need to understand more about what matters to the people who will buy and use the things we design. This demands some knowledge of their activities and their thoughts and feelings about these; related goals, aspirations, rituals and values; the personal, social and cultural context; meanings attached to different features and objects. This issue is far more complex than a single research method can address.

Second, people's experiences occur within a physical and socio-cultural context and through time. These contextual, dynamic, multi-sensory, spatial and temporal dimensions stretch the limits of traditional modelling tools to explore and communicate what it will be like to interact with the things we design. We need richer representations to fully explore this terrain.

Third, this work involves ever more diverse groups of colleagues and clients. A design team is likely to involve professionals from multiple disciplines such as interaction design, industrial design, engineering, architecture, human factors, business, marketing and branding. Each brings a unique perspective on the issues and an individual approach to solving them. To work effectively it is important to find ways to develop a common vision of what the team is trying to bring into being.

Finally, design work concerned with *"experience"* often involves decision-making at higher levels within a client organisation than is usually the case in traditional product design projects. Designers need ways to grasp and reflect executives' strategic goals for their company and its brand — their intended outward expression of the company. Ultimately the challenge is to find compelling ways to convince decision-makers that current customer experiences and perceptions could be enhanced through introduction of specific new design expressions of their offerings.

The next section briefly outlines some developments in design practice that begin to address these new demands.

WHAT THIS MEANS IN PRACTICE

So, designers today work with more people with different perspectives on more complex design problems. The eventual human, customer or user experience is a focus that everyone can rally around, at least conceptually, see Figure 1.

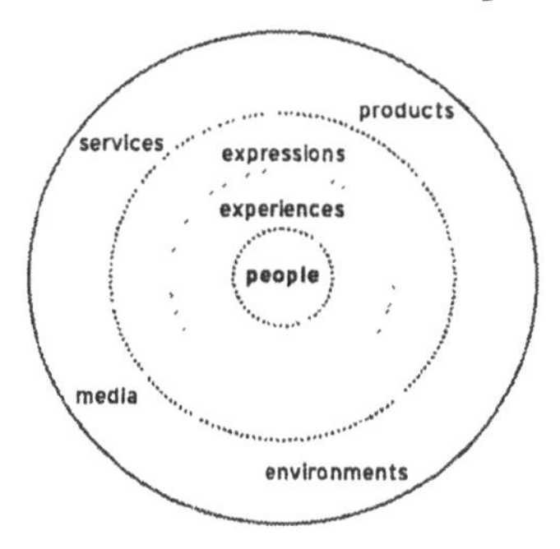

Figure 1 People's experience as an organising focus for design expressions of multiple kinds.

But this is an abstract concept. The real power of design is in creating tangible expressions of ideas; this is the key to addressing the demands of working in this new territory. The more that representations of insights, evidence, design ideas and final concepts can be made experiential in some way, the more all parties — team members, clients, decision makers and users — will be able to grasp how the representations relate to other people's experience.

A few years ago Bill Moggridge of IDEO invented a tautology to encourage communication of ideas through tangible media: "The only way to experience an experience is to experience it." Such thinking led Buchenau and Fulton Suri (2000) to promote the use of prototyping methods that allow designers, users and clients to try things out for themselves rather than witness demonstration or description. Such *"experience prototyping"* is relevant here, but there are also other useful activities that are becoming more integrated into design practice at different points. Several also overlap in their usefulness to address multiple demands.

Discovering what matters

Contextual observations and users interviews have become standard practice in design. To explore more emotional concerns, projective methods that involve asking participants to create something from their own experience are proving to have inspirational and informational power for design. These include methods such as cultural probes of a user group (Gaver *et al.*, 1999), collage making, personal narratives, photo-journal assignments and cognitive map-making around the design topic. Rather than analysed data, it is often the raw images of real people, places and things, the maps or collages themselves and the unedited personal stories that best capture important insights in ways that design and client teams can relate to. There is a personal connection to these tangible artefacts that stirs empathy, so that we sense viscerally what matters to the people we study. Retaining the raw evidence as images, stories and video for later communications and persuasion is important too.

Developing common vision

Another advantage of these rich personal representations is their ability to help create shared experience and reference points within a design team. This is a powerful asset, providing the foundation for a common point of view among professionals of various different disciplines. There is value in taking this a step further, involving design team members and clients directly in attending carefully to their own personal experience of a relevant situation. Situations might be naturally occurring (as in recording behaviour and experience around "using water and plumbing fixtures at home") or contrived (as in role-playing "emergency-room patient" and making the entire journey, except the actual injury and treatment!). The experience might be recorded in written or spoken form or through video- or photo-documentary. Such exercises are especially valuable at the start of a project as a way to establish mutual respect for each other's perspectives and begin to form an initial team point-of-view.

Creating richer representations

Experience and emotional responses are dynamic in the sense that they occur in time and space, through the flow of some kind of activity. As a team begins to explore important experiential qualities, it's important to go beyond words and static representations. One way is to connect important qualities to other real-life experiences that represent analogous behaviours. For example, in considering how to enhance customers' experiences in a branded coffee house, exploring activities like visiting a village market or a high street would inspire us to consider dynamic and multi-sensory qualities that many traditional models lack.

Storyboards and user scenarios, as simple sketches or video, capture dynamic dimensions too and are well-embedded in design practice. Taken to an experiential level, scenarios can be walked-through and acted out with simple props. This offers the opportunity for real-time discovery by participants, whether users, designers or clients, and allows for well-grounded idea generation and refinement.

Convincing decision makers

Key decision-makers at high levels within client companies are often focussed upon *"big picture"* issues and may have limited interest in details. But if design is to fulfil its potential as a business advantage, it is important to begin by exploring with them strategic directions, values and how they wish the company to be known. A valuable role for designers is to turn these visions into tangible form as early as possible, working directly with clients. It can be helpful to have clients visually map their company against others within similar industries along dimensions such as progressive/traditional, informal/formal. Another method is to create metaphorical descriptions for the company, imagining it as a particular kind of car or country and listing the equivalent qualities. Other methods might involve them in predicting next year's headlines or creating rough video-advertisements from a future imaginary campaign. All these methods begin to give tangible expression to the way a company wishes to be perceived and the kind of emotions it wants to evoke through its products and services.

Early agreement about these issues allows us to assess how well a company's intentions align with customers' perceptions as we discover what really matters to them. It is also a great starting point for the eventual story we tell about these discoveries and proposed design solutions. Beginning the story with the company's vision, and unfolding it with rich experiential material and evidence will bring executives along on the design journey in a dramatic revelatory way that simply presenting the design solution cannot.

CONCLUSIONS

In summary, the opportunity is for designers to exercise greater influence on the designed world, building out from people's current and desired experiences to support a company's brand through design expressions — whether products, services, environments, media or hybrid offerings. One of the profession's major strengths is to make it possible for people to experience things in a tangible way — whether these are abstract ideas or actual things that do not yet exist. This ability needs to be exploited to its full, in developing new forms of prototyping and communicating with team members and clients and others in client organisations.

REFERENCES

Buchenau, M., and Fulton Suri, J., 2000, Experience prototyping. In *Proceedings of DIS 2000*, (New York: ACM Press), pp. 424 – 433.

Gaver, B., Dunne, T., and Pacenti, E., 1999, Cultural probes. *Interactions*, **6**, 1, pp. 21 – 29.

Pine II, B.J., and Gilmore, J.H. 1998. Welcome to the experience economy. *Harvard Business Review*, July-August, pp. 97 – 105.

Sutton, R., 1992, Feelings about a Disneyland visit: photography and the reconstruction of bygone emotions. *Journal of Management Inquiry*, 1, pp. 278–287.

The comprehension shift HMI of the future – designers of the future

Michael Robinson
Fiat, Italy

INTRODUCTION

Modern technology is advancing at a tremendous rate, leaving most consumers confused and frustrated. The automobile context offers arguably the greatest (and oldest) battlefield in user's love-hate relationship with the machine world. A battle that historically embraces not only technical issues (like next-generation feature races) but also rich, cultural (local versus global ethnic values), social (status, lifestyle), psychological (stress damage, emotional involvement), even health issues (ergonomics, well-being). In order to design sophisticated products like automobiles in this new digital age, car designers must widen their intellectual scope past industrially feasible, aesthetic beauty to include these unusual aspects called "emotional awareness".

One of the major concerns however in the automobile world today is safety, but safety signifies survival, and when survival becomes a priority, then quality of life has very little opportunity of success. According to the NHSTA, last year in the U.S., 6 million accidents were reported to the police, which means that the true number is closer to double or triple if you count all the unreported, supermarket fender benders. Not all crashes cause mortalities fortunately. Four million were simply PDOs (Property Damage Only) while 3 million people were injured and 41,821 men, women, and children lost their lives in highway accidents in the U.S. alone, for an average of one every 13 minutes. Social costs, including medical care, car repair, insurance costs and lost earnings add up to an estimated total of $150 billion dollars in the States every year! So why do accidents happen? Is it all a matter of cell phones distracting drivers, or is there something else?

COMPREHENSION-COMPLEXITY CRISIS

The automobile is easily the most complex object in our daily lives. Jet fighters may be more complex but our partners and children do not use them regularly. Personal computers are just as diffuse but do not kill people when they crash. Our latest luxury sedan has 52 onboard computers, offering dot.com communication complexity added onto existing mobility complexity. As complexity continues its steady climb in our automobiles, so-called "easy-to-use" interfaces are false illusions for both users and designers. Technological advancements often hide life-threatening consequences due to the fact that machine complexity has recently surpassed human comprehension, inverting a "total control" scenario here-to-fore taken for granted.

THE COMPREHENSION SHIFT

One possible solution could be the implementation of another complex system such as human emotions in order to match and neutralise the dangers of increasing complexity in the automobile context. The mechanisms of human emotions are virtually unknown by most people and certainly uncontrolled, unexploited, but soon will become our primary reference in human-machine interfaces. Between today's 100% manual control and tomorrow's 100% full-automatic control (hands-free driving), we will pass through a hybrid driving period which will allow scientists and designers to evolve the necessary technology while allowing the general public to learn to accept talking with machines.

In the infamous sci-fi film, 2001, when HAL asked Dave to show him his sketches, "You're really improving, Dave", any one of us would have replied, "Don't patronise me HAL!" The difference is that we all treat computers as indispensable but (rather stupid) rational support systems, but soon they, as other machines in our lives, will become proactive (rather intelligent) emotionally aware, support services. While most artificial intelligence developers unfortunately have dismissed "affective" computing as something they (the computers) will self-learn by correspondence on the job, I tend to agree with Marvin Minsky who recognises a general lack of understanding of human "common sense", which aims towards what could called AEI or Artificial Emotional Intelligence.

EMOTIONAL AWARENESS

Researchers are coming closer and closer to recognising and monitoring human emotions by comparing redundant physiological signals. MIT Media Lab's Rosalind Picard says: "If you can't measure it, you can't manage it." The hot question today is: "Now what?" Human-to-Human Relationships, which will soon be widely used by HMI developers as behaviour references, offering appropriate situational attitudes in cars: as if designing a factory-installed, trusted family chauffeur. Unfortunately, corporate HMI experts have little or no training in cognitive sciences and are actually quite ignorant in such matters. Eventually, this obligatory shift will sweep the field, producing concepts like Profound User Databases (PUDs), capable of personalising every machine (car, PC, etc.) through shared experience collection, taking full advantage of the technology enabling, digital revolution.

VISION

Today, car designers are starting to look into emotional interfaces aimed at achieving that HMI nirvana called "well-being", asking benchmark questions like: "Does this car really understand me?" The next step will be industry-wide applications of mood sensors for cognitive navigation assistance, and finally, around 2050 (roughly 8 car-generations from now), driving on public roads will become illegal, entering into the Zero Accidents Era, completely eliminating highway accidents and mortalities. Imagine no more steering wheels, pedals, air bags, bumpers, or overweight, crash-worthy superstructures. Cars will take on totally different parameters for beauty. These oversimplified steps embrace the full spectrum of modern AI development and consequently represent giant leaps in technology.

TRANSFORMATIONS

Cars are the most complex object in our daily lives, but continue killing people due to human error. Up until a short time ago, humans were considered by car developers as *intruders* in the technological auto environment, whereas today, they are slowly becoming integrated into the overall system, underlining a giant gap in industrial expertise, as we discover that purely automotive aspects in car development amount to only half of the overall picture. A similar transformation is taking place subsequently in the car designer world due to overwhelming obstacles forcing them into sales-reducing compromises. Here again, integration is the easy buzz word but requires much more than platform-style development groups. Real integration means putting a little designer in every development team member's head, spreading creativity throughout the corporation, while exchanging technical, economical and cultural wealth among colleagues.

CARRY BACK APPLICATIONS

Stepping outside the automotive context in search for novel HMI solutions has led us backwards, Human-Human Relationships, which not only offer promising new directions for human safety and simplicity, but raise intriguing questions in the world of Emotional Awareness in purely interpersonal, non-machine oriented relationships. Present, technical comprehension will slowly begin to shift towards unexplored facets of human behaviour, which will take full advantage of the uncountable, hidden potential in human emotions. Reform implications will set new quality-of-life standards for the future, in areas such as general society, corporate management, learning, products, even humans, even designers, if we can only learn to approach them from this unorthodox viewpoint.

EMOTIONAL AWARENESS IN SOCIETY

Our masculine society is slowly gravitating towards a more feminine approach (nothing to do with suffragettes or feminist movements). The intuitive care and concern mothers have for their children is exactly the type of empathy all humans should have for one another. Reading needs, projecting comprehension and loving care are all behavioural milestones of the future.

Emotional awareness in corporations

today	**tomorrow**
patriarch oriented social values	matriarch oriented social values
capitalism guides the western world	spiritualism guides the eastern world
masculine approach (regardless of sex)	feminine approach (regardless of sex)
mandatory rationality & functionality	emotional & interpersonal mentality
forced aggressiveness & control	sensitive & helpful cooperative & self
obligatory ambition & expectations	fulfilling atmosphere
victims of stress & frustration	psycho-physiological well-being & balance

Giant, aggressive, multinational conglomerates are becoming the ideal solution for survival in times of uncertainty, offering safety in numbers instead of ideas. Within those environments, individuals are pressured and corralled into levels of stress previously unthinkable. New management styles will reward those that create value in human resources as well as financial.

Emotional awareness in education

today	**tomorrow**
power leadership	creative leadership
king-of-the-hill ambitions	backstage motivators
union labour	social magnets
employee gripe factor	employee attraction factor
embedded bureaucracy	embedded services
stagnant culture	efficient courtesy norms
strategy enforcement	strategic Intuition
mass control	research & development as a social skill
creating value for shareholders	creating value for shareholders
strictly financial	first human, then financial

How many children love their high school subjects, their textbooks or their teachers? Tomorrow's schools will be more like *information candy stores* where young minds just can not get enough knowledge, creating a sort of *intellectual obesity* through virtual field trips which will replace books with Disney style information magnets assuring fun learning through expert's eyes. Top-down, force-fed learning will be replaced by juvenile curiosity, accelerating brain performance through sheer desire.

Emotional awareness in products

today	**tomorrow**
classist system	worldwide learning
intellectual racism	intellectual equality
90% students hate learning	school becomes kids favourite place
learning is boring (stale toast)	learning is exciting (embedded interest)
top-down learning	bottom-up learning
force-fed (fois gras) info input	curiosity creates hunger for knowledge
classroom disruption	emotional learning base
widens the communication gap	reduces the communication gap
performance anxiety	self-control
bad emotional habits	self-confidence (emotional intelligence)
correct behaviour	correct behaviour
forced control	info candy shops

Future Ralph Naders will collect consumer, interface harassment cases from humiliated users that could not turn off their VCR. Products will offer user seduction or *embedded courtesy*.

Emotional awareness in humans

today	**tomorrow**
user humiliation	user seduction
insult factor	attraction factor
marketing incentives	gratification incentives
human gullibility tests	human-machine erogenous zones
abuse and abandon	proactive partnership
gadget race	understands user
rational coherence	emotional coherence
form follows function	feelings awareness sensors
objects refuse suggestions	profound user database
ignorant and stubborn machines	onboard "trusted butler"
too many switches	subliminal switches
encyclopaedia size users manuals	"I think therefore I am"

Individuals will be motivated to recognise their generally accepted lack of emotional awareness and take full possession of their hidden potential, pushing HHR to more profound levels.

Emotional awareness in designers

today	tomorrow
emotional illiteracy	emotional intelligence
90% unaware of mental potential	finally playing with a full deck
rational bias rewarded	multiple intelligence mix rewarded
overpowering social models	balance becomes new reference
short term goals	long term goals
lack of life plans	looking farther downstream
simple AI thought transactions	higher AEI forms of thinking
big blue wins again	telepathy replaces cell phones
performance orientation	interpersonal relationship orientation
in families, friends, jobs, attitudes	in people, environments, roles, rewards

Car designers are facing a global transformation phase due to corporate inadequacies and technological disappointments. In order to produce a winning design, not only in aesthetic terms, but above all, from commercial, conceptual, even historical viewpoints, designers must overcome such an impossible number of bureaucratic roadblocks, that many give up before they begin sketching. Most milestones cars throughout history are considered exceptions rather than rules. Those gorgeous cars recognised for their irresistible sex appeal are often impossible to drive due to strange iDrive style gadgets, which disorient and alienate would-be buyers. Sexy fender fantasies are slowly morphing into sexy interface opportunities, inviting design schools to rethink their coursework and design teams to reorganise their priorities.

Creative process

today	tomorrow
feasibility - aesthetics compromise	emotional awareness
Industrial incompetence	New collaboration code
innovators = rule breakers	creativity expansion
creative minds viewed negatively	broadband mind transfer
objects precede concepts	concepts guide objects
Inertia guides most designers	proactive ideation strategy
art & technical background	cognitive science & sci-fi
design schools teach a,b and cs	New areas of designer competency
no history	culturally prepared designers
cultural preparation non-existent	deeper intellectual preparation
HMI ignorance	emotional awareness
unsketchable technology applications	digital HMI opportunities

CONCLUSIONS

Today, automotive HMI teams are discovering that the comprehension-to-complexity crisis can be inverted by benchmarking Human-to-Human Relationship (HHR) references, concentrating on how humans get along with one another. Social, cognitive, and other behavioural sciences will guide future product development, providing the missing emotional link in the technological revolution. Digital-Human-Machine Interface (DHMI) opportunities will redirect AI development towards Artificial "Emotional" Intelligence (AEI) thanks to the softer, interpersonal aspects of HHR. As a result, carry-back DHMI applications in HHR scenarios will eventually offer revolutionary, alternative role models for reforming society, industry, schools, and finally humans in a continuous, inside/outside-the-box development cycle. This metaphorical approach to creativity helps us all stop and recognise how little time we dedicate to courageous thinking: an essential element in times of uncertainty.

EXPERIENCE DRIVEN DESIGN

Slightly pregnant

Daniel Formosa
Daniel Formosa Design, USA

RESEARCH AND DEVELOPMENT OF PACKAGING AND INSTRUCTIONAL MATERIALS FOR ORAL CONTRACEPTIVES

More than 10 million females in the United States rely on oral contraceptives. The pill is a highly effective contraceptive, but only when taken correctly. In that regard, oral contraceptives are different from many other medications - the consequences of missed birth control pills can have life-altering effects. Comparing it with aspirin, one mother, not planning on having more children, said with a smile, "Missing a birth control pill *can* result in a headache, but you'll have it for about 17 years."

This paper discusses a range of issues addressed by a team of designers employing research and design techniques to improve patient compliance for oral contraceptives. The team was responsible for developing and conducting research, gaining first hand knowledge and experience on issues pertaining to oral contraceptives and patient compliance. Two related projects were addressed by the team: 1) the design of the pill dispenser and 2) the development of packaging, patient instructions and educational materials. In finalising the physical design, an extensive colour study was undertaken for the dispenser to help understand the impact of colour on perception, attitude and compliance. The methodologies employed by the team varied throughout the project according to the needs. The common bond was the goal of improving patient compliance and the exploration of how various aspects of design can help achieve this goal. This paper will focus on the design of the pill dispenser.

SEXUAL ACTIVITY IN THE UNITED STATES

Teenagers

Levels of sexual activity in teenagers do not vary considerably among the United States, Great Britain, Canada, Sweden and France. Fifty to sixty percent of females have had sex by age 18. By 20 the number increases to approximately 80%.

Differences between the countries can be found, however. US teenagers have the highest rates of pregnancy, childbearing and abortion. They are also less likely to use the pill or other hormonal methods of contraception. A sexually active teenager who does not use contraceptives has a 90% chance of becoming pregnant within a year. Of the 2.7 million teenagers using contraceptives, almost half use the pill.

Compared with other countries, fourteen-year-olds in the US are more likely to have sexual intercourse. The 1.8 million 14-year-old females incur more than

22,000 pregnancies each year. Of the 5.6 million females aged 15 to 17 years there are over 300,000 pregnancies each year. Almost one third of these results in legal abortions. One-sixth results in miscarriages. The remaining 50% are births (Henshaw, 2001).

Adults and teenagers

Approximately half of the 6.3 million pregnancies in the United States each year are unintended, and for either adults or teens, an unintended pregnancy is a life altering experience.

Forty two million women in the United States, approximately 70% of women of reproductive age, are sexually active but do not want to become pregnant. A sexually active female who does not want to have more than two children will need to use some form of birth control for 20 or more years. Because the pill is an "easily reversible" method, it is the method most widely used by women in their twenties.

Birth control "Failure Rates"

Various agencies in the US use the rather unemotional term "failure rate" when evaluating the effectiveness of various methods of birth control. Statistics compare failure rates of "perfect use" patients, who comply flawlessly with the regimen, and "average use", the real-world effectiveness. Among the reversible methods - a category that excludes procedures such as vasectomy or tubal sterilisation - the birth control pill, used perfectly, is surpassed only by the implant in effectiveness. Theoretical "perfect use" of oral contraceptives shows a failure rate of just 0.1%. In practice the failure rate is closer to 6%, a considerable difference.

DESIGN AND PATIENT COMPLIANCE

The history of oral contraceptives

In 1957 scientists at Ortho, a major US pharmaceutical company, began research into hormone treatments that could prevent pregnancy. In 1960 the US Food and Drug Administration approved the pill for use. In 1963 Ortho introduced the first Dialpak™ dispenser. By 1965 the birth control pill became the most prominent, readily reversible method of contraception.

Compliance has always been a factor. The Dialpak™ design addressed usability issues and compliance. It is essential that oral contraceptives be taken one per day, at approximately the same time each day. Hormonal requirements change week by week, in accordance with the menstrual cycle, and dosage is critical. A 28-pill (4 week) package is common, although pills are also offered in 21-pill packages. The last 7 pills in the 28-day pack are placebos, included to help patients maintain their daily routine. Ortho's pill dispenser was unique in that it arranged the pills in a circle, labelled according to day of the week. The pills were attached to a dial that must be indexed, one click for each day, before a pill could be dispensed. At a glance the patient is able to keep track of her daily dosage with a clear indication of whether a pill was taken or possibly forgotten.

Design goals

Based on the original design a new generation of packaging was being conceived. The goal was to further improve compliance. Adherence to the regimen of one pill a day is difficult for some patients, putting them at risk. The "drop-out" rate, the number of patients who for various reasons stop taking the pill, is high.

Another design consideration was environmental friendliness. Instead of discarding the entire pill dispenser every 28 days as previous practice, a refillable "cassette" system was being developed. In addition to the readily apparent ecological benefits, this would allow a nicer, more durable dispenser to be produced. The drawback would be an extra step for the patient to insert the refill.

While an improved dispenser was the original goal, and a challenge in itself, the design team that was ultimately assembled at Ortho-McNeil took on many different aspects of design. They addressed not only the aesthetics of the dispenser, but usability, perceptions, product graphics, instructions and information presented to patients as well as professionals.

Hitting the streets

If birth control pills are 99.9% effective in *perfect use* but only 94% effective in practice, then the solution would be to immerse the team in knowledge about the 5.9% of patients having problems.

Immersion meant hitting the streets. In preparation the design team met with teens, adults, physicians and health care providers, visited family planning clinics and chemists, and looked at other methods of contraception available to the patients. In addition to the extensive information contained in medical journals, the team searched articles in popular magazines to understand exactly what patients, or would-be patients, are reading. We also found that no matter how credible the "official" source, women relied heavily on information from friends. This source was depended upon more than we would consider a comfortable amount, since we found a lot of misinformation being circulated.

The team conducted visits to clinics in tough high-risk urban neighbourhoods, focus groups with doctors, chemists and patients, one-on-one interviews and usability tests to uncover difficulties encountered by women on the pill. Perceptual research measured emotional responses to the proposed designs.

ISSUES ADDRESSED BY DESIGN

Design of the pill dispenser

The original Dialpak™ tablet dispenser, although innovative, was in need of an update. It was clearly an offering from a pharmaceutical company. While this was, of course, the situation, there are obvious disadvantages. Many women would rather keep their use of the pill private. This is true of adults and teens alike. At the same time we wanted to encourage the idea of portability, to ensure that the pill is at hand when required. An obvious, easily identifiable, and pharmaceutical-looking dispenser discouraged keeping the pills within easy access.

The new design encompasses a smaller, more rounded shape, easily interpreted as a cosmetic compact rather than a birth control pill dispenser. The ability to fit in the pocket was important, therefore the reduction in size and the elimination of sharp corners was crucial.

Another "elimination" was the logo. In effort to be less identifiable, the pharmaceutical looking Ortho logo was removed from the outside of the dispenser. The new design accepts a drop-in refillable cassette, and is recyclable. Its 28-day regimen allows new patients to use a "Day 1" start, beginning immediately after the menstrual period, a procedure more harmonious with the female cycle and offering more immediate protection.

Colour perception and compliance

As the design was nearing completion the team turned its attention to colour. While this may seem somewhat superficial to the final design (colour can be less critical in many other products), colour was deemed to be a crucial factor in this product. There are a number of attributes that needed to be communicated visually, and some things that needed to be hidden from view. The team explored ways in which appearance and colour could impact the success of the final design. Preliminary investigations disclosed that "personal" was a highly desirable attribute. Patients viewed the entire topic of contraception as personal - my body, my lifestyle, my private life, my choice, me as a female. The team set out to find a colour that was perceived as "personal" by the greatest number of women. The team needed a colour that would alienate the fewest patients.

Investigations began with a spectrum of colours, including primaries, pastels, metallic finishes and neutrals. Attributes were presented as semantic differentials - posing questions such as "does this colour make the pill dispenser discreet or obvious?" in a way that allowed to users to rate and discuss their perceptions (Oppenheim, 1966). The responses were statistically correlated to their personal preferences. This technique allowed the design team to predict which attributes would be preferable, identifying directions to be further explored. Therefore (again using "discreet" as an example) even if we did not have a "very discreet" colour in our current range, the analyses predicted that this was essential to the patient's personal preference. If we found a colour that was perceived to be "discreet" - whatever that may be - we would be successful. By revealing statistical correlations among various design attributes, the team identified the emotions responsible for personal preferences in design, and to our ultimate goal, those that could help improve oral contraceptive compliance. The final answer was a very warm but very neutral colour.

The colour research leads to another innovation. With any single colour it was possible to be "very" personal to some patients at the exclusion of others, who had different personal preferences. Therefore a personalised *selection* of colours and patterns would be the ideal way to reach our goal. Unfortunately it is only possible to offer one version for mass distribution, keeping costs low and simplifying distribution to the chemists. But on special order a variety of colours and finishes are possible. The idea of personalised pill dispensers was proposed. Our goal was to find six colours and patterns that spanned the greatest number of patients. Colours that would evoke "love it" or "hate it" responses were possible, since the "hate it" group would have five other options. From more than 40 proposed designs, and through a series of quantitative studies into emotional responses that enlisted hundreds of respondents around the US, the team identified six colours and patterns that put the largest number of patients into the "love it" category. These are currently offered as Ortho Personal Paks™, with finishes ranging from gold accent to a leopard pattern.

Research and development of patient instructions

Instructions and educational materials were critical components in the system. Communicating how to start taking the pill, what to expect, what to do if you miss one, two or three pills, side effects, preconceptions and misconceptions greatly affect patient compliance. This work entailed a significant amount of research and development, but is only briefly mentioned here to present a complete overview of the design team's involvement.

Implications for design methodology

Key to the success of this project is the fact that the design team focused on the patients, and specifically on the segment of patients that are most at risk. To meet their goals the design team became directly involved in research and considered all aspects of design in addressing oral contraceptives and patient compliance.

REFERENCES

Henshaw, S.K., 2001, *US teenage pregnancy statistics,* (New York: The Alan Guttmacher Institute).

Oppenheim, A.N., 1966, *Questionnaire design and attitude measurement,* (New York: Basic Books Inc.).

Fit and hit: two experience driven design strategies and their application in real life

Matthijs van Dijk, Rene Konings
KVD, The Netherlands

INTRODUCTION

KVD is a design agency focusing on product and product support, based in Amsterdam.

Design requires and involves a redefinition of design starting points at the conceptual level. It is not about cosmetic redesign. Designing is the investigation of the kernel of things; to challenge what may appear to be fixed or a given and discover new ways of interpretation. This is what will be referred to as 'RE-FRAMING' When thinking about new products it is easy to become fixated on its hardware. The manifestation of a product is one of the tools to express its' identity. The origin however, is embodied in its relationship with the user. It is about the qualitative interaction with the user: the product *experience*. Since we live in a society of continuous change, many products do not/no longer evoke the intended effect with its users: they are/become 'misfits' in time. For products to 'fit' to their context, one needs to understand future needs, desires as well as concerns of users. Using the product design strategy FIT and the product support strategy HIT, fulfils this need.

The essence of these approaches is a definition of the *experience* between a stimulus - be it a product or product support - and the user. The product experience is driven by its specific context; domain & time dependent. The product support experience is driven by concerns people have with a specific product. 'RE-FRAMING' is the essential activity required making that step ahead. It is an investment in your future that requires time. To practice what we preach, we will demonstrate our approach by the case of 'Colourbox' for Akzo Nobel (The Netherlands / Sweden).

RE-FRAMING

Product design is about how people experience products; the qualitative relation between the user (U) and the product (P). The physical appearance is nothing but a means to trigger experiences. Framing new experiences is far more important than the actual design of the product itself.

But what kind of strategy is needed to be able to design that experience? In cooperation with the Technical University of Delft, a framework was developed, based on the interaction between the user and a product (Hekkert and van Dijk,

2000). The elegance of this interaction is that it tells something about the needs concerns and desires of the user as well as about the qualitative characteristics of the product.

FIT product development strategy

To ensure a perfect fit of new product experiences in real life, the interaction is directed by its specific - domain and time related - context. The context exists of e.g. sociological, demographical, technical, political as well as psychological factors (Hekkert and van Dijk, 2001). Out of this interaction new product characteristics can be distilled. These characteristics will act as the starting point to the design process. This is called FIT: an evolutionary, Darwinistic, way of designing (Figure 1).

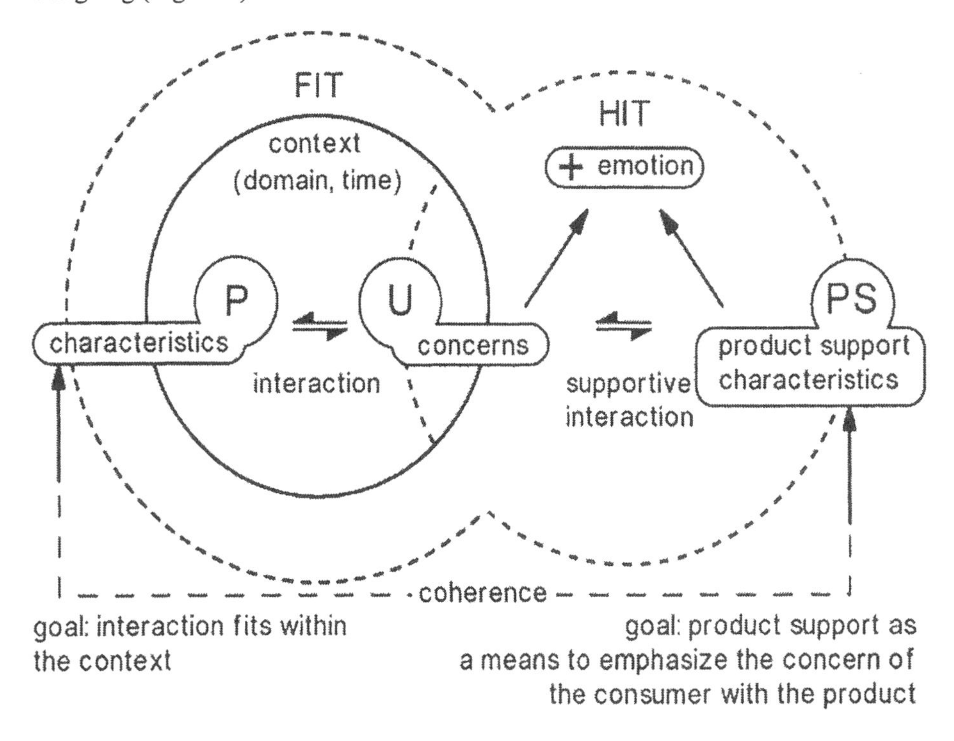

Figure 1 FIT and HIT framework (Copyright Hekkert/van Dijk/Desmet/KVD 2002).

HIT product support development strategy

To get a grip on designing product support there is a second order strategy connected to FIT. It reflects the relationship between product support (PS) and the product (P). To realize this coherence, concerns are distilled out of the interaction defined in the FIT phase. These concerns are the driving forces to develop another interaction, one between the product support (PS) and the user (U). The interaction, between user and product support on one end, and concern on the other, will elicit an emotion (Desmet and Hekkert, 2002). This emotion should be a positive one (Figure 1). Out of the new interaction, qualitative product support-characteristics are being developed with which the design process can now be continued.

Designing qualitative product and communication characteristics, we call RE-FRAMING. Experience often shows qualitative starting points - from market research or client personnel - to be either inappropriate and/or insufficient. Clients and users 'live' in the present and are therefore biased.

COLOURBOX CASE STUDY

Introduction

Sample fans are often used when the colour of a coating is to be selected. In the car repair branch, each supplier of coating materials edits per car brand or per number of car brands one or more colour fans regularly. Consequently, a car repair workshop has an extensive colour fan library. Since the available colour range per car brand usually changes frequently, such colour fans are generally loose-leaf systems, which can be updated when necessary. For this purpose, the available fans comprise of a pin, which can be disassembled consisting of a headed externally threaded pin, and a hollow internally threaded, also headed bush.

A drawback of the existing colour fans is that they cannot be stored in a well-ordered, conveniently arranged way. Also the disassembling of the pin when new loose leaves are to be added or when old ones are to be removed is time consuming. The heads of the pins protrude and may cause damage, e.g. if the fan is laid upon a car to be repaired. Further, the prior art fans easily fan out unintentionally when gripped or carried (Zonneveld, 1999).

Assignment

Design a colour sample-fan/storage-device system. 'Rough', 'strong' and 'basic' in accordance with the physical surroundings of a car repair workshop, may best describe the characteristics of the system. The quantitative requirements of the new system have to be defined by the design agency together with Akzo Nobel Corporate Colours. Each complete colourbox-system consists of 1 storage device unit and 8 sample fans. Refer to Figure 2 for the designing process.

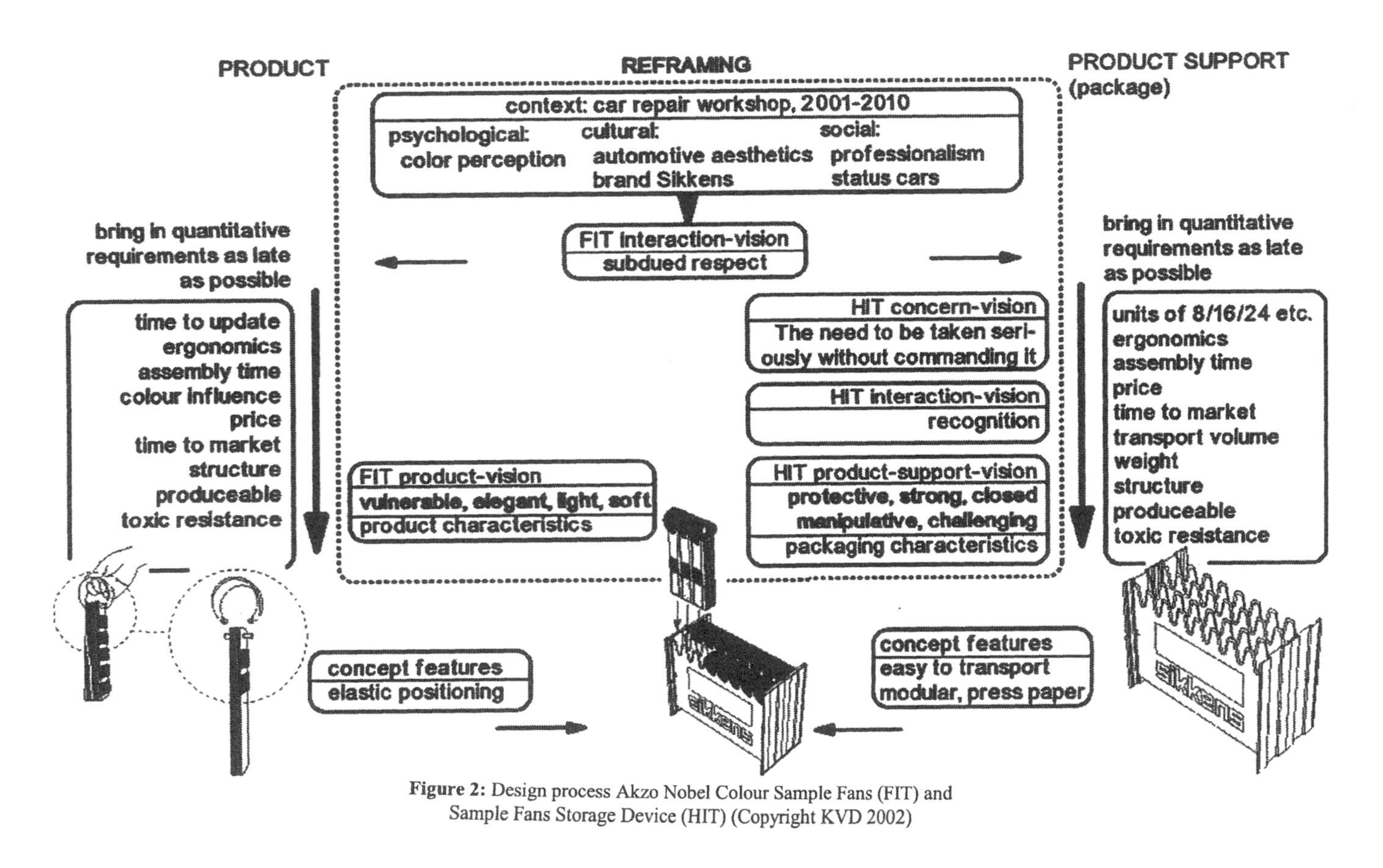

Figure 2: Design process Akzo Nobel Colour Sample Fans (FIT) and Sample Fans Storage Device (HIT) (Copyright KVD 2002)

CONCLUSIONS

Often a design briefing is based on preoccupations of the client. Using the Reframing Techniques FIT and HIT showed that completes other characteristics were needed to accomplish the performance of the Colourboxes with the users. At the start of the design process we therefore recommend to only involve industrial designers.

In the end the client was surprised by the new concept of a colourbox. After some time they understood what now makes the product so special.

REFERENCES

Desmet, P.M.A., & Hekkert, P.,2002, *The basis of product emotions*. In *Pleasure with products: Beyond usability*, edited by Green, W. and Jordan, P., (London: Taylor & Francis), pp. 61-68.

Hekkert, P. and van Dijk, M., 2000, *ViP: Visie in Productontwikkeling* (Vip: Vision in Product design), Internal report, Delft University of Technology.

Hekkert, P. and van Dijk, M., 2001, *Designing from context: Foundations andapplications of the ViP approach*. In *Designing in Context, Proceedings of DTRS 5*, edited by Lloyd, P. and Christiaans, H., (Delft: DUP Science), pp. 383-394.

Zonneveld, H., 1999, *Sample fan and storage device for sample fans*, In *European Patent*, Patent Application No. 99204332.3, Case number ACO 2746 PDEP.

Bridging the emotional gap

Lee Crossley
PDD Ltd, UK

INTRODUCTION

The recent demand is for design teams to make provision for the full experience of the user whilst creating a broadening strategic offering. Gaining an understanding of people for product design is an increasingly difficult challenge. Understanding the more emotional aspects of experience is difficult enough, but eliciting them within sensitive topic areas is quite a challenge. This paper describes an initial stage of a collaborative design research project that immersed a design team into the world of youths for product ideation. PDD, a design innovation consultancy, worked together with youths in the UK and Japan to understand male grooming experiences leading to emotionally focused opportunities.

In designing for experiences there appears to be a gap in the methodologies used for understanding people and their emotions; and between the roles and mindsets within design teams. This paper attempts to demonstrate an approach to increase understanding of these issues within the timescales of a consultancy.

Brief industry overview

Design consultancies are moving beyond the role of outside contractors designing products to client specifications. Instead, they are evolving into collaborators in developing new strategic knowledge that fosters groundbreaking innovations (Sethia, 2001). Clients may now ask 'what should we be creating knowing who we are and what we do?' This is in correlation with people increasingly wanting to spend money on products and services that can help improve their emotional well-being. Design is now less about creating artefacts and more about creating and staging a new compelling story for people to experience. Understanding the stories and roles people would like to play in the future is essential before we consider which products will support these new roles. Stories and tales speak directly to the heart rather than the brain. Businesses need to imagine their offerings; the way good novelists imagine their stories, and the way directors build and craft films (Jensen, 1999). The clients primary objective in the case study was to develop a strategic, emotionally driven connection between their brand and the male youth grooming market.

The world of male grooming is in transition. Men are increasingly caring more about how they look, how they are perceived and more importantly how they feel about themselves. Traditionally, male grooming products have been designed and promoted on a functions and features standpoint: they are technically excellent but do not connect with people on an emotional level. However, today grooming is more about pleasure, and the more sensual aspects of experience. The typical story used to communicate male grooming products is one of masculinity, power and performance. The shift in consumer behaviour had signalled the need for design that understood evolving males feelings, insecurities and desires. Our research aim is to understand how emotional values are generated during everyday, functional activity.

DESIGN AND STRATEGIC 'DESIGN RESEARCH'

The original research question focused on what is 'cool', and what attracts male youth in an emotional way. There are a number of ways the project could have been approached. Firstly taking into account the core design team were mostly young and male, then surely they had a pretty good idea of what was cool, needed and desired for the target audience? Designers start to think about design ideas as soon as they get the brief - bringing with them their own subjective ideas of what would emotionally trigger the user. Often the design team look at existing objects and technology, and build up lifestyle images such as pictures of extreme sports, snow boarding, skate boarding etc. The danger here is to rely on stereotypes and reduce the user to abstract generalities.

The design team can only rely on their own experience. They do not, necessarily, have awareness of knowledge of other cultures or how individuals think and feel about this highly personal topic that is not often discussed openly between men. Without this knowledge design teams may be able to discover what aesthetically attracts this group, but the look of the object is only a part of the story, you need to get beneath the stereotypes. In contrast this project took an alternative approach in redefining the question through design research.

An interdisciplinary approach was taken in the 3-month project involving a concurrent exploration of new insights about technology and its human relevance; new understanding of how people experience things - living, learning, working and playing; and new ways of thinking about business goals and structures that utilised design based processes and communication tools.

THE APPROACH - IMMERSION IN CONTEXT

Research is a creative exploration, but is rarely seen that way. Design research crafts problems in the same way that the design process crafts solutions. This new vision for design research is focused on rapidly immersing the design team into the lives, hearts and minds of people in a short space of time. The challenge for this project was to get young men inspired to tell us their own stories and express their emotions about a mundane functional activity. The next section will describe how this was achieved. Creating a dialogue and encouraging story telling can be affected by cultural differences. The key to user research is building a good rapport and a relationship, both with the user, and with the design team. The research team worked at two levels firstly, a core close-knit team containing two designers and one design-researcher that stayed with the project from start to finish. Secondly a larger multi-disciplinary review team was involved at key stages. The core team was immersed in different cultures and sub-cultures in the UK and Japan over a period of three weeks.

The approach used to explore teenagers grooming worlds was based on a triangulation of tools and methods approaching the subject from different angles. These aimed to discover the past, present and probe possible future grooming experiences and aspirations. Literature in this area does not usually offer much practical advice on integrating a methodology into fast paced-design projects. The first part of discovery was to gain deep insight into people's life context, which involved building a picture of people and their world.

Life context

Understanding people and their life context includes investigating attitudes, aspirations and values that determine how individuals interpret objects, situations, and interactions they meet in their day-to-day activities. By using a technique called character modelling where the team and sometimes the user has a kit with questions, cameras and collages, they are able to frame and understand the lifestyle of the person they are creating for, before the team explores what people do. In this project this stage involved immersion: spending time with family and friends, going shopping, eating meals, to understand their friendships, networks and social activities. This was coupled with building a contextual map, using video, still photos, maps, and cultural items, creating a picture for the team of the person's environment, their town, home, personal space and work to understand how they organise their lives and the stuff around them. Touring participant houses, including their bedrooms, introduces the designers to rich information about people's personalities, their values, lifestyles and possessions. The bedrooms also acted as a personal comfortable forum for them to talk about and express themselves and their desires. The value of understanding life context is that the characters and objects with which people surround themselves are indicators of how they stage and perceive themselves.

Undertaking shopping trips gave us insights into what currently inspires youths for a number of different genres: music, fashion, entertainment, food etc. It quickly became clear that male youth, and people in general seek experiences. It is the experience of discrete communication that fuels the desire to SMS message, but not necessarily to own a phone. Their goal is to look attractive, but not necessarily to own a shaver. As Jensen states "The crown is the object that inspires an individual to wish to be king: the dream is not to own a crown, it is to be king"(Jensen 1999).

Figure 1 Teenager describing his most meaningful objects.
Observing the grooming process in the bathroom

The key to the first discovery stage was to understand youth's life stages; it was the emotional aspects of 'stages not ages' that was an important issue for the team.

Grooming experiences

To understand the present we needed to appreciate the past. Discussing past experiences and exchanging grooming stories with users and their friends and family enabled us to generate an holistic picture of when, how, where and why. Talking to different people in the social circle enabled different views and different characters of the user to be expressed. Many products relating to hygiene and health experiences fall into the 'taboo' category in many cultures because they are related to personal and private activities. Discussing the 'soft' issues of product experience around sensitive topics is outside most males' comfort zone. A number of previous research projects undertaken at PDD had shown there was no language surrounding this subject: males don't talk to other males about grooming. Young males come about the whole activity in an adhoc manner, not sure between them how they started and what was the best technique. The need for education was apparent.

Observation and reflection

Once a relationship had been built with the participants- 'hanging-out' and discussing personal objects in their personal space, it was less of a big step for the youths to let us observe their grooming routine. The value of observing the grooming process was key. Products used, the body language throughout the whole event were captured on video. This would have been more uncomfortable if we had not built up a relationship and understanding beforehand. The value of immersion into a variety of cultures illuminates the global commonalities but more importantly highlights local influences. The core grooming need for young men across cultures appeared to be social acceptance, in a multitude of forms, whether its through complete removal of embarrassing unwanted hair or creating a new image, being attractive to girls, or getting into bars. The emerging issue was that the current products and communication campaigns were not resonating with the emotional needs of the youth market, due to a lack of understanding of the life stages of young males. The opportunity for resonance was for the brand that understood the emotional life stages of young groomers and as a result generated new stories and contexts for them to experience.

DREAMS AND OPPORTUNITIES

After being immersed in the minds and hearts of youth culture, and understanding what roles these men want to perform at what times, the design team were able to create snapshots of experiences. At PDD, if a member of the design team cannot experience people's life context first hand, they have access to an advanced system that digitally archives all the video, visual imagery, notes, transcripts and interpretations captured from the field. The importance at this stage is for the research to inspire and not just inform, through visual maps, storyboards, videos and oral storytelling keeping the 'experiences alive' for discussion and debate between key stakeholders. From the 'pictures of experience' the design team are inspired to generate visions of potential new ones based on 'real' people to help focus their minds for conceptual discovery. From this, ideas were co-created with key stakeholders including youth from the first stage and 'new to the project' participants. The value of co-creating with everyday people allows access to desires and dreams that are not easily articulated in words or actions; they also give validation to the teams ideas and interpretations of their worlds. Creative kits that enable people to explore potential scenarios, opportunities, and activities help bridge the gap for the design team from what is, to what could be. This second phase, which has been written in a follow-up paper, involved using more generative methods to explore people's ideas and desires that led to re-defined design concepts.

DISCUSSION AND CONCLUSIONS

Traditional design/research roles and boundaries are blurring, partial adoption of each others roles builds empathy within a design team. The new initiative has called for a new breed of designer that undertakes a large part of their time as a design researcher. This role focuses on the link between traditional discovery research and conceptual development, evolving and developing new generative methodologies. The new designer-researcher is concerned with constructing and communicating pictures of experience and creating experience scenarios, in a language that is accessible to the creative process. Finding out 'what's cool' is not necessarily the route to innovative creations, and products that will sell. There has been more design than ever in the last few years, and much of it failed. The launch of the Imac has inspired an era of design excess. Robert Brunner of Pentagram states: "So many companies said they wanted to 'Imac' a new laptop or telephone, they saw [it] as a metaphor for colourful, exciting, successful design" (Parks, 2002). Copycatting current 'cool' styles is not enough, and does not touch on 'what cool is'. The case study outlined in this paper could have started off differently, we could have created 'cool' grooming products, but this would not have created a 'cool' experience.

Design research today can help to redefine the big questions. Involving youths in the development process enabled us to redefine this question and re-evaluate what the design team thought was 'cool'. This subsequently had a big influence on what has come out of the conceptual stage. Designing for a broader strategic experience requires understanding every aspect of experience including difficult to access emotions. Designers need better techniques, but nothing can replace first-hand experience.

REFERENCES

Jensen, R., 1999, *The dream society: How the coming shift from information to imagination will transform your business*, (New York: McGraw Hill).

Parks, B, 2002, Cool design wont save a dud product, March 11, 2002, http://www.business2.com/articles/web/0,1653,38671,00.html

Sethia, N (2001) Generating and exploiting interdisciplinary knowledge in design: Product development and innovation in the New Economy, *2001 IDSA Conference on Design Education 'Designing Your Life'*. Available at http://www.idsa.org.

Positive space

Roshi Givechi, Velma L Velázquez
IDEO, USA

INTRODUCTION

We explore motivations that underlie product experience and should be considered when designing products with greater appeal. Our focus is rooted in people's use of and co-existence with products, as opposed to reactions to the brand image.

HUMAN NATURE

People have an innate capacity to experience and respond to their surroundings. It is our nature to experience through multiple channels, dynamically and holistically. It is also our nature to make and use objects, and our selection and use of them reflects our goals, skills, and preferences.

Within the realm of design, people's experiences with products and services are also holistic. Beyond selecting a product that meets a functional need, its appeal matters to us: Does it feel good to touch? Is it prestigious? Do I believe in it? Is it controversial or risqué? Does it signal that we are part of a hi-tech world?

THE POSITIVE SPACE

When experiencing products, people's emotions are triggered through the senses, through knowledge or expectations, as well as through an established connection with a product over time. Our experience with a product evokes an overall emotion or impression. Our association or reaction to it fosters the emotional response we develop toward a product. It is everything about the product, that we call its positive space, which can become more significant than the product itself.

The positive space represents the aura of a product, the sum of its physical attributes plus its intangible essence—or the meaning it hosts for each of its users. It's the space where what the product brings and what people bring meet and interact, forming a new and ideally strong relationship. The intangible essence of the positive space emerges in relation to the user, so it is personal and depends on the person's sensory experiences, memories, moods, preferences, skills, goals, lifestyle, rituals.... It is fluid, it can change with the context of use (environment, time of the day, other people involved). The essence also evolves over time, continually shaped by experiences shared between people and products.

CREATING POTENTIAL IN THE POSITIVE SPACE

When choosing products, people's motivations hold a significant emotional component. Although people choose products for practical reasons to meet basic needs, we believe that people choose products that suit who they are. That is, products with the "right" positive space.

Better designs create potential in the positive space to provoke reactions from people. To design with the intent to satisfy function alone is no longer enough. Better design makes space for a feeling of identity, achievement, inspiration, and joy... While immediate reflexes toward a product can seduce us, it is the overarching experience constructed around a product, and its integration into our lives that enable us to fall in love.

People's relationships with products are as complex as many of the products themselves, compounding the overall connection between the two. However, through our experience in the industry, three areas emerge for creating stronger potential in the positive space: empowerment, delight, and connection.

Empowerment

Successful design empowers, satisfies the arguments of philosophical beliefs, of value systems, and mental constructs. Products that empower extend human abilities and establish a feeling of satisfaction and accomplishment.

Useful and usable products such as Personal Digital Assistants (PDA's) like the Palm V, or the Sony CLIÉ meet these criteria. We might not have known we needed these tools, but once begin to use them, they become indispensable. With the complexities of modern life, they save and organize our information and schedules that we cannot humanly memorize, all in a compact, slender and portable form. They fit into our lifestyles, enable us to feel modern, in control, and "on top of things."

In the same way that PDAs package management tasks with panache, so can headsets. In fast food restaurants, drive-thru attendants are equipped with headsets, enabling them to multitask and move around: they can listen for orders, communicate back to people, and have their hands free for handling money, food delivery, etc. Interestingly, the headsets also give these particular employees prestige and a sense of importance—a welcome "rock-star" appeal—rooted in the association with techno-gadgetry as well as reflecting a higher level of responsibility to co-workers and patrons alike. Capability conveys empowerment.

Electric and hybrid vehicles arm us with a different kind of empowerment—through personal satisfaction of supporting a cause we believe in. For example, choosing products that work to preserve our environment and our global community makes us feel that our beliefs are acknowledged, and enables us a means of living by our ideals every day, even if on a small scale.

DELIGHT

Successful design delights, inspires the senses, surprises, is fun. Delight brings back the child in us, and invites us to play.

Both physically and conceptually, the music box embodies many of these characteristics. Whether it's the fact that the box acquires a purpose beyond storage to contain a miniature dance performance, or the actual music that triggers a memory, the music box tends to evoke a smile. While the music box's magical appeal sparked its popularity and resonance with people, fewer products today seem to offer the simple delights. Perhaps it is product complexity that moves us further away from the pleasurable potentials. Nonetheless, many are trying to trigger a more positive response.

Apple Computer's Powerbook sleep mode indicator exemplifies delight through its behavior. The small LED slit emits a green glow that slowly brightens and dims, mimicking the rhythm of breath as if the computer itself were resting like a person. The little light that personifies the machine is truly endearing...and it's just a piece of plastic! The sleep indicator may not be a pivotal design consideration, yet Apple made an effort to go beyond the baseline and offered us something more—something that inspires our senses.

In a similar way, IDEO Japan designed a CD Player for Muji stores that meets similar criteria. The design affords us something, tempting us to pull on the string suspended below the player, which initiates the CD to spin and music to play. The minimalist design and elementary interaction distinguish it from other players well.

Just as the design of the Muji CD player exceeds mere function, information appliances can too. In NCR's 3 Goldfish and a Computer, intended to provide information about NCR to people waiting in their lobby, giant knob-like buttons trigger the display of digital information on a plasma screen. The information graphics follow the random movement of live goldfish, swimming merrily in the aquarium below the screen. The unique link between technology and nature not only complements their environment, but the intersection of the two provides an element of unpredictability that makes its use and presence more memorable.

CONNECTION

Successful design connects, fulfils the need for identity, association, and community belonging. Through it we feel the comfort and support of a larger group, the freedom to connect with our own identity, or we feel both.

The LOMO cameras depict a phenomenon where the purchase of the product marks one's entry into the LOMO Society, a community of people who share or want to subscribe to a common attitude. New members receive a colorful book of Lomo photographs taken by ordinary people; the camera kit includes LOMO's Top 10 rules, revealing the free-spirited and more spontaneous "shoot from the hip" direction that it promotes so well. Lomographers are encouraged to establish their own identity in the community by sending in their personal favorites, keep connected through the LOMO website, and travel to various LOMO exhibits around the world. Ironically, the various cameras' qualities aren't even that predictable or necessarily good, but the imperfections fit in perfectly with the low-tech appeal and set the foundation for its group of avid collectors.

While the fondness of such objects allows people to loosely connect, cell phones connect people more literally. Their basic functionality and portability allows people the freedom to remain active and still reachable. Nokia cell phones in particular were early to transcend the practical aspect of their product and offer some options for personalization through colour, something other than black or silver. How revolutionary! They provide access to a niche community through their reputation, and are sensitive to the needs for personalization—enabling people to express more of their own personality within the world of cell phone users.

From a project in Latin America, we learned the value a shampoo bottle from a well-known brand can have for low-income consumers. Beyond the practicality of washing one's hair or the aesthetics of the bottle itself, what mattered was that someone could afford this particular brand. The feeling of prestige and link to a higher social status that came from the bottle inspired people to display it as decoration in their humble homes–a symbol and reminder of this association.

HOW DO WE DESIGN FOR THE POSITIVE SPACE?

Now that we have highlighted some elements that make for a stronger positive space, how can we use this information as designers? Below we discuss some of the lessons we have learned.

LOOK CLOSER, FOCUS ON QUALITY

Where do we look? If we consider our definition of positive space, it is everything about the product as well as the connection between a product and the individual that becomes a part of the product meaning. We should understand what the product means to the individual, through his/her verbal and non-verbal expressions, in order to extract cues that inform design. By drawing our focus as practitioners on the subtleties that catch people's attention, we will get much closer at seeing and sensing what can inspire the "right" positive space. By looking closer, we will become better at finding out what matters to people.

We believe that fewer, deeper personal conversations with people provide richer, higher quality insights than gathering data from a larger group at a shallower level. By observing people do things in the context of their environments and providing probes to spawn conversation, we can discover more. Get closer to the people you are designing for. Spend quality time with them. The subtleties based in real experiences cannot be detected from afar. Once we extract insights, it is our responsibility to make new connections between products and emotions.

Make New Connections

If we leverage the more subtle aspects, or look at how the intersection between two disparate products or services can set the stage for fresh responses, we can better design for the positive space. Concept projects, self-initiated internal projects typically based on an agreed theme, offer designers the opportunity to experiment with these new connections and design more freely. Whether the themes include new applications for materials, workspaces or visions of future technologies, the embodiment of the ideas in a clean visual form gives the concepts a sense of real possibility. The concepts themselves provide a way to inspire each other and hopefully the community as well, and consequently serve as a new point of discussion around which designers and product makers can communicate.

Speak the Emotional Language

To better design for emotion, we need to better understand people's emotions in relation to their products. In our practice, we have found that we can interpret emotions more clearly by connecting with people directly, sensing what they express at an intuitive level. Emotions are rich and complex, and a direct experience that allows the essence of emotion to be understood first-hand by designers provides better texture for design.

Prototype the Positive Space

We need to learn about the emotional subtleties that can add to a product offering by prototyping specifically for the positive space. This means prototyping a step beyond functionality to evaluate emotion and discover raw, gut-level reactions. There is value in seeing potential users react to prototypes, as there is also value in allowing designers to sense their own reactions to the design, and gauge whether the experience satisfies them. By prototyping the positive space, we are able to feel the difference by having the closest connection to the matter at hand. Because of real-world design tradeoffs, we recognise the need to negotiate the development of product attributes. However, we hope the focus on the positive space will only enhance what surrounds us and engages us every day.

CONCLUSION

It is our duty to design with emotion in mind. By learning from the relationships people have with objects and their surroundings, we can harness aspects of those connections and promote richer, more meaningful experiences.

Light, emotion and design

Phillip G Mead
AIA Texas Tech University, USA

INTRODUCTION

Recent medical light research suggests a number of new design opportunities for architects, interior and industrial designers. Bright light boxes that approximate outdoor light levels are now used to treat those with depression, eating and sleep disorders. However, the unfortunate side effect is a headache due to glare. As an architect, the author has worked on daylight design options that are less glaring and provide more interesting views than bright light boxes. However, daylighting is limited to certain locations in a building and subject to short winter days. This provides opportunities for both interior and industrial designers to extend daylight hours which allows therapy to be more integrally woven into everyday routines.

Background

The interplay between light and design has been long and fertile. Ancient temple design tracked solar movement to worship sun gods, mark time and dramatize rituals. The Roman and Renaissance architects Vitruvius and Alberti both wrote about window placement to maximize light and warm up interiors (Morgan, 1914; Rykwyrt, 1988). Not until the mid 19th century did medical design advocates like Florence Nightingale (1860) begin to promote the healthful and emotionally uplifting qualities of light in houses and hospitals. She noted that patients on the sunny side of a hospital had higher spirits than those located in darker areas. This observation prompted her pavilion hospital plans consisting of narrow patient wings so light could enter from two sides (Nightingale, 1863). Light and design progressed further when Dr. Finson in 1903 was given the Nobel Prize for proving that sunlight could help cure tuberculosis (Holick, 1998) resulting in hospitals with sunbathing balconies. This was also the dawn of modern architecture from which Frank Lloyd Wright and Le Corbusier began to promote the health qualities of light and air. But the incorporation of light in architecture for health reasons appears to fade after the institutionalisation of a more productive "fat" compact hospital plan with its maze of interior rooms that shorten walking distances for nurses and doctors (Gerllach-Spriggs *et al*, 1998) Additionally, the dissemination of sulphanilamide and penicillin in the 40's replaced the more logistically challenging outdoor light therapies (Becker, 1985). In the mid 70's, energy conservation, and building practices further restricted light and window use in buildings. However, today a new interest in healthy design is rekindled by the affects of sick building syndrome and daylight studies of increased student performance in schools (Heschong, 1999).

LIGHT RESEARCH AND THERAPY TODAY

In 1984 Dr. Norman Rosenthal and his team of researchers from the National Institute of Mental Health, first identified and successfully treated winter depression or "Seasonal Affective Disorder" with high light levels. To counter depression, the research team focused on the use of bright light to suppress the neurotransmitter melatonin, which aids in sleep but causes sluggishness in the winter and the morning. To many, the high intensity light feels like a mild electric pulse (Rosenthal, 1998). Rosenthal *et al.*'s (1987) later studies demonstrated how the lack of winter light increased the appetite for carbohydrates which may compensate for a lack of the "happy" neurotransmitter serotonin (Rosenthal, 1998). Serotonin, a precursor to melatonin, is the target of anti-depression drugs like Prozac and Zoloft.

Parallel to winter depression is Dr. Daniel Kripke's research on sleep patterns and light. His team of researchers at the University of California San Diego showed a relationship between sleep disorders, low light and the dysfunction of the body's internal timekeeper also known as circadian rhythms. In both winter depression and sleep disorders, bright light suppresses the brain's excess of melatonin, which is highest during mid sleep. However, for those who have difficulty waking-up, melatonin levels remain high in the morning resulting in sluggish feelings for several hours after waking.

One possible explanation for the affect of low light levels on depression and sleep disorders may be evolutionary. Until recently, humans spent the majority of the day outdoors with light ranging between 500 to 20,000 lux, but today, as the American Lung Association (2002) estimates, Americans spend 90 percent of their time inside which on average is roughly 10 times darker. According to Dr. Kripke (2001) humans function normally in the wake/sleep cycle when exposed daily for an hour to 1500 to 2500 lux of light However, this exposure is rarely reached even in mild climates like San Diego (Espiritu *et al.*, 1994). Additionally, today's low indoor light standards focus on energy conservation and visual comfort which is three to fifteen times lower than Kripke's minimum recommended levels. (100 lux for homes, 300-lux for classrooms and computer stations to 500-lux for offices (Stein and Reynolds, 2000).

Light boxes and their limitations

Today's high 10,000 lux light boxes are used for 20 minute intervals while older therapies used 2500 lux for an hour. 10,000 lux therapy is quick, but unfortunately glaring because the contrast between the lamp and its background can be a magnitude of 1000. Additionally, the lamps are bulky so they are limited to tables and desks, which restrict use to a few activities. Routines like waking from sleep, bathing, dressing, meal preparation, or reclined reading cannot be easily adapted to the boxes. Additionally, the light box has no view, unlike a window, which can offer a rich variety of views with high light intensities.

DAYLIGHTING POSSIBILITIES AND LIMITATIONS

It would appear that sitting next to a window would produce enough light to suppress melatonin, but light measurements show otherwise. (Mead, 2001) However, there are other ways to accommodate high intensities of light for therapy. The best solution is to be outside which can have far higher light levels than light boxes. However, the outside is often uncomfortable, so to counter these, small peninsula rooms surrounded on most sides by windows and skylights can bring the outside in.

(Mead, 2001). However, furniture placement in these areas must be oriented so that the eyes face the outside and not a dark interior. Outside views also offer a broader range of spatial sequences than the inside (Ludlow, 1976). If scenes are of nature, stress reduction may result as found by Roger Ulrich (1984) of Texas A&M who discovered that postoperative surgery patients who viewed trees from a hospital window, improved recovery time and lowered doses of pain medication over patients who viewed a brick wall.

However, daylighting solutions are limited because high levels of natural light are mostly confined to spaces within a few feet of a window and seasonal variations or time of day are not constant. The farther one is positioned away from the window or skylight, the more drastically the light levels drop (Brown *et al.,* 1992). For example, moving from a position of two feet from a window to that of four feet can drastically reduce the amount of daylight from 60 to 80 percent. Blinds may also be drawn because privacy and sun glare are a common problem.

Timing is another problem. Since the Romans, east-facing windows have been touted as an important orientation for bedrooms...presumably to aid wakefulness in the morning. Additionally, east facing rooms in mental hospitals have been found to shorten stays for patients suffering anxiety and depression (Benedetti *et al.,* 2001). Since light therapy is most needed in the morning to suppress melatonin, the eastern orientation is sensible, but in winter, when waking can take place one to two hours before sunrise, an eastern orientation is moot. Currently, dawn simulating alarm clocks are used to gently awaken, but the light lacks sufficient intensity to feel fully alert.

INTERIOR DESIGN POSSIBILITIES

If a typical morning routine takes place in the bedroom, bathroom, and kitchen, it would make sense to locate an *alternate* bright light switch in these areas. Since all three rooms are relatively small, 2500 lux of bright indirect light could bounce enough light on the walls and ceiling to reduce glare. In a shower or bath, the small confines of a 3' x 5' space could reach much higher light levels than 2500 lux allowing both the water and light to gently awaken the bather. To soften the severity of a sudden shock of bright light, a timing device within the switch could slowly turn-up the light.

In the office, extra bright light could accompany work stations for employees with winter depression or for those who feel sluggish in the morning. Since break areas do not have the worry of glare as much as computer stations, these rooms could be lit to high outdoor levels. French doors to the outside could encourage outdoor use where extra high levels of light and fresh air can recharge the body.

Currently St. Gjorans Hospital in Stockholm Sweden employs light therapy rooms that emit 1500 to 2500 lux of indirect lighting. The rooms are glare-free because hidden florescent tubes are aimed at the ceiling. However, the set-up is fairly elaborate because patients as well as furniture are draped in white sheets to ensure a maximum of light reflectance. This prompts a better solution from industrial designers.

INDUSTRIAL DESIGN POSSIBILITIES

Today's light boxes and visors give the advantage of portability, but glare is a problem. Like St. Gjorans hospital, glare reduction might lie with the integration of light sources with walls and ceilings. If high intensity lamps could shine both on the background (walls and ceilings) as well as the eyes, a marked reduction in glare could result. If such a lamp were placed in the corners of a small room such as a bedroom or bathroom, then the light could easily reflect off several surfaces and soften the contrast between lamp and background considerably.

CONCLUSION

In architecture, interior and industrial design, the integration of lighting with walls and ceilings is a well established principle. Additionally, the position of windows, light fixtures and switches should be strategically placed and woven into daily routines. However, a proper balance must be struck in the timing of when to use bright and dim light. If all rooms are brightly lit at all times (including the evening) prohibitive energy costs and an inappropriate shifting of circadian rhythms could result. This necessitates the careful identification of strategic rooms and the placement of alternative light switches. As with most planning, finesse is a key attribute when integrating light into daily living.

REFERENCES

American Lung Association web page: www.alaw.org/air_quality.

Becker, R., 1985, *The body electric*, (New York: Morrow).

Benedetti, F., Colombo, C., Barbini, B., Campori, E., and Smeraldi, E., 2001, Morning sunlight reduces length of hospitalization in bipolar depression. *Journal of Affective Disorders*, **62**, pp. 221-223.

Brown, G.Z., Haglud, B., Loveland, J., Reynolds, J., and Ubbelohde, M., 1992, *Inside out design procedures for passive environmental technologies*, (New York: John Wiley and Sons).

Espiritu, R., Kripke, D., Ancoli-Israel, S., Mowen, M., Mason, W., Fell, R., Klauber, M., and Kaplan, O., 1994, Low illumination experienced by San

Diego adults: Association with Atypical Depressive Symptoms. *Society of Biological Psychiatry,* **35**, pp 403-407.

Gerlach-Spriggs, N., Kaufman, R., and Warner, S., 1998, *Restorative gardens*, (New Haven: Yale University Press).

Heschong, L., 1999, *Daylighting in schools: An investigation into the relationship between daylighting and human performance*, Report submitted to: The Pacific Gas and Electric Company, August 20.

Holick, M., 1998, Biologic effects of light: Historical and new perspectives, In *Biological Effects of Light 98,* edited by Holick, M. (Norwell, MA.: Kluwer Academic Publishers), p. 15.

Kripke, D., 2001, Phone interview by author, recorded on notepad, Lubbock Texas, May 25.

Ludlow, A., 1976, The function of windows in buildings. *Lighting Research and Technology,* **8**, pp. 57-68.

Mead, P., 2001, Assessment of design configurations for the therapeutic use of daylight, In *Biological Effects of Light 2001,* edited by Holick, M. (Norwell, MA.: Kluwer Academic Publishers), pp. 75-82.

Nightingale, F., 1860, *Notes on nursing,* (New York: Appleton).

Nightingale, F., 1863, *Notes on hospitals,* 3rd ed, (London: Longman and Green).

Rosenthal, N., 1998, *Winter blues*, (New York: Guileford Press).

Rosenthal, N., Genhart, M., Jacobsen, F., Skwerer, R., and Wehr, T., 1987, Disturbances of appetite and weight regulation in seasonal affective disorder. *Annals of the New York Academy of Sciences,* **449**, pp. 216-230.

Rykwert, J., 1988, *Leon Batista Alberti: The art of building in ten books of architecture* (Cambridge, Mass: MIT Press).

Stein, B. and Reynolds, J., 2000, *Mechanical and electrical equipment for buildings*, (New York: John Wiley and Sons).

Ulrich, R., 1984, View through a window may influence recovery from surgery. *Science*, **224** , pp. 420-21.

Textile fields and workplace emotions

Judith Mottram
Loughborough University, UK

INTRODUCTION

The development of the design solution for textile screen and panel coverings for the Axiom Office Furniture Programme is described and the relevance of existing knowledge of colour theory and visual perception is considered. The design process made a conscious attempt to establish the potential for emotional correspondence between end users and the product. Consideration of theories of perception confirms that the design solution may achieve this objective.

THE AXIOM DESIGN BRIEF

The Axiom programme was designed to provide a solution for larger, open-plan office environments as well as the private office. Early discussions with the design consultant identified issues that were to define the aesthetic parameters of the textile component. It was considered important to enable workers to feel 'at home' at their workstations, and to minimise the intrusion of distractions. In keeping with the rest of the programme, the textiles were conceived as an 'added value' product that would touch the senses to reinforce appreciation of functional efficacy.

It was agreed that the textile designs would attempt to harness recognition and memory sub-consciously, to encourage end-users to establish a relationship with their workplace. An extensive colour palette was required, with a range of patterns that would be distinctive in the market. Colour and pattern ranges of competitive products were predominantly focused on primary or 'true' colour, of high saturation and dark tones, with small pattern repeats with distinct tonal contrasts. A twenty-four plain colour palette with four patterns was specified, organised in inter-mixable but distinctive subsets. The product was also subject to particular technical and mechanical specifications. An opportunity was seen to contribute to environmental gain, by controlling the degree of light absorption through colour selection, and by supporting economical fabric cutting through patterns that respected the mechanical properties of the substrates. The patterns had to read 'upside down', where the fabric wrapped up and over the desk screens, and support proximity of computer based work. The product was also required to meet economic targets, flammability standards and 'wear' tests.

THE DESIGN SOLUTION

During the course of the project there was interplay of empirical and theoretical knowledge, intuition and expediency. The following account identifies where existing knowledge relates to design decisions and explores the implications for the potential success of the outcome.

The central axis of the colour palette was conceived as a series of 'rumbling' greys. This absence of colour was seen as a neutral backbone for the palette, to mirror the dominant colours of the urban environment, and to reflect the materials used in the furniture. In the background, the notion of 'medium grey' as a standard was present. This grey, located at equal distance on the grey scale from pure black to pure white, is used in photography as a normative technical standard and has been linked to the use of 'blurring' in contemporary painting (Lang, 1994). The sense that the colour could be like looking at 'nothing' was seen to support the requirement that the palette should not distract the worker from the task at hand, and the reference to distant atmospheric colour was seen to aid optical relaxation. Mid to light tones and unsaturated colours were to be employed for similar reasons. Added benefits of this were the minimisation of light absorption and clear market differentiation. Experiments on office colour preference and task errors (Kwallek, 1996) appear to have established that users generally make misguided judgements on environmental factors that may support productivity.

The palette was based on subtractive colour mixing from a small number of pigments, to produce 'broken', subdued colours. Discrimination from the straightforward black and white mixes of competitor ranges was the motive for a mixed grey, but it was thought possible that discernment of constituent colours might allow for recognition of other associations, and support inter-relationships between the various hues within the palette.

The selection of pigments for mixing the hues included both earth colours (pigments derived from natural sources) and synthetic colours. The intention was for the pigments to reference both the natural world as well as the materials and processes of the technological work environment. Through an iterative mixing process, working with artists' gouache paints, seven pigments were selected: Red Earth, Yellow Ochre, Cyan Blue, Lemon Yellow, Spectrum Violet, Black and White. The cyan and lemon (inorganic synthetic colours) are two of the subtractive primary colours of the CMYK model (Cyan, Magenta, Yellow and Black as used in the print industry), while the red and yellow earth colours (iron oxides) are emphatically of nature. The particular violet used is reminiscent of the purple-dominant blue that Castelli (1992) cites as the 'habitual, persistent and familiar chromatic stimulus' of television.

The extent to which pigment recognition can be determined within mixed hues is unclear, but the human visual system can discriminate millions of colour sensations. Modern ethno-linguistics, physiological optics and experimental psychology have all contributed to our current understanding of the mechanisms of colour perception (Gage, 1999). However, the relative importance of association and sensation is less clear, with chromatherapy and synaesthesia proposing models that do not accord with recently reported genetic differences in colour vision (Riley, 1995). As Riley noted, 'with research pointing to complete individuality in perception, what theoretical models have a chance?'

The colour mixing to establish the Axiom palette exploited the principle of complementarity in the subtractive mixing of secondary and tertiary hues. Whilst the pigments selected reflected the notion of a primary set, each of the generic basic colours was undercut by the addition of its complementary. This procedure effectively utilised Chevreul's laws of simultaneous and successive contrast, providing a restoration of balance from the perception of one colour by the admixture of its opposite on the colour circle (Chevreul, 1864). The objective was to generate a subdued and supportive palette, inherently interchangeable due to the subliminal recognition of the constituent pigments.

Interchangeability was also designed-in through structuring the palette into four main colours, each presented in five different tonal values. Gage (1999) notes that the value-content of hues had been considered to be the primary determinant of harmonious juxtaposition by Ostwald and others, with complementaries of equal value on the grey scale being most pleasing. The Axiom tonal steps do not necessarily reflect Ostwald's intervals of 3, 4, 6, 8 or 12, but this principle of tonal harmony may also be reflected in the predominance of rod cells in the eye and in ethno-linguistic studies of colour names, where light/dark colour terms are the first to appear.

The principles used in developing the patterns included acknowledgement of the need for the eye to rest on infinity at regular intervals when working at the computer screen, the need for task focus, and the objective to make the worker feel a relationship with the environment. The use of abstracted imagery from nature was seen to address optical relaxation and environmental identification, with the strong horizontal orientation of the screens and panels reinforcing the suggestion of landscape vistas. The distracting propensity to track pattern repeats was considered to automatically reinforce awareness of relative distance, thus breaking the allusion to landscape space. To compensate for this, it was determined that the repeat scale should counter easy identification. The 'Holy Grail' at this point was to generate a design that was recognisable at five metres but dissolved into a neutral blur at one metre from the observer.

Familiar but generalised images were needed to balance the removal of the city and the office from the experience of nature. Images of sky, water and vegetation were selected for development on the basis that they provided stable, calming and restful memories of the fundamental attributes of the natural world, and may provide a counterpoint to the multiplicity of words and images within the contemporary office environment.

Photographic reference material was manipulated using 2D imaging software, then exported into Jacquard weave software for further development. The initial selection of source material concentrated upon images that did not suggest a reading of top or bottom, responding to the specific requirements for the textile covering to go up and over the Axiom screens.

Marketing literature produced by the manufacturing company describes the product as giving an 'opportunity to tell a story on the journey through the work place', with meaning 'given from a distance', dissolving 'into abstracted sensation at close proximity'. But the question remains; do the image fields contribute to emotional attunement of users?

VISUAL PERCEPTION AND EMOTIONAL ATTUNEMENT

Various aspects of visual perception theory suggest that the design solution might achieve its objectives, but such theory was not consciously employed during the design process. Gordon's (1997) survey of visual perception indicates that the role of invariants and ratios in the construction of meaning may be of particular relevance. Barry (1997) clarifies the role of unconscious emotional reactions to visual stimuli in relation to perceptual activity, concluding that while allowing response to subliminal messages, parallel cognitive rationalisation develops visual intelligence. These primarily ecological approaches to optics appear to provide useful models with which to consider this project.

Gordon notes Gibson's stress on perceiving as an active process in a complex environment, and considers the emphasis on 'invariants' as his most important contribution to understanding direct perception. As regularities in the relationships of light and texture in the optic flow as we move, or higher-order patterns of change, invariants allow surface and object properties to be extracted. Effectively, invariants enable us to know, at a pre-cognitive stage, about the seen environment. By providing surfaces with visual characteristics associated with the patterns of the natural world, it is possible that the Axiom designs could remind the user of the invariants operational when negotiating space outside the office. We know what it is like to see water close by – we understand the scale of the images relative to us, and how they change as we move past. On the basis of work building on Gibson's theories, it has been suggested that invariant responses were acquired during human evolution, as opposed to learnt in early life. Patterns of stimulation afforded by landscape experience are thus likely to be deeply rooted. By referencing these, location recognition may be stronger than provided by patterns with less association or less accord with light, colour and texture in space. A significant body of work on environmental features that positively influence human functioning has been linked by Heerwagen (1999) to Wilson's (1984) concept of 'Biophilia'. If the invariants offered within the environment reference visual features that are innately attractive to and echo human affiliation with nature, a sense of well being is more likely to be achieved.

Detection of invariant information within the environment brings with it the opportunity for meaning to be ascribed, by virtue of what the sight 'affords' the observer. So if there is perception of the invariant of unmoving clouds, does that mean the viewer could know that it is not windy? What associations might be afforded by such perception? Could calmness, stasis, and safety be suggested?

The direct perception of ratios is central to the concept of invariants and affordances, of viewer to object and object to ground and distance. This suggests that, when vision stimulates association, and the associated form has a fundamental invariant or affordance relationship, the scale and ratios inherent in the first object need to be in step with those of the association. Whilst this concept was not consciously employed during the design process, there was a feeling that there was a recognisable 'correct scale' for the images.

That the images developed into the textiles designs were drawn from nature, with an undetectable pattern repeat, also relates to fractals. Identical repetition is not common in the natural world, but the self-similarity, or fractal development, of the patterns of nature is thought to be conducive to emotional and cognitive functioning (Heerwagen, 1999). These patterns are also generally characterised by the lack of figure/ground distinctions. Gestalt theories of visual perception consider the figure to be attended to more readily than the ground and, although the screens and panel could be seen as figures within the physical environment of the office, they do become grounds at workstation level.

CONCLUSION

A theoretical basis to claims that the Axiom textiles might affect end-user emotions has been demonstrated, although the design process drew on the knowledge base of the design team in a largely unconscious way. Design decisions depended on the assumption that the responses of users would be similar to those of the designers, reflecting shared levels of visual intelligence. It could be suggested that emotion is utilised in all design that consciously addresses specific problems. Theories of visual perception have not established purely physiological or cognitive explanations of visual phenomena; direct emotion and associations are seen to be closely entangled. If we take design seriously, as more than superficial adjustment, we cannot design without emotion. Knowledge is operational, but lived experience may be more so.

REFERENCES

Barry, A.M.S., 1997, *Visual intelligence*, (Albany: SUNY Press).

Castelli, C., 1992, The theory of pallor. In *Notes on color*, edited by Radice, B., (Milan: Abet Laminati), pp. 61-90.

Chevruel, E., 1864, *Des coleurs et de leurs application aux arts industriels*, (Paris: J. B. Baillière et fils).

Gage, J., 1999, *Colour and meaning*, (London: Thames & Hudson).

Gordon, I.E., 1997, *Theories of visual perception*, (Chichester: Wiley).

Heerwagen, J., 1999, Towards a general theory of the human factors of sustainability. In *Proceedings of AIA-USGBC conference: Mainstreaming Green,* Chattanooga, TN, USA, pp. 3-6.

Kwallek, N., 1996, Effects of nine monochromatic interiors on clerical tasks and worker mood, *Colour Research and Application*, **21** (6), pp. 448-458.

Lang, L., 1994, The photographers hand. In *Gerhard Richter*, edited by Lagdira, J., (Paris: Editions Dis Voir).

Riley, C.A., 1995, *Color codes*, (Hanover: University Press of New England).

Wilson, E.O., 1984, *Biophilia*, (Cambridge: Harvard University Press).

From a socially intelligent robot concept to an ad: eliciting audience participation throughout the graphic design process

A Bennett, S Restivo
Rensselaer Polytechnic Institute, USA

INTRODUCTION

A dynamic communication artefact—like a socially intelligent robot (SIR)—that is being designed for mass consumption warrants a participatory design process with a methodological infrastructure that has the potential to engage mass audiences—from conception to production and distribution. The question then is: in which context can we elicit and measure consumer response as an appropriate indicator of the mass audience's positive emotions towards a mass-produced communication artefact (i.e. a consumer product)? One way to elicit a positive emotional response from a consumer to a given product is through the use of a simple ad placed in a commercial venue that is accessible to mass audiences. Traditionally, we use advertising campaigns to market and to sell products to a mass audience. Increased sales of the product directly measure the ingenuity and effectiveness of the advertising campaign. From increased sales, we can also infer that if the audience has been persuaded to buy, then we have indirectly measured the audience's positive emotional response to the graphic design of the communication artefact. However, when we use advertising campaigns just to increase sales for a product that has already reached a final manufactured form, we do not take advantage of advertising's full potential in the design process. Advertising campaigns can be used from conception to production and distribution to gather audience input and gauge the audience's emotional response to the idea of a product before it reaches a prototypical form, to an idea that has reached a prototypical form, and to a final form that has already been manufactured. This paper presents a methodology for eliciting audience participation throughout the creative design process by use of the most basic commercial marketing convention—the ad.

INTERACTIVE AESTHETICS

A design can be conceived in terms of information. In fact design is information. That designs are socially constructed may seem more or less transparent. It is equally the case that information is socially constructed (Restivo and Steinhauer, 2001). Understanding that designs are social in this deep sense suggests that we could be more self-conscious about and committed to the social process of designing. Indeed, usability testing in the fields of technical communication and information technology, for instance, do already engage select members of a target audience in the design process after a prototype has been conceived and designed. However, to make design more social, audience input needs to be gathered much sooner in the design process, from conception. If a product is manufactured for mass consumption, then input from the mass audience is appropriate and necessary. The question then becomes: how can we elicit participation from mass audiences throughout the design process?

Engaging a mass audience in the design process—from conception to production and distribution—may be done through the application of Interactive Aesthetics (Steinhauer 2000). IA states that the aesthetics that we used to design communication artefacts may be used to elicit audience input and participation in otherwise remote design processes if they are made to be interactive. A communication artefact may be socially conceived and designed (by a mass audience) if interactive aesthetics is applied throughout the design process. For instance, an electronic ad may be used to elicit new ideas for products during the initial phase of product conception. It may also be used to elicit the advantages and disadvantages or likes and dislikes regarding an already-manufactured consumer product. In the latter example, the electronic ad can generate feedback via email discussions in regard to the audience's positive or negative emotional reaction to an already-manufactured object. This information can then inform the conception phase of a participatory design process.

Interactive Aesthetics brings the theory and practice of the graphic design discipline into interdisciplinary design processes (from the conception phase) where it is not typically found. When Dwiggins, coined the term graphic design in 1922, he narrowly defined it as a profession that brings structured order and visual form to printed communications (Meggs, 1992). Helfand (2001) defines graphic design broadly as the art of visualizing ideas. However, since all who practice graphic design today are not necessarily artists (or even graphic designers), a modification of this definition of graphic design (that the American Institute of Graphic Design has adopted) is in order. Graphic design is the act of visually translating ideas in order to inform, to mesmerize, to persuade, to instruct, or to elicit input from a target audience. As a result of broadening the definition of the term graphic design, the range of objects that qualify as communication artefacts broadens to encompass not only the conventional graphic design objects (posters, stationery, annual reports) but also a wide range of interdisciplinary unconventional graphic design objects (websites, products, computer software applications, and even robots) that are vehicles for transporting information to mass audiences. n the application of IA during the design of conventional and unconventional consumer objects, there are many opportunities to elicit audience participation throughout the communication design process.

In our on-going iteration of IA, we now conjecture that aesthetic experiences are rhythmic. They are about breathing and the ever-present and on-going vibrations of the (social) body. That is, they are about postures (Schumacher, 1989), and postures are rhythmic. There are already hints of this conjecture in the writings of John Dewey and Lev Vygotsky (see the discussion in Valsiner and van der Veer, 2000: 216-17). This suggests that designs are rhythmic, dynamic, evolving signs, icons, and indexes. This would make the connection between designs and emotions profoundly intimate. Emotions are the rhythms of resolving oppositions or tensions that exist in between humans and humans and also between humans and natural and manufactured artefacts. A design is an element in such an emotional circuit. It is iconic to the extent that it represents what it describes, what it "does" (and then, is this equally true of all designs?). Perhaps we need the concept of graphic de-sign here. As a sign in a relationship of interpretative and aesthetic causative dynamics with the viewer/audience, it is an index. Every time a graphic de-sign is viewed, it may be something different. This is not the case every time for every individual (though this has to be the case in an immeasurable sense) –but it is the case for every target or distinctive audience representing a class, race, sex, age, or gender category. In part, we are inspired in these directions by a reading of the Prague Lingusitic Circle and the Russian semiologists. Applying this concept to the graphic de-sign of a particular object helps to make clear that artefacts in general are information systems (Restivo and Bennett, 2000).

Emotions are generated by synchronizing bodily and speech rhythms (rhythmic synchronization), and the movement of interacting actors into a smooth series of connected ritual interactions (entrainment). The challenge for SIRs engineers – and for the graphic de-sign process as we are reconceptualising it - is to make possible not only the micro-coordination of movements between humans and robots but more intimately between humans and robots as graphic de-signs. Robot and human must be able to get into the same rhythm so that they can anticipate the next beat, move, or stress (rhythmic entrainment; Collins, 1988: 188-208). We now need to amend, or link Breazeal's (2001: 2121) claim about humans being profoundly social to the claim that humans are the most socially rhythmic of all species. Our brain waves fluctuate, our muscles vibrate, our hearts beat, and we breathe rhythmically. Speech is an especially rhythmic activity. Getting into the flow of an interaction means making a series of adjustments that coordinate rhythms. The significance of this for SIRs R&D is that humans can "keep up the beat" by anticipating rather than by pure reactivity (McClelland, 1985). Furthermore, synchronisation is pleasurable. It makes social life possible. Social life is built around social solidarity rituals, and interaction rituals and ritual chains are the micro-level manifestations of our species survival specialisations in social rhythms.

INTERACTIVE AESTHETICS APPLIED

Herein lies a methodology for the application of Interactive Aesthetics to multidisciplinary and interdisciplinary unconventional projects. For example, the design of social and humanoid robots emerged as a major area of research during the 1990s. Breazeal (MIT Media Lab) is one of the pioneers in this field, best known for her development of Kismet, the social and emotional robot. Kismet, an emotional robotic bust, is a robot prototype designed express emotions using facial gestures much the way a real child expresses emotions. Conceived, designed and constructed by Breazeal, the Kismet prototype communicates non-verbally and expresses such emotions as sadness, surprise and anger.

We have been working independently and at times collaboratively with Breazeal on issues, problems, and questions that Breazeal raises in her work on

Kismet. Our previous presentations at conferences co-organised with Breazeal, for example, discloses some of the communication design issues and problems that roboticists face who are intent on mass-producting machines that can express and respond to emotions. In our previous work we address specific questions about the design of socially intelligent robots (Bennett, 2001; Restivo & Bennett, 2000). For example: Why build, or try to build a socially intelligent robot (SIR)? What are the ethical and value issues surrounding the very idea of SIR? What are the design implications of a social theory or emotions applied to social robotics? How can interactivity between the user and a socially intelligent robot prototype inform the creative design of the final form? Is audience-input necessary in the creative design process of a SIR? Now we ask: How can an ad be designed to elicit audience participation in the creative design process of an SIR?

In general, when one design an ad, one visualises verbal information through the use of aesthetics in order to: 1) attract the attention of the viewer in order to persuade him/her to linger for more information; 2) give the viewer a memorable experience. The act of visually translating verbal data entails employing emotive aesthetics – images, graphics, colours, typefaces – from one's reservoir of cultural experiences and knowledge. Herein lies a conundrum. Because we live, communicate in a multicultural global society, it is indeed likely that one can only communicate to an audience that one understands preferable through experience. Therefore, in order to speak to a mass audience that is multicultural, one may need to gather input from the mass audience. We – as designers of future products – can no longer continue to design products that we believe to be necessary for a mass audience that comprises cultures different from our own. We must get feedback from the audience. In the special case of the Kismet prototype that is on its way to members of the mass audience to participate in the design of socially intelligent robots. A well-designed advertisement strategically placed within a commercial venue/context like the World Wide Web has the potential to persuade the audience to participate in the current design process of a dynamic communication artefact. First, we can use an electronic banner ad (see Figure 1 for template of a banner ad) that sends the emotionally aroused viewer to a web page for more information. Then, on the web page another ad like the one in Figure 1 could ask the viewer to participate and to interpret the Kismet prototype. In the template of the second ad, the graphic image is detached from its native environments – the MIT Media Lab. It is stripped of context. Centred in a white space, each viewer much rely on their past experiences, memory, or knowledge to interpret its aesthetics and implied function. This experimental advertising research project (located at the www.rpi.edu/~bennett) proves that the interpretation of visual language – the way that the designer intends – depends on the audience's cultural experiences. It will inform (and inspire) interdisciplinary creative design projects where researchers from non-design disciplines collaborate with a graphic designer and the target audience on the creation of mass-producible communication artefacts.

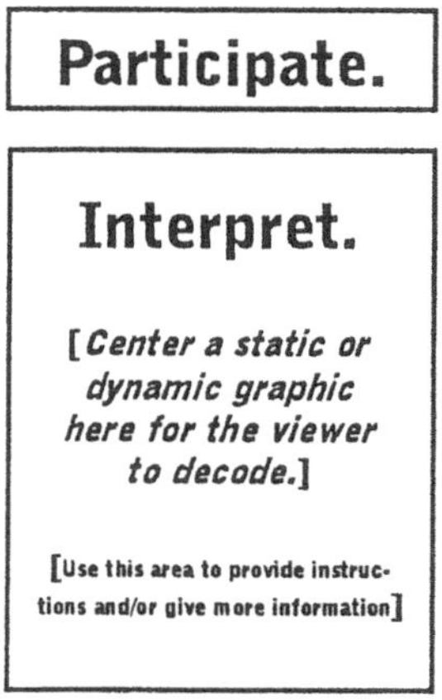

Participate.

Interpret.

[*Center a static or dynamic graphic here for the viewer to decode.*]

[Use this area to provide instructions and/or give more information]

Figure 1 Template for electronic banner ad and ad for webpage that elicits audience input

REFERENCES

Bennett, A., Creatively Designing Socially Intelligent Robots. In *Proceedings of the IEEE International Conference on Systems, Man, and Cybernetics*, Tucson, Arizona (October 7-10, 2001), pp. 2118-20.

Breazeal, C., Socially Intelligent Robots: Research, Development, and Applications. In *Proceedings of the IEEE International Conference on Systems, Man, and Cybernetics*, Tucson, Arizona (October 7-10, 2001), pp. 2121-26.

Collins, R., 1988, *Theoretical Sociology* (New York: Harcourt Brace Jovanovich).

Helfand, J., 2001, Screen: Essays on Graphic Design, New Media, and Visual Culture (New York: Princeton Architectural Press).

Restivo, S., Bringing Up and Booting Up: Social Theory and the Emergence of Socially Intelligent Robots. In *Proceedings of the IEEE International Conference on Systems, Man, and Cybernetics*, Tucson, Arizona (October 7-10, 2001), pp. 2110-17.

Restivo, S. and Steinhauer, A., 2000, Toward a Socio-Visual Theory of Information and Information Technology. In *Proceedings of the IEEE International Symposium on Technology and Society*, Rome, Italy (September 6-8, 2000), pp. 169-175.

Schumacher, J., 1989, *Human Posture* (Albany: SUNY Press).

Steinhauer, A., 2000, Transcending Space and Culture: Interactive Aesthetics that Facilitate Remote Participation in Graphic Design Processes. In Adjunct Proceedings of CoDesigning 2000, edited by Scrivener, S.A.R., Ball, L.J., and Woodcock, A., (Coventry: Coventry University), pp. 155-160.

Valsiner, J. and van der Veer, R., 2000 *The Social Mind* (Cambridge: Cambridge University Press).

This article has originally appeared in the Journal of Design Research 2002, Vol 2. Issue 2, http://jdr.tudelft.nl

The branded hotel: an educational experience

Ed C R Hollis and Paul A Rodgers
Napier University, UK

INTRODUCTION

Brands and trademarks have existed for almost as long as organised trade itself. However in recent years, particularly in Western culture, consumers use brands as a powerful way of defining and expressing their own identities. Brands have a cerebral dimension which is the reputation they enjoy in the minds of the consumer and, therefore, they must engender trust, loyalty, and quality if they are ultimately to be successful (Williams, 2000).

In today's market, it is no longer sufficient to design functional products or services. Consumers demand pleasurable and/or exciting experiences. In this climate, the relationship between brand and product is blurred. As Steve Hughes, director of PSD, puts it: *"Now people are recognising that the brand is the product and the product is the brand"* (Redhead, 2001).

Like the product-brand shift, the hotel has changed as a medium. The most successful hotels now address the desire for a total experience – be it the fantasies in the deserts of Las Vegas, the urbane wit of Starck's interiors for Ian Schrager, or even the still expanding Disney empire. The hotel is no longer a building type, but the manifestation of a desired experience or imaginative world (Moore, 1999).

Today, designers are faced with the task of treading the thin line between the total experience and the themed experience. In respect of this trend, the authors set a design brief to 24 pairs of third year undergraduate Design Futures and Interior Architecture students of Napier University. The pairs were chosen randomly by lots and the gender ratio was split evenly at 50:50. The brief, "The Branded Hotel" scheme, based on Sall and Colin's RCA project, set out to explore the relationship of branding and design in several ways (Sall and Colin, 2001). The site chosen for the location of the hotel was the former General Post Office (GPO) building in George Square, Glasgow.

Firstly, the student pairs were asked to reflect on what qualities their selected brand communicates, how these qualities are communicated to consumers, and how might these two dimensional qualities translate to the function of a 3D physical hotel experience. These values were then applied to the design of a "branded hotel", starting with the spaces of entry and rest. That is, the reception and the bedroom.

Secondly, students were asked to design a "branded object" for the bedroom and consider how the collision of an everyday hotel object (*e.g.* kettle, iron, trouser press) and their chosen brand might extend the brand's two dimensional values into three-dimensional materiality.

Thirdly, the students were requested to represent (physically) in a three dimensional model of the site landscape, the two dimensional values of the brand they had been allocated. They were asked to consider how this may change aspects of the landscape such as the hotel's façade, the site, the traffic, the circulation around the buildings, and the experience of approaching and entering the hotel.

In summary, the role of the designer has radically altered over the consumer/leisure revolution of the last twenty years, and design education, traditionally focused on form, function, and construction as autonomous exercises, needs to respond to. If designers do not engage with the imagineers of total experiences, they will be usurped by them. This student project is one such attempt to both engage and challenge the status quo, and to prepare students for the real tasks that they will face ahead.

WHAT BRAND AM I?

We buy brands, once our basic needs have been satisfied, because we are tribal by nature and enjoy acquiring and showing products off. Products, be they trainers, jeans, jewellery or underwear act as markers of tribal belonging. Brands are a way through the maze of products which promise certain qualities both physical and psychological (Lawrence, 2001).

Figure 1 What brand am I? student examples.

With this in mind, the "Branded Hotel" project kicked off by asking each student to interrogate their partner's branded profile by asking what brands are they wearing? What brands are they smoking? What brands are they listening to? The rationale behind this exercise was to explore the reasoning that students use when making consumer choices and to understand what their choice of brands said about them as individuals.

The student pairs were asked to consider what they thought their partner was trying to say with their choice of brands on show (*i.e.* jeans, trainers and cigarettes). They were then asked to draw or photograph each other and use it to communicate their brands and what they said about them (Figure 1).

BRAND VALUES

The next part of the project required student pairs to clearly define the values of the specific brand that they had been assigned. They were asked to present an overall description of the brand and consider what qualities the brand communicates? How are the values of the brand communicated? How might these two-dimensional values manifest themselves into the function of a hotel? Figure 2 shows the response of one of the student pairings involved in the project.

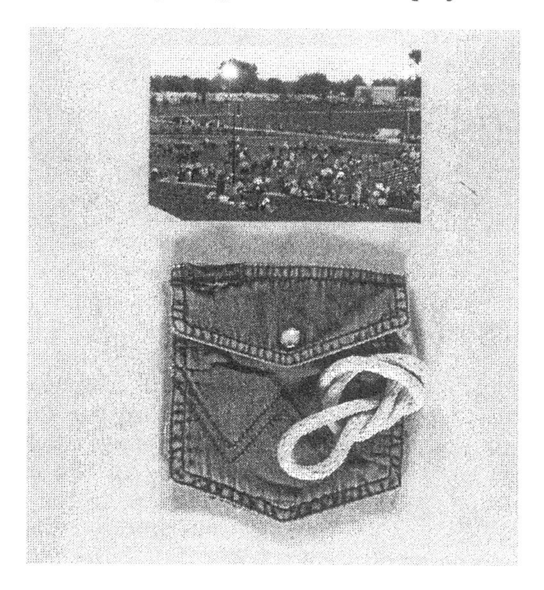

Figure 2 Brand values presentation board example (*WRANGLER*).

ENTRY AND REST

The next part of the "Branded Hotel" project witnessed the break up of the Design Futures and Interior Architecture student pairs. Individual Interior Architecture and Design Futures students were each given specific tasks to complete relating to the design of their "Branded Hotel" (Figure 3). However, collaboration and communication during these individual tasks was encouraged.

The Interior Architecture students were asked to design the reception of the hotel, and asked to consider how the reception expresses the brand of the hotel? What (and how) does it imply about the hotel within? How does the reception address the street outside? The Design Futures students were given the task of designing one hotel bedroom and bathroom of their "Branded Hotel". They were asked to reflect on how the room with bathroom expresses the brand of the hotel? What does it imply about a night in the hotel? How does the bathroom relate to the reception? Finally in student pairs they were asked to negotiate the planning of the room and the reception to make a coherent experience which expresses the brand of the hotel overall.

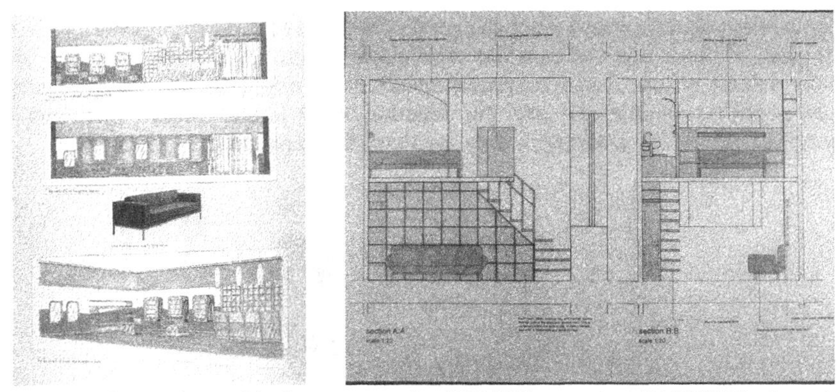

Figure 3 Reception and bedroom examples (*VIRGIN reception, EASYJET bedroom*).

HOTEL AND OBJECT

The next stage of the project was again conducted independently. The Interior Architecture students were set the challenge of designing the "Branded Hotel" containing 15 bedrooms, a restaurant, a bar, a reception (which had already been designed), and appropriate service spaces within the site designated to you. Students were asked to provide plans and sections of the hotel as a whole, concentrating on organisation and clarity of circulation and considering how these express the brand of the hotel? Contrastingly, the Design Futures students were requested to design a "Branded Object" which might be used in the hotel r oom such as a mini kettle, a hair dryer, or a trouser press. The goal being to represent, physically, in a three-dimensional model the two-dimensional values of the brand they had been allocated. Figure four shows an example of two of the "Branded Objects" designed by the Design Futures students.

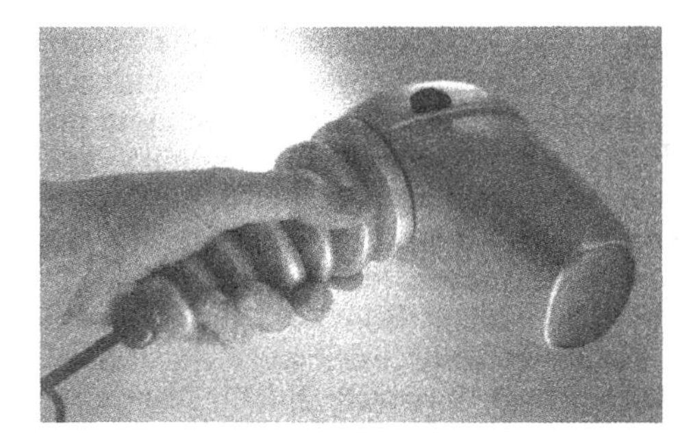

Figure 4 Branded object examples (*NIKE hair dryer, MINI picnic hamper*).

LANDSCAPE AND THE JOURNEY

The final part of the brief asked the students, again working individually on independent projects, to design a journey from the starting point of the reception area to the end point of the bedroom (Interior Architecture students). Here, the students were asked to convey the conceptual framework of the brand values of the hotel in a physical form (*i.e.* plans and sections of the hotel).

The Design Futures students were set the task of amalgamating their particular brand and its values with the immediate landscape surrounding the Hotel at George Square, Glasgow. The purpose of this task being to extend the brand's values into new territory (*i.e.* the façade, site logistics, and local context), considering how the two-dimensional brand values materialise three-dimensionally? And how the landscape, the local setting and infrastructure might be affected by the manifestation of the brand's values? Figure 5 illustrates two examples of the "Branded Landscapes" designed by the Design Futures students.

Figure 5 Branded landscape examples (*GUINNESS, HILFIGER*).

In summary, both sets of students found the fifteen-week project to be both interesting and demanding. The gradual build up of the project starting from two dimensional brand values through to representing the "Branded Hotel" in three dimensions initially posed difficulties for some students. However, as the results show the students eventually produced very exciting and successful solutions to the challenges posed. The students have shown that they have engaged and challenged the traditional design education focus on form, function, and construction and embraced the notion of "total experience" in their proposals which should prepare them well for the challenges that they will face in the future.

REFERENCES

Lawrence, F., 9 July, 2001, Me and my labels. *The Guardian G2*, pp. 8-9.

Moore, R. (Ed.), 1999, *Vertigo: The Strange New World of the Contemporary City*.(London: Laurence King).

Redhead, D., 2001, *The Power of 10*. (London: Laurence King).

Sall, C. and Colin, K., 2001, City symmetry. *Blueprint*, **138**, pp. 76-79.

Williams, G., 2000, *Branded?* (London: V&A Publications).

Emotive communication using mobile IC devices

Roland Keller
Siemens ICM MP P TI3, Germany

INTRODUCTION

In the 14/2002 edition of "Interacting with computers" (Elsevier Edition), Gilbert Cockton writes that the statement "Cogito ergo sum" – "I think therefore I am" by the seventeenth century French philosopher Descartes should be amended to read "I feel therefore I am". This is on the basis that human existence is no longer proved by our intellectual prowess but by our emotional aptitude. Cockton uses this to appeal to developers and designers to take more account of the emotional aspect of human-machine interactions. Nor is Cockton alone in voicing this opinion. At present, computer systems interact with users in ways that do not allow for the complexities of naturalistic social interaction (Scheirer *et al.*, 2002). There is an increasing awareness of the importance of emotions in human communication interaction (HCI).

No one doubts that a user interface on a mobile phone, for instance, should be designed to take users' emotions into account in human-machine interaction. In this case that means: When language is used by a person to interact with a machine, the rules for person-to-person communication apply (Rubinstein & Hersh, 1984). Reeves and Nass (1996) argue that human-machine interaction is inherently natural and social, so that the rules of human-human interaction apply to human-machine interaction (Picard & Klein, 2002). According to this, the same rules apply to communication between people and machines as to communication between people. So HCI designers endeavouring to teach machines how to deal with emotions can learn from what is known about human communication in order to put more emotions in the human computer interface. The question arises as to what specific aim is being pursued by human-machine interaction and what the purpose is of investigations in the human-machine field.

We can assume that human-machine interaction is not an aim in itself but serves a superior purpose: to improve, or at least support, communication between people. It is not interaction between people and machines that is of existential significance to humans but the interpersonal exchange. So after Descartes' "I think therefore I am" and Cockton's "I feel therefore I am", the statement "I communicate therefore I am" appears to gain more credence. For we must not lose sight of the fact that machines are there to assist us. To assist us in what way? What is a human's need out of life but to develop, live, and flourish, individually and as part of a larger society? Moreover, as these needs are identified, yet are known to go unmet due to a paucity of human support, what role can and should computational technology play in helping to address and even satisfy those needs? (Picard & Klein, 2002).

Construction machines are used to construct, cars serve to move us from one place to another, and a communication device is used for communicating. Communication devices – and hence also the HCI – are therefore aids to communication between people. And we endeavour to design these aids to allow the best possible direct human-human communication that is not linked either to time or to specific locations.

PROCESSES IN HUMAN-HUMAN COMMUNICATION

Communication can be defined as a process of understanding between people, between humans and machines, or between machines. Communication serves to exchange information on either a unilateral or a reciprocal basis. Prerequisites for communication are a sender who generates the information, a transmission system for transporting the data, and a recipient who receives the information. A mutually intelligible language is also an essential requirement for communication. The content of a piece of information does not correspond to that given it by the sender but to the response it triggers in the recipient.

A process defined by communication theory takes place before any information can trigger a response in the recipient. Given a mutually intelligible language between sender and recipient, the recipient will perceive the information using the five senses we humans have (sight, hearing, smell, taste, touch). Perceived information is filtered by the recipient. The filters employed include experience, education, individual values, and personal character. The information is then assessed on the basis of the emotional world felt by the recipient. Only now can the recipient react and in turn become a sender. As this process takes place within the recipient and is beyond the sender's control, all senders can do is convey their message in such a way that, on the one hand, as many senses as possible are appropriately stimulated in the recipient and, on the other hand, that they make as much additional information available to the recipient as possible. It is important for the recipient when interpreting information to be aware of the particular emotions that have been instilled in the information by the sender. Irony, for instance, is very difficult to convey using a solely text-based medium as certain additional information is missing. One example: the sentence "I can't do that" can be an apology, a reproach, a question or an appeal, or it can be meant ironically, depending on stress and intonation. This accounts for the very quick appearance in the world of text chatting of emotionally expressive symbols (emoticons) to extend the otherwise limited scope for expression. The written word can be given an ironic coding by using the "winking eye" emoticon, as in e.g. "I can't do that ;-)".

In direct human-human communication, the possibility of emotionally charging information is determined by the human physiological premise and the user's psychological state. But if location or time differences between communicating parties mean they use technical equipment as a conveyance, the limits of emotional enriching are set chiefly by the means employed: the communication device.

We are already familiar with one monosensory aid that lets us convey additional auditory information perfectly, namely the telephone: it allows acoustic signals to be sent in a way that enables the addressee to make a relatively unambiguous assessment of the emotional aspect of the information received. Additional information can be attached in the variable stressing of the basic information, for example. As the telephone is restricted to the audio channel, it seems obvious to add another one, such as video. Video telephones would make it possible to convey both auditory and visual additional information. However, for various economic reasons the video telephone has never managed to establish itself as a mass market product – although communications theory would indicate this to be a logical extension of the telephone.

TYPES OF HUMAN-HUMAN COMMUNICATION

What has been said about communication processes shows that the recipient of information has to rely on receiving it with additional, emotive information. What now remains to be examined are the possibilities the sender has of conveying additional, emotive information. Independently of the sender-recipient model there are several variable parameters in communication. For example, we have to distinguish whether people are communicating with or without an aid (type and nature) and know who is communicating with whom (number of persons involved). Furthermore, information can be received immediately after being sent or with a delay (synchronous, asynchronous) or after a change of location (mobile). The sender's state of awareness is also an important factor (implicit, explicit).

Type and nature	- Direct, immediate (without any aids: I'm talking to you) - Indirect, through a medium (with an aid: I'm phoning you)
Number of persons	- One (human-machine: I'm dialling a phone number) - Two (between two people: I'm talking to you) - A group (among several persons: I'm talking to each of you)
Time	- Synchronous (the recipient receives the message the same time it is sent: we are on the phone to each other) - A synchronous (the recipient receives the message with a delay: you retrieve messages from your voice mailbox)
Awareness	- Implicit (unconsciously: pause for effect while talking.) - Explicit (consciously: clearing your throat while talking.)

As by definition it precludes the use of aids, direct communication will be set aside in the following. Besides, human-machine communication is not the main focus of what we are considering here for the reasons given. In indirect, interpersonal communication the remaining parameters can be freely combined.

The implicit conveying of emotions has so far only been applied to niche areas (psychiatry, for example). Users' fear of being under surveillance has impeded not-medical or not-legal applications for the conveying of emotions implicitly. It is also expected that concepts for conveying emotions synchronously will be transferable to asynchronous conveying. So the obvious approach is to concentrate on the explicit conveying of emotions (additional information) in indirect (here using IC devices), synchronous communication between two people. It should also be possible to use the solutions for mobile applications.

EXAMPLES AND PROSPECTS

Research has long been conducted into the possibility of relocating communication to virtual space and imaging reality there on a 1:1 basis. It means attempts are also being made to convey emotions such as facial expressions, attitude, gestures, etc. It involves representing persons existing in reality with fictional figures (avatars) in virtual space. Being freely configurable, these synthetic representations can be partly distinguished from the people who actually exist through different characteristics or modes of behaviour.

It is an approach involving indirect, synchronous, two-person communication with additional information being released explicitly: the conveying of emotions has to be consciously released by the sender before these emotions can be reproduced in virtual reality. But the relocation of interpersonal communication to virtual space is hindered by both the very sizeable portion of various sensors and effectors and the number and extent of necessary technical devices that are not capable of mobile use. The requirements placed on research are therefore to develop solutions that will enable users to convey additional, emotional information on a mobile basis making use of as many senses as possible. U sers should be able to manage with as few sensors as possible, with all the necessary elements being integrated in one device. It must be noted here that every sensory organ can perceive additional information and so convey emotions.

Example 1 (human-machine)

As mentioned earlier, developers are now more sensitised to emotions in human-machine communication and some concepts have already been turned to practical account. In most of the concepts that have been implemented, emotions are incorporated in the form of personalizing or individualizing the device. Below are some examples arranged according to sensory organ:

Audio - Ring tones that can be produced individually or selected according to personal preference. Ringer signals can be user-defined.
- Context-sensitive advisory tones. Advisory tones can be produced differently depending on the ambient conditions.
- Speech quality. The speech quality of the mailbox function and of audio navigation is geared to the user.

Optic - Interchangeable faceplates. Changeable covers to suit personal taste.
- Display quality. Realistic imaging by high quality display.
- Screen saver. Graphics displayed by the device in sleep mode.
- Wallpaper. Graphics displayed in the background.
- Visual advisory signals. Signals that can replace and/or supplement the ring tones depending on the particular situation.

Haptic - Device material. The material contributes to the devices emotive value.
- Input elements. By their position, material, and form, the input elements affect how the device is emotionally rated.

Olfactory -Aromatic ringing. Ringer tone that can be smelt.

Example 2 (human-human)

The real challenge is to find answers to the question: How can emotions be conveyed with the aid of devices? In contrast to approaches relating to human-machine communication, there are very few concepts for explicitly conveying emotions in indirect, synchronous communication between two people that have been implemented and can also be employed on a mobile basis. Below are a few examples arranged according to sensory organ:

Audio - Voice-defamiliarising filter. Filter allowing a voice to be changed.
- Background acoustics. Sounds and melodies played in the background of a conversation.

Optic - Mood display. Display of a person's moods using colours.
- Emoticon. Graphic elements composed of keyboard characters that are used in text-based communication (chat text).

Haptic - Vibration signals. Movable element that reproduces the sender's mood.
- Travel-path feedback: the further an input element is moved from its initial position, the greater the effort needed to advance it further.

Olfactory - Smell signals. Ringer tones in the form of an aroma.

CONCLUSIONS

Based on the assumption that interpersonal communication will continue to have a very high rating in our society in the near future, it can be stated that emotional information added to that which is being conveyed serves as an aid to understanding it (decoding or assembling information fragments). This additional information, which can make use of all human sensory organs, is critical for the recipient's error-free interpretation of the information. A consequent requirement placed on innovative devices of the future is to incorporate more perception channels in communication and so give the sender of information an opportunity to enrich it with additional, emotional information.

We know of two different approaches to effecting innovation: technology- and user-driven. In the area of emotional communication, or the explicit conveying of emotions in indirect, synchronous communication between two people, the state-of-the-art makes a sufficient number of useable technologies available. But there is a lack of innovative concepts.

Product and interface designers have put more effort over recent years into incorporating the aspect of emotionality in human-machine interaction. However, human-machine interaction is only an aid to interpersonal communication, and the question "How can emotions be conveyed with the aid of devices?" remains unanswered.

REFERENCES

Cockton, G., 2002, Yes/no or maybe – Further evaluation of an interface for brain-injured individuals. *Interacting with Computers*, **14**, pp. 341-358.

Fernandez, R., Klein J., Picard, R. W., and. Scheirer, J. 2002, Computers that recognise and respond to users emotion. *Interacting with computers*, **14**, pp. 93-118.

Hersch, H.M. and Rubinstein, R., 1984, *The human factor,* (Digital Press).

Klein J. and Picard, R. W., 2002, This computer responds to users frustration. *Interacting with computers*, **14**, pp. 119-140.

Nass, C. and Reeves B., 1996, *The media equation: How people treat computers, television and new media like real people and places*, (Cambridge: Cambridge University Press).

GENERATIVE TOOLS

Generative tools for context mapping: tuning the tools

Pieter Jan Stappers
Delft University of Technology, The Netherlands

Elizabeth B-N Sanders
Ohio State University and SonicRim, USA

INTRODUCTION: GENERATIVE TOOLS

Recent directions in design require designers to become more and more aware of the user's experience, emotion, the situation of product use, and social and cultural influences. Designers need insight in the diverse contexts surrounding a product's use, and especially within the field of participatory design, a number of techniques have emerged to more widely explore the user's life than had been customary in traditional, function-centred design.

Among these techniques are the cultural probes (Gaver *et al.*, 1999) and the generative tools pioneered by SonicRim (Sanders, 1999; 2000; Sanders and William, 2002), in both of which respondents are asked to make designerly artefacts to express aspects of their situation, their life, their worries and joys, etc. For instance, respondents are given a 'toolkit' of words and images and asked to make a collage expressing good and bad aspects of their home or work situation. These collages are then used for inspiration by the design team and (in SonicRim's generative tools) the respondents also present their collages to each other. These presentations carry much information that may not be directly apparent from the collages. The collages and (especially) the transcripts of the presentations are analysed using elementary statistical methods, such as counting the co-occurrence of images and words. More sophisticated analyses, such as using multidimensional scaling to reveal the patterns in chosen images and words, can also be performed.

In the design development process, generative methods such as collaging can be used together with other methods in a converging perspectives approach (Sanders, 2000) that draws simultaneously from three perspectives: marketing research *("what people say")*, applied anthropology *("what people do")* and participatory design *("what people make")*. When all three perspectives are explored simultaneously, we can understand the experience domains of the people we are serving through design. When we bring these people through guided discovery and give them the generative *make* tools, we have set the stage for them to express their own creative ideas.

Generative techniques mentioned above are extremely rich sources of information. However, the statistical work that is involved in analysing the sessions is laborious and mind-numbing. In joint research of TU Delft and SonicRim we try to validate assumptions about how toolkits should be constructed, and to optimise the way in which the resulting data are analysed, both in making it less cumbersome and more rich in analysing the structure of collages and presentations.

We describe a series of small experiments in which respondents were asked to make a collage expressing their 'home' experience (a task for which data exists now from Europe, the USA, and Asia). Between the experimental conditions we varied (1) toolkit imagery (pictures versus abstract shapes) to test the assumption that pictures lead people to express emotions and memories, abstract shapes lead to diagrams expressing processes; (2) structure of the artefact (picture collages versus verbal mindmaps) to gauge the influence of pictorial information on the resulting artefacts and presentations; (3) the medium (pasting pictures and words on paper versus arranging a collage using computer software), in order to see whether computer tools lead to richer or less rich artefacts and presentations. The last question is of great practical importance, as using computers has the promise of facilitating the statistical analysis in many ways, but runs the risk of stifling people's creativity, as is often found in creative design tools (e.g., Stappers & Hennessey, 1999).

EXPERIMENT

The experimental sessions took the form of a condensed-style generative tools session. Participants were students of Delft University of Technology who did not have collage-making techniques as part of their curricula (i.e., not students of Industrial Design Engineering or Architecture, for whom collage-making is a formalized skill). In the sessions, each participant made an artefact to express their "home experience: past, present and ideal", then presented his or her creation to the other participants. One week before the session, participants received a small diary workbook with questions and exercises about their current living situation; this was done to sensitise them to the topic, i.e., to set them reflecting about the home experience before the session. Session leaders were students of Industrial Design Engineering carrying out the sub-experiments as part of an introductory research methods course.

Overall experimental design

The series of experiments compared groups of participants creating and presenting expressive artefacts in one central, and three differential conditions. Because the central condition was so important, two sessions were held for this condition. For each of the differential conditions, only a single session was held.

The differences between the conditions for the independent variable lay in the material they used to make the collages. In the analysis we looked at the use that was made of these materials in the artefacts, the form of these artefacts, and most importantly, the content of the presentations. Figure 1 gives an overview of the design of the experiment; Figure 2 shows examples of artefacts created in the sessions.

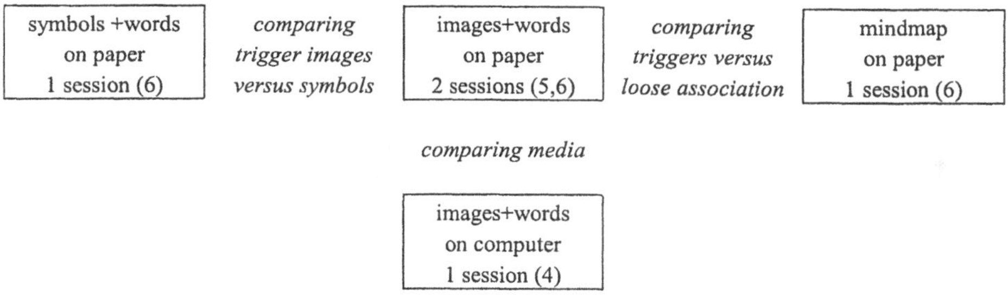

symbols +words
on paper
1 session (6)

comparing trigger images versus symbols

images+words
on paper
2 sessions (5,6)

comparing triggers versus loose association

mindmap
on paper
1 session (6)

comparing media

images+words
on computer
1 session (4)

Figure 1 Conditions in the experiment, and the number of sessions that were held in each condition. The numbers in brackets behind the sessions are the number of participants in each session.

In the central condition, participants received 125 images and 108 words, which served to trigger associations. These triggers were given to them on sheets of A3 size paper. The images were chosen to cover feelings, things, ideals, etc. that people might associate with the home experience. Participants made their image and word collages on an A2 size sheet of paper. They were also given a set of large markers with which they could add lines, words, or drawings. In the differential conditions, variants of these materials and instructions were given to the participants.

In the 'computer' condition, the same triggers were used, but instead of working on paper, participants used a custom-made computer program. This program resembled the central condition as far as possible. Also, the program was designed to pose as little 'interface' distractions as possible. Participants could navigate between their collage and pages of words or images by mouse-clicking tab fields on the top of the screen; they could import words or images into their collage by clicking these images or words; then, in their collage screen, they could rearrange the words and images. Unlike in the other conditions, the program did not permit participants to add new words or draw lines into the collage. This feature was left out in order to keep the program simple to use.

In the 'symbol' condition, the trigger images were replaced by brightly coloured symbol shapes, such as hands, circles, arrows, hearts, each large enough that something could be written inside them.

In the 'mindmap' condition, participants received an A2 sheet of paper on which an organic, hollow triangle ending in three branches was already sketched in. They had to choose a starting image that best represented their idea of 'home', and then 'grow' the mindmap by adding associations as branches. No further trigger words and images were given. Because participants were expected to be unfamiliar with mindmaps, they were shown an example of a complete mindmap about a different topic, taken from Buzan and Buzan (2000), and were given a sheet of helpful hints, such as 'make it pretty', 'branch out', 'use colour', 'draw small pictures', and 'ask yourself: who? what? where? why? These hint sheets were previously used by Keller (personal communication).

	Example artefact	Observations (as compared to the central condition)
central	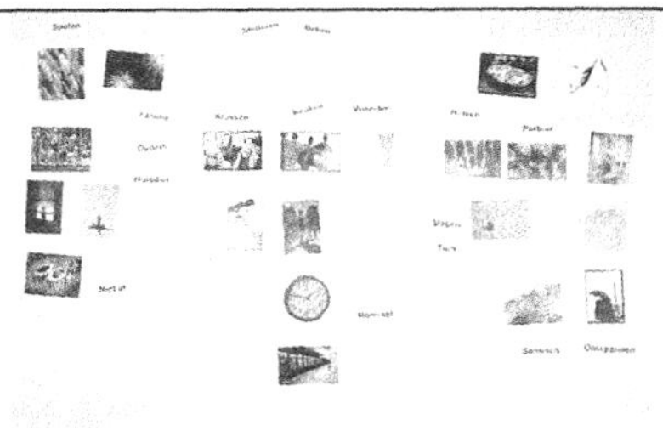	General observations were that in all conditions different compositions occurred, where diagonal, three vertical columns, triangles and large clouds were the most common. (The mindmaps form a special case, where all compositions were branching out)
computer	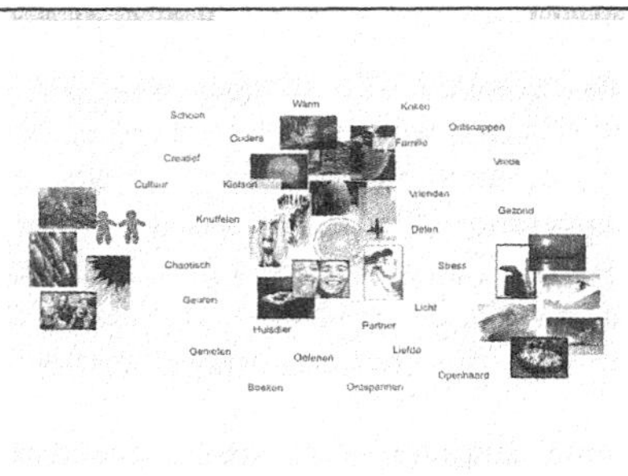	Participants take shorter time to create the artefact, because the possibilities for aesthetic improvements are limited. The presentations take as long as with paper. Artefacts are simpler, more crisp, but artefacts and presentations are not less expressive or rich in depth and width.
symbol		Presentations take longer, and appear to have the same breadth (number of topics addressed) as the image+word condition, but more depth (number of statements made). The presentations are more structured, but also less anecdotal.
mindmap	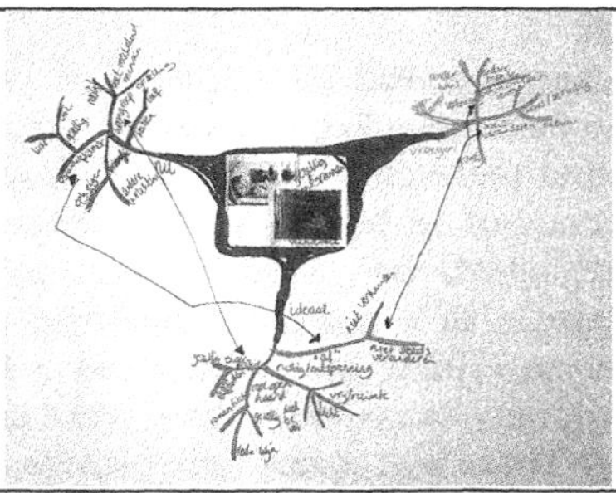	More instruction and encouragement is needed (participants have to create all content themselves). Similar topics were addressed. More links between topics were discussed (rather than individual elements).

Figure 2 One sample collage and observations for each condition.

RESULTS, DISCUSSION AND CONCLUSIONS

Both artefacts and presentations were studied. For the artefacts, the number of triggers and the compositions of the artefacts were compared. For the presentations, we measured the length (how long did the participant talk), the breadth (how many different topics did the participant address), and the depth (how informative were the statements about these topics). Because of the exploratory nature of this study, and the length available here, we do not present details of the analyses, but summarise our general impressions in Figure 2.

The overall results were encouraging. We have seen that non-designers can express themselves creatively using a variety of tools. There were no winners or losers among the conditions; each captured unique and useful insight into people's lives and expectations for their future. The symbol+word collages are best carried out after the picture+word collages (as is the current practice): the former help participants to group and summarise ideas received from the more emotionally coloured associations that they produced in the latter. Mindmaps can be a practical way to either start or summarise, and require little preparation. Finally, the computer condition is especially promising, as it showed that digital media could be integrated in a creative process. The key will be to learn the best possible applications of each and to continue to explore more efficient analysis methods. In further work, we also intend to look at possibilities to make more use of the computer-recorded data (position, time data of construction) and create a tool that enables analysers to immerse themselves by playing with visualised relations.

REFERENCES

Buzan, T. and Buzan, B., 2000, *The mind map book*, (London: BBC Worldwide).

Gaver, B., Dunne, T.and Pacenti, E.,1999, Cultural probes. *ACM Interactions*, January+February, pp. 21-29.

Sanders, E.B.N., 1999, Postdesign and participatory culture, In *Proceedings of useful and critical: The position of research in design* (Helsinki: University of Art and Design), pp. 87-92.

Sanders, E.B.N., 2000, Generative tools for co-designing, In *Collaborative Design*, edited by Scrivener, Ball and Woodcock, (London: Springer-Verlag), pp.3-12.

Sanders, E.B.N. and William, C.T., 2002, Harnessing people's creativity: Ideation and expression through visual communication, In *Focus groups: Supporting effective product development,* (London: Taylor and Francis), in press.

Stappers, P.J., and Hennessey, J. M., 1999, Towards electronic napkins and beermats: Computer support for visual ideation skills. In *Visual representations and interpretations,* edited by Paton, R.C. & Neilson, E. (Springer: Berlin), pp. 220-225.

Can personality categorisation inform the design of products and interfaces?

Michael Goatman
University of Hertfordshire, UK

INTRODUCTION

This study is part of the pursuit of bringing technology towards people. It lies in the increase of accessibility in the operation of digitally based equipment. The digital world has provided an enormous reservoir of facility for people to use and the designer's job in this context is to provide an access to this facility, which maximises the satisfaction as well as the achievement of the user.

The emerging software based systems provide a broadening access to facility, but with it the emergence of the word "compliance", the need to understand and comply with the system in order to be effective in its operation. It is the position of this paper that this is not a direction that will prove ultimately desirable, that technology should be the servant to people, not the other way round, and therefore the idea that technology should be made to comply to the user is an important area of development for the future.

This is not an isolated view. Josephson (1992) suggested that the 1990s was the accessing decade where the perceived need was speed of access to information. He suggests that the next decade is the decade of sensing, with intelligent agents in media being used to customise to individual taste. An example of this is in walking up to your door, the door can sense that you live there and open for you.

The focus of the paper is towards the area of interactive products, those that present information for retrieval and manipulation. It deals with the way in which we as individuals naturally choose to carry out this mental activity, and suggests that we do so in different ways.

THE PROPOSAL OF THE STUDY

The study suggests that:

- We process information in different ways
- These ways can be identified and categorised
- Methods of presenting information in the design of a product called interfaces can be identified and categorised
- Interfaces may potentially be allocated to individuals for their most comfortable and effective experience.

This suggestion poses a number of questions:

- Can individual ways of thinking be identified and categorised?
- Can individual interface methods be identified and categorised?
- Can a way be found which can apply a person's individual method of thinking for the adaptation of interface?

The application of the idea

The application of the idea is envisaged primarily in two areas:

- A guide for designers and marketers in the appropriate allocation of formats within the Product Creation Process. An established basis of knowledge could be provided for application within development teams.
- The possibility of an intelligent interface application that would allow a computer package to address an individual and provide an alternative of interface to the underlying software structure that would give the most comfortable and efficient operation.

THE COMPONENTS OF THE EXPERIMENT

It became the intention of the study therefore to develop a platform through which these three elements could be quantified, and relationships between them explored. The proposal was to institute an empirical test in which participants would experience each component, and provide response regarding their preferences. The results could then be examined for comparison.

Categorisation of ways of thinking

The question regarding the ways in which people prefer mentally to process information was identified as an aspect of personality, the study of which is an established practice within psychology. The categorisation of personality for clinical or professional purposes is extensively researched. The study examined the processes and methods adopted for the purpose of personality analysis, and identified the Myers Briggs Type Indicator (MBTI), as making a distinction relating appropriately to the design context of the study. Briggs Myers (1993) cites that the MBTI is based on the assertion of preferences, and focuses on fairly general characteristics at the level of 'how your mind works' rather than specific talents or expressions of strength. The approach is to assume that to prefer behaving in a certain way is not to prefer to behave in the opposite way. The MBTI categorises preferences within four pairs of opposites. These are:

Extraversion (E) and Introversion (I) | Sensing (S) and Intuition (N)
Thinking (T) and Feeling (F) | Judging (J) and Perceiving (P)

This provides a categorisation of 16 type profiles identified by a four letter coding. Data is gathered by a questionnaire consisting of around 90 multiple-choice questions that takes about half an hour to complete. The outcomes of the MBTI are expressed in descriptive words, and always in a positive interpretation. This description of preferred actions provides a medium that can relate to expressions of the other test components.

Short route identification of personality

For the identification process to be effective a much more immediate access to the person's personality profile would need to be identified. A proposal for this was suggested that stemmed from an exercise given to students on a BA(Hons) Product Design Degree. The students were required to design a garden in a given rectangle within a limited time frame. This was used to demonstrate how different formats could be used to present the same information. The resulting formats however were noticeable for two reasons:

- A similarly identifiable selection of formats was produced whenever the exercise was administered.
- The formats being produced by individual students were identifiable, and to some extent predictable by the tutor who was used to working with each student, with the way in which the student approached the organisation of their studies.

This second aspect gave rise to considerable interest in the exercise as an analysis of individual mental characteristics.

The idea is therefore proposed that evidence present in a simple drawing, created in response to the requirement to design a garden, may provide evidence of certain aspects of the characteristics of an individual's personality, including the way that they most naturally organise information. A method is therefore suggested that allows a software applicant to be provided with an optimum interface scenario through the input of a simple drawing.

Categorisation of drawings

The drawings produced from the garden exercise could be identified into six categories that are illustrated below. They are defined by the most dominant features of the layout structure.

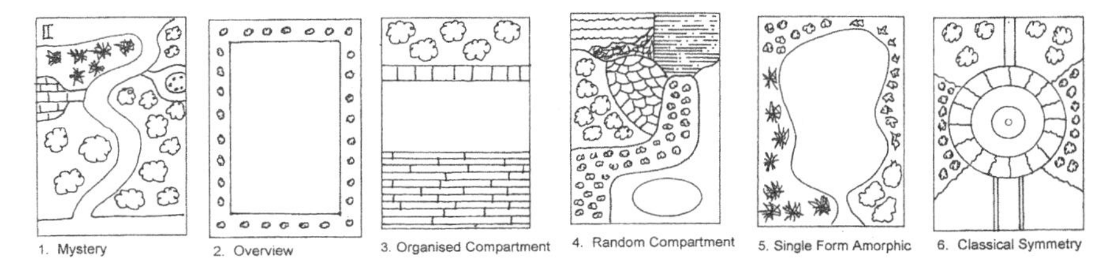

Figure 1 Categories of Garden Drawing identified by graphic feature.

Categorisation of design preference

The chosen medium of identification of design preference is that of experience. Tyler Blake, speaking at a conference given by Philips Corporate Industrial Design in The Netherlands in 1986 stated the premise that product design is the design of experiences, and that the design of mental behaviour relates to software interfaces in the same way that design of physical behaviour relates to the design of hard products.

A description of each of the six categories of garden drawing was made as a depiction of the experience that the layout suggested. This was expanded to a more defined description that was identifiable with the experience of using an established format of interface. Conceptual examples of the six related interface formats are illustrated below

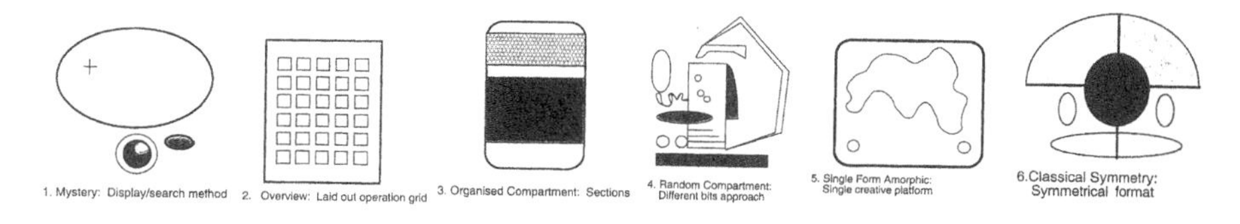

Figure 2 Generic examples of product interfaces demonstrating an experiential relationship with the six garden drawings.

THE EMPIRICAL TEST

An empirical test was then conducted in which each participant experienced the three exercises below.

1. The graphic garden layout exercise.
2. The standard Myers Briggs Type Indicator.
3. A computer screen interface exercise carrying out a comparable task on six separately designed interfaces.

This gave a categorisation for comparison with 16-6-6 elements. Results from each individual's participation were recorded and the data examined for relationships between the results.

The outcome of the empirical test

Evidence that the garden drawing with the winding pathway, garden category 1, showed a relationship with E extravert types was demonstrated in a number of aspects of the results.

Certain factors were experienced in the test that limited its effectiveness in demonstrating relationships between many of the elements. These included:

- The test participants were made up from volunteers, and results suggested a tendency for certain personality types above others to respond to such an invitation. This gave a skew to the test segment.
- Individuals relate to a computer interface format that is most familiar to them, and therefore do not approach each element equally.
- The test segment of 43 participants was less than enough to produce statistically significant comparisons within this categorisation.

THE CONTINUATION OF THE STUDY

It is the intention to continue the experiment with developments employing simpler taxonomies of comparison, focussing on single relationships such as those between the extravert personality and a garden drawing showing a dominant feature of a winding path. Methods of categorisation of design preference remain an ongoing area of investigation.

If evidence can be established to show statistical significance regarding relationships between these components, a platform may be formed for use as a guide to product design briefing, and to the creation of preferred interfaces that can be provided automatically by an intelligent software system.

The study is seen as an exercise at the early stage of an important future endeavour to make technology products sensitive and adaptive to their human users.

REFERENCES

Blake, T., 1986, Paper presented at *The Future of User Interface Conference*, Philips Corporate Design, Eindhoven, The Netherlands.

Josephson, H., 1992, Interface metaphors. Paper presented at *The language of interactivity Conference*, Sydney, Australia.
Available at:
http://www.afc.gov.au/resources/online/afc_loi/presentations/hal+j.html

Briggs Myers, I., 1993, *Introduction to type*, (Oxford: Oxford University Press).

Gender designs: aspects of gender as found in the design of perfume bottles

Katrin Wellmann, Ralph Bruder, Karen Oltersdorf
University of Essen, Germany

GENDER-SPECIFIC DESIGN SEEN AS A CULTURAL PHENOMENON

The aim of this paper is the analysis of gender-specific design, a cultural phenomenon which is a product as well as a constituting element of the culturally valid concept of gender. When we speak of emotions we always speak of somebody having or seeking certain emotions. This somebody is – usually – either male or female. The single person's emotions are under the influence of human experience, of socialisation and history of his or her body (see Johnston, 1999, and Tiger, 2000). These in turn are influenced by everything that belongs to our environment – including products.

So the question arises: Which concepts are these products based on – these artificial bodies – that are directed at either of the sexes? Which are the bodies' features, their semantics – what do they symbolise and how recognisable or effective are they in transporting their meanings? These questions seem worth asking since everyone's personal identity is, amongst other things, made up of gender-specific attributions which are co-defined by product or packaging design.

There are evidently several products on the different markets that are gender-specific. This expression alludes to products that target only one of the sexes, or those that are explicitly declared as 'unisex', targeting both sexes. We want to know: What is it that makes people recognise products as directed at either women or men? Which stereotypical features does a product need to be perceived 'correctly'?

Empirical Research

For this analysis, perfume bottles serve as a good example of gender-specific design, because they are, in most cases, directed at either of the sexes or explicitly at both. Moreover, one can concentrate on their aesthetic features, which dominate their design, because there are not too many functions to be fulfilled by the bottles. Another reason is the intimacy of these products - they occupy the users' personal space, their bathrooms, and the contents of perfume bottles are applied to the body. The emotional importance of perfume bottles in the purchase of a fragrance becomes clear when we take into consideration the results of blindfold tests. It was observed (Machatschke, 2000) that "only forty percent of perfume users recognise their favourite fragrance without the bottle."

For these reasons, we decided to evaluate a total of 27 perfume bottles. Three of these were unisex products. Then there were 24 partner products, that is 12 pairs, each consisting of one perfume for women and one for men, of the same brand or under the same product name. Our work was directed by the following questions:

1. Which features are 'typical' of products directed at either women or men?
2. In which features do partner products differ from each other?
3. Which concept of the relationship between the sexes is the basis for the design of partner products?
4. Which concept is the design of unisex products based on?

First, we analysed formal design elements (properties such as shape, dimensions and proportions, surface textures, colours and graphical designs) to find those that were most common in products of the same orientation. (Orientation means products for women, products for men or unisex products.) In order to unify our analysis, we assigned attributes to four elements of the bottles: head, neck, shoulders and body.

Second, we analysed similarities and differences between partner products in order to understand their relationship.

Lastly, we conducted an empirical study with forty students (equal numbers of men and women), first showing them the bottles in grey shades without graphics, then coloured without graphics and then coloured with graphics. They were asked to answer two questions: "Which orientation does this product have?" and: "Do you know the product's name?" The latter question was intended to elucidate those attributions that were influenced by already existing knowledge about the product. This empirical research enabled us to find out which products were easily recognisable as being directed towards men, women or both. One could also see from the answers given at which stage the product was recognised, e.g.: Was it recognised in grey scale? In this case, it meant it was the bottle's shape and not its colour which transported the essential information.

1. Which features are "typical" of products directed at either women or men?

We came to the conclusion that stereotypically feminine features of the bottles were round, warm, light/lucid, soft/delicate, golden, waisted body and sloping shoulders. Representative of feminine features is the product *pleasures* by *Estée Lauder* for women. Three quarters of the twelve products for women in the test were identified correctly by the test subjects. Some products for women show only a few of the mentioned feminine features – and even some masculine ones – but were nevertheless recognised correctly.

Stereotypically masculine features of the bottles were angular, straight, cold/cool, dark, silver, black, short neck or no neck and heavy base. Many of these are found in the product for men by *Helmut Lang*. In contrast to the results when testing products for women, only half of the products for men in the test were identified correctly. Each of the recognised products possesses at least five of the previously mentioned stereotypical masculine features and rarely one or two feminine ones. It becomes obvious that the impression of masculinity could only be achieved when the bottles represented stereotypically masculine features and left no room for interpretation. As soon as the subjects became insecure in defining the product's orientation, their decisions tended to be in favour of the feminine.

This does not surprise in a culture which developed from patriarchy, since every basic distinction (like the one between male and female) automatically involves the definition of what inner content needs to be differentiated from what outer rest (see Luhmann, 1996). In our culture it is defined what is masculine or male. All of the remaining features belong to femininity. In brief, everything that is not specifically masculine is automatically labelled feminine.

2. In which features do partner products differ from each other?
3. Which concept of the relationship between the sexes is the basis for the design of partner products?

Seen in comparison and relation to each other, partner products can be separated into three groups. The cases within the first group, called congeneric ("equal but different"), show almost identical features in the partners. They differ only in a very few small, yet stereotypical, distinctions. For example, the products *Elle* and *Lui* by *Emporio Armani* are coloured differently: *Elle* is golden, *Lui* is black. In addition, the bottle for women has a small indentation, whereas the one for men has a small protrusion.

Products of the second group, called related ("different but equal"), are similar but vary in details. This can be a variation in proportions, colours, stoppers, graphics – the kinship is still visible. It seems as if there was one underlying theme, twice interpreted. Examples of this category are the partner products by *Bulgari*.

The third group, called extraneous (socks and shoes), consists of partner products whose formal contiguity is difficult to discern. There are suggestions of a common origin, e.g. the just perceivable stone-like surface in the two products *1881* by *Cerruti*. Products following this concept can rarely be recognised as partners without thorough scrutiny.

Within all of these three concepts for relationships between the masculine and the feminine, the differentiation is achieved by the use of bipolar stereotypes. This means that the product for women is almost always smaller, rounder, lighter – the product for men bigger, more angular and so on.

By defining the missing results – what exactly did we *not* find? – we notice the one-sidedness of our findings. What still fails to exist is a concept that unifies both sexes. A simple inversion of existing stereotypes – as found in some products – does not completely satisfy.

4. Which concept is the design of unisex products based on?

Features of unisex products are divided into neutral, feminine and masculine features. Features are called neutral if their appearance is represented equally in products for men and women (features such as opaque or large head). The feminine features of unisex products are mostly rounded, light and sloping shoulders, whereas the masculine features are straight, cool, short neck, heavy base, black and silver. The masculine side therefore prevails significantly. Designers or producers of these products seem to suppose that the impression should be more masculine in order to be approved of by both male and female consumers.

Conclusion

Consequently, no real 'unisexuality' could be detected. On the one hand, there is the predominance of masculine features signalling a male-dominated combination sexuality. On the other hand, there is an inclination to align unisex products with the concept of functionality - the impression of asexuality seems intended. Subjects frequently put forward functional associations when asked to evaluate the unisex products. Since functionality or objectivity are also generally regarded as stereotypically masculine characteristics, the impression of masculinity is reinforced by this.

When empirically researching and analysing gender-specific products, we find stereotypically feminine features in products for women and stereotypically masculine features in those for men. We find that products which are intended for men seem to "require" more stereotypically well-defined features than those intended for women. Likewise, certain patterns are detectable in the way the relationship of partner products is presented. Furthermore, we can identify the underlying concept of so-called unisexuality.

Ernst (1999) points out that "the traditional co-ordinate system of gender relations and the allocation of 'masculinity' and 'femininity' seems to have started wavering. However, in the reality of society the social disparity due to gender is still significant." The results of the analysis of gender-specific products such as perfume bottles show the continuing existence of traditional values concerning male and female. If we presume that products do influence their consumers emotionally, touching and often affecting their make-up of personal and role identity – then we also see the consequences such traditions have. To fit into their assigned roles, women and men commonly use gender-reinforcing or gender-forming external features like clothing. For example, men typically wear dark suits with shoulder pads (angular) and a tie as a straight line – women waisted, light dresses and shoes with high heels (small base).

So which emotions do we as consumers seek, when trying to match our choice of products with stereotypical expectations? A possible answer could be: security in our identity. Design of gender-specific products helps create and support role-images. In order to add zest to our product worlds, alternative paths are waiting to be trod. Stereotypical representation on the traditional bipolar model alone leaves a lot to be desired.

REFERENCES

Ernst, S., 1999, *Geschlechterverhältnis und Führungspositionen. Eine figurationssoziologische Analyse der Stereotypenkonstruktion*, (Opladen/ Wiesbaden: Westdeutscher Verlag), translated by Katrin Wellmann.
Johnston, V. S., 1999, *Why We Feel. The Science Of Human Emotions*, (Cambridge, Massachusetts: Perseus Books), p. 88.
Luhmann, N., 1996, *Protest: Systemtheorie und soziale Bewegungen,* (Frankfurt/Main: Suhrkamp), pp. 107.
Machatschke, M., 2000, Der Mann, dem die lila Kuh aufs Wort gehorcht. In *Die Welt*, www.welt.de, (October 11th, 2000), translated by Katrin Wellmann.
Tiger, L., 2000, *The Pursuit of Pleasure*, (New Brunswick, New Jersey: Transaction Publishers), p. 53.

How to create Linus's blanket

Yoko Kaizuka
University of Central Lancashire, UK

BACKGROUND

Feeling Foreign

I am Japanese but currently living in the North West of England. When I first came to the UK, I used to wonder why British girls wore unnecessarily tight fitting clothes. They appeared to me as if they had bought the wrong sizes. Now I get used to seeing that as the taste of British girls. Something ordinary for British people appeared so strange to me at the start of my stay here.
The effect of humour, preference of colours, shapes, textures etc., are very hard to articulate or explain logically and are derived from our received cultures. They hugely affect our daily life and design activity. Since I came over to UK, I have always felt vaguely uncomfortable - as though I were wearing the wrong spectacles! Expressing subtle feelings was very difficult for me because of my different cultural background. 'Although basic emotions, i.e. Joy, Distress, Anger, Fear, Surprise, Disgust, are universal and innate, emotional expressions are not like words, which differ from culture to culture; they are closer to breathing, which is just part of human nature' (Evans 2001). Linguistic expressions of emotion vary from culture to culture, and it was interpretations of language that created my feelings of unease. For example, I knew the meaning of the expression "Oh my God!" I understood that a person was expressing astonishment, but it took time for me to sympathise with it because the word always trapped me into thinking 'which god does the person mean?'

In certain cases, emotion exists although there is no name for it. 'Amae' is a Japanese noun that describes a very complex emotion. In Takeo Doi's 'Anatomy of Dependence' (Doi 1971) he showed many examples that this emotion amae, does exist not only amongst Japanese but also amongst his American patients. However there is no word for it, in the English language. Amae is often rather simplistically translated as 'dependence', but Evans described it as 'a kind of joy, a feeling of acceptance, of belonging, of being valued by a group of people' (Evans 2001).

The emotion amae is very complex and ambiguous; we cannot express it as 'my amae feeling (e.g. for Jane) consists of 30% affection, 17% sympathy, 12% security and 8% hate'. So I started trying to find a user and product relationship that relates to amae in order to explain it to other people. An ideal icon of this relationship exists in Schulz's cartoon 'Peanuts': it is Linus and his blanket. Linus is a boy who always carries a security blanket with him. His blanket is such a low-tech product; it may be made of acrylic or pure cashmere; just an ordinary household product but it brings Linus comfortable feelings, a sense of security and of belonging. 'Linus's blanket' is suitable as a visual metaphor to represent amae, but is open-ended and not scientific or logical.

Ordinary, everyday routine activities offer positive stimuli, for instance the smell of clean bed linen, a nice cup of tea. The way to find comfort depends on each person and their personal preference and cultural background and so in the future '...design will have to pay far more attention to the domains of human interaction and culture theory than to the technological issues with which it has thus far been engaged'. (Friedman 2002)

A SURVEY

My MPhil project is concerned with the examination of visual methods relevant for research into consumer needs and new product development. I aim to explore the possibility of developing further techniques that may have greater sensitivity to the emotional factors relevant to design. In order to identify the issues contained in the notion of Linus's blanket and to aid future work, I decided to undertake action research in the form of workshops and a questionnaire. The workshops took place in February 2002, with groups of design students at University of Central Lancashire, from a variety of sub-disciplines.

As a pre-workshop activity, participants were asked to submit a picture of an object that they considered to be 'good design' and to describe their reasons in a short paragraph. They were asked to wear their favourite top and denim jeans to the workshop. I took photos of each individual to form a visual record of their involvement, and also of their personal bag /carryall. I thought that their bags and the objects inside would indicate each person's essential kit for daily life. This aspect of the workshop was aimed at projecting the participants as 'Linus' and to record outward expressions of amae in their everyday lives.

The questionnaire asked about aspects of personal comfort, security and feelings embodied in the concept of the Linus's blanket. It aimed to record personal interpretations of the relationship between 'Linus and his blanket' (as an ideal product-user relationship). As part of the workshop I asked the participants to design a product that was stimulated by their understanding of the Linus's blanket concept. This aimed to compare objective and subjective definitions of the blanket's emotional properties. The participants were 1st Year Product Design Students (21 all male, age group 18-20) and 2nd Year Furniture and Textile students (11 females, age group 18-40, 4 males, age group 18-24). The day's programme was as follows:

1. Participants were asked to propose an object that would offer comfort and re-assurance in terms of personal need. The focus of this proposal could be himself or herself or another clearly identified character.
2. Participants were given an explanation of 'Linus's blanket' including some visual examples, background information and asked to complete a questionnaire.
3. Participants were asked to re-design or modify their proposed comforter in response to information received about Linus's blanket.
4. Participants were asked to work in groups to produce a comforter for suggested users to include a written specification and visual references.

Students were asked to select and consider one of the following user groups:

- A single professional living and working in a foreign country
- Themselves, but imagined as retired at the age of 65
- A young teenager in the year 2010
- A technophobe in a technological society
- A disaster relief worker
- A family

RESULTS

Pre-workshop activity: most chose as their examples of '*good design*' objects relating to their own practice area but which were beyond their own personal experience, such as a luxury sports car, or other famous 'designer' products. In their descriptions, they discussed styling and aesthetic appeal rather than practical working function.

Favourite top and jeans: participants wore clothing influenced by current market trends. Most of their tops clearly displayed logotypes or other graphics. One student wore a worn out rugby shirt. In the questionnaire, he explained that the shirt came from the university's rugby team and that he is captain and had won the university league. Another wore his favourite football team shirt. I was interested to see the varieties of denim that would be on show. The choice of denim is culturally complex and displays subtle fashion and lifestyle preferences. Graphic display is not as overt in the case of denim jeans, it is more expressive and complex, but each person's denim jeans looked subtly different and communicated quietly their owner's style and personality.

Bags/Carryalls: most carried wallets, mobile phones and several of the product design students also had mini-disc players. Many of their mobile phones had customised logos, plastic covers and ring tones. One student had an asthma inhaler. The majority of the bags were worn rather than carried: worn as backpacks or slung across the chest. The contents of the carryalls gave an insight into each person's life stage and showed their minimum essential tools for a day away from home. (This stage of the workshop was particularly intriguing for me as a foreign researcher, and I was able to cross reference to my Japanese experience.)

Workshop part 1- empathising with Linus and 'his' blanket: students drew various kinds of things. These drawings could be categorised as 'Aspirational Comforters': e.g. a sports car, famous designer chair. 'Physical Comforters': e.g. chocolate, beer. 'Escapist Comforters': e.g. a time machine and other science fiction type ideas. Most of the craft orientated students' reactions related more to physical comfort. All of the science fiction, futurist ideas came from product design students. Some students were very attached to physical comfort ideas but seemed to lack the objectivity required to develop ideas further. One textiles student drew chocolate and a bath twice as her comforter rather than designing her own object. I asked her what she would produce herself in term of a comforter. She answered that it would be a 'sensual textile'. This response can still be categorised as a physical comforter, but unlike her other answers, it stimulated various different and personalised imagery.

Questionnaire- myself and 'my' blanket: participants were asked to describe themselves in terms of their own personal preference of smell, sounds and textures. Some saw their equivalent of Linus's blanket as their pet or memories. Keeping in contact with friends was also considered important.

Workshop part 2-Designing 'their' blanket: many stated in the questionnaire that their emotional attachments and relaxation related to personal memories. However 9 out of the 12 groups drew electronic gadgets. Telecommunication tools, involving storage of personal visual and audio memories, were suggested as important aspects of the function; for instance, a baseball cap with a built in music stereo system and mobile phone; a digital photo/video album with built in mobile phone function. Few attempts were made to encompass aspects of a '*simple*' thing from their past: i.e. a blanket or a teddy bear.

The most resolved concept was a product designed for family use. The suggested user scenario related to a child feeling lonely at kindergarten. The product was proposed as an egg-like form, made of a squeezable flexible material. He/she can squeeze the egg and a signal goes to the other members of the family. The family member can receive the signal and squeeze the machine thus sending a vibration with colour illumination to the child's machine. By transmitting these vibrations between themselves, a family can simulate the feeling of holding hands or of touching. I discovered that the group who developed this idea had known each other longer than the other groups. They were more open with each other and communicated more effectively, using words and drawings, whereas the less familiarised groups were more self-conscious and distanced, and were not able to cover the same amount of ground in the time available.

When considering the needs of third age users one group cynically designed a sofa with built in TV remote control as they considered TV watching to be the major activity of the over 65's, but did not give any attention as to how the shape of the sofa might relate to physical comfort. From my perspective this group's work exemplified the difficulties of considering an unknown other's needs and how society often lacks optimism for the elderly.

SURVEY CONCLUSION:

When I structured the survey, I included Linus as an open-ended question in order that it might serve as an ideas parking lot. I aimed that the concept of Linus's blanket could be more effectively communicated when expressed through the combination of pictures and words. Participants supplied personal opinions via the questionnaire regarding their own emotional comforters and by projecting these onto various existing products through design drawings. They were able to relate in an abstract way to their personal memories. The workshop successfully created conscious personal connections with the amae emotion, but the channelling of these feelings by designing for others proved difficult and narrow in its expression. I had hoped to free myself and the participants from cultural bias, but when asked to draw product ideas, participants displayed their 'consumer culture' by presenting only electronic, technology-orientated ideas, and by ignoring or forgetting the simple pleasures and sensations involved in every day activities.

Although the emotion amae, which can be stimulated by Linus's comforter, is universally understood and experienced, it lies in the unconscious and therefore expression in terms of the design process is blocked and difficult.

PERSONAL INSIGHTS

The participants expressed their personalities and lifestyle though their clothing and belongings and during the activities in the first part of the workshop. For instance, if the wearer of the rugby shirt had not explained his personal connection to it, then those feelings would have gone unnoticed. He stated that he felt that his equivalents of Linus's blanket are his friends; his relationship with the rugby shirt also expresses this. Users build emotional and physical relationships with products, just like Linus. The asthma sufferer may not feel amae emotion for her inhaler. He/she needs it for *physical* security, which is not the same as feelings of amae, which bring the added complexities of *emotional* security.

Participants told me that personal memories of being with families and friends and being able to communicate with others are crucial for them to feel emotionally comfortable. These feelings exist without the use of technological devices. Simple materials offer physical comfort and the stimuli to recall good memories and bring amae emotion directly to users. I feel that products that provoke amae, have a propensity to absorb the user's memories and connect the user with other people. 'Simple' materials, such as denim, seem to have 'more space' for a user to build the complexities of amae feelings while simultaneously experiencing those of physical ease. The ageing process of some materials, their sympathetic look and feel, help people to experience them as 'human' and part of their own life. Although denim garments are mass-produced, they serve as devices to convey aesthetic, social, cultural and tactile stimuli. Like many textile products, denim is culturally complex and it is both an emotional *and* physical material. As it's colour fades away, the original shape deforms, it becomes part of the user's life and grows older alongside the user. It becomes a witness of personal history. I think that the history and personality of materials such as denim have much to bring to a fuller understanding of product-user relationships.

In the future, human interaction will be sustained by technologies able to store visuals, sounds, and memories; interdependence will be supported by telecommunication. However these technologies may lack the full dimension of amae, which exists through a synthesis of emotional and physical comfort. Product designers need to learn more about the complexities of the emotional properties of so called 'simple' materials. I advocate that 'Linus's blanket' can be used as a visual icon to represent complex emotions that are otherwise very difficult to express through language alone. By referring to such an icon designers can be stimulated to access and consider the fundamental human feelings of amae in their work.

REFERENCES

Friedman, K., 2002, Towards an integrative design discipline. In *Creating Breakthrough Ideas: The Collaboration of Anthropologists and Designers in the Product Development Industry*, edited by B. Byrne and S. E. Squires, (Westport, Conn.: Greenwood Publishing Group) (in press), p. 11.
Doi, T., 1971, *Anatomy of Dependence*, (Tokyo, Japan: Kodansha).
Evans, D., 2001, *Emotion*, (New York, USA: Oxford University Press) pp. 6-7.

Science and design – two sides of creating a product experience

Nicole Eikelenberg, Sytze Kalisvaart, Marc van der Zande, Frans Lefeber and Huub Ehlhardt
TNO Industrial Technology
The Netherlands

INTRODUCTION

Up till now designers interpreted designing comfortable products as realising the absence of discomfort in a product by correctly applying human factors knowledge. Ergonomics are traditionally focused on minimising discomfort and health complaints, which are necessary but not sufficient. True designing with comfort as a major focus point sets the need for an approach that looks at (comfort) expectations and emotional aspects as well.

The concept of comfort

Helander and Zhang (1997) found that in the case of chairs, 'Comfort' and 'Discomfort' are two fundamentally different concepts. 'Comfort' is associated with cognitive, emotional but also physical aspects and related to 'Well being', whereas 'Discomfort' is associated with physical aspects alone. Thus, comfort and discomfort are not two extremes of the same continuum. Next to physical comfort, the term comfort is associated with a positive emotional experience.

Traditionally, designers have considered the issue of emotion and experience as an implicit part of their design task. Depending on the project and the client, this 'soft side' of the product design was either discussed with a client or completely left to the designer's 'magic'.

Now that consumers have become used to products no longer causing physical discomfort, a high level of physical comfort and also emotion and experience have gained importance in marketing and sales. This applies not only to consumer products, but increasingly also to professional products like safety shoes and hand tools. Therefore, requirements for the manageability of these issues in the product development process are set and a scientific basis is sought.

A designer might develop a perfectly comfortable sleeping mattress but if the product does not give a heightened comfort expectation, it will not sell. The challenge for the designer is to design a mattress that raises a certain comfort expectation equivalent to the actual comfort level.

SCIENCE: COGNITIVE MODEL OF COMFORT AND SATISFACTION

A cognitive model for comfort and consumer satisfaction was developed (Figure 1) to assist designers in their DfCCS work and to provide a scientific basis. The model can help as a framework for common understanding of how comfort and satisfaction are achieved and which factors influence the result. It can also play an important role in communicating comfort and satisfaction issues with product companies.

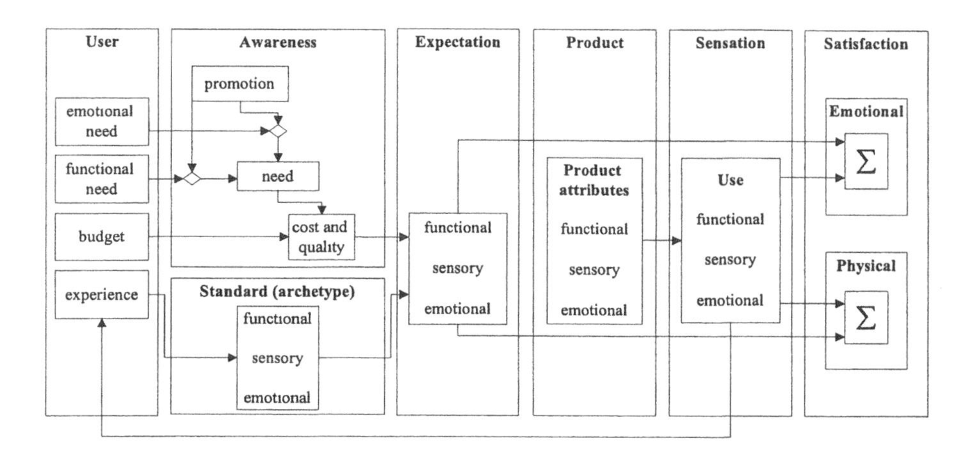

Figure 1 Cognitive model of comfort and consumer satisfaction

The following notions served as starting point for the model:

- We use Helander and Zhang's definition of 'Comfort' where comfort is a function of physical comfort (human factors) and emotional experience. Discomfort we define as the absence of physical comfort.
- Because of the evolution of the meaning of comfort from purely physical to the Helander and Zhang definition, we avoided the term comfort in the cognitive model.
- Emotional and physical expectations (before purchase) and actual emotional and physical sensations (during use) determine consumer satisfaction. For example, a user might buy a chair that looks comfortable but upon use, he discovers the smooth material constantly lets him slide forward. The user is not satisfied.
- Comfort is always appreciated in comparison with an individual standard expected level of comfort and satisfaction (a reference level).

DESIGN FOR COMFORT AND CONSUMER SATISFACTION

In the development of TNO's Design for Comfort and Consumer Satisfaction tool the following steps were taken:

- Analysis of the approach product developers would adopt intuitively;
- Analyses of the design process flow for identifying tool opportunities;
- Development of tool concepts;
- Analysis of the effectiveness of the tool concepts;
- Adjustment and integration of the tool concepts into a DfCCS-tool

Currently some of the tool concepts are being analysed on their effectiveness.

Analyses of the intuitive approach

We gave three teams of four TNO product developers the assignment to redesign a product previously developed at TNO for Comfort and Consumer Satisfaction. The questions were:

- Determine which aspects you find important for Comfort and Consumer Satisfaction of this product;
- Try to improve the Comfort and Consumer Satisfaction of this product;
- Decide how you will continue after this session.

We found that some of the designers include experience aspects in the use of the term 'Comfort', for example 'it looks heavy'. One team treats 'Comfort' as more than 'Prevention of discomfort' alone, another team treats it as 'Prevention of discomfort' alone. It is stated that 'Comfort' is mostly tactile, whereas 'Experience' and 'Consumer Satisfaction' are also visual.

Though not part of the assignment, the teams generated many comments and questions on functionality. Apparently, they do not want to handle comfort and satisfaction in isolation from functionality.

The intuitive design process described above appeared fairly unstructured. Therefore, a positive intervention could be to provide an approach that suggests the aspects to be covered when doing DfCCS.

Designers have no problem dealing with the vagueness of 'Consumer Satisfaction' and 'Experience'. Moreover, they consider the satisfaction aspect as one of their core competencies. Seldom would they involve cognitive psychologists or other specialists in handling this issue, unless the business importance and uncertainty about comfort and consumer satisfaction were very large. Apparently, these designers are not inclined to involve researchers for satisfaction. For comfort, specific requests for ergonomic information were made.

INTEGRATION OF SCIENCE AND DESIGN: THE DFCCS-TOOLS

Based on the cognitive model, the analysis of the design process and the intuitive approach of designers in handling comfort as a design aspect tool, tool opportunities have been identified. In the problem definition phase of a design project mainly facilitating tools for comfort-goal setting, identification of relevant comfort aspects and prioritising of these aspects are needed. In the design phase of the project mainly expert knowledge and prediction methods of achieved comfort levels (and expectation levels) are needed. The central tool for the problem definition phase is discussed below.

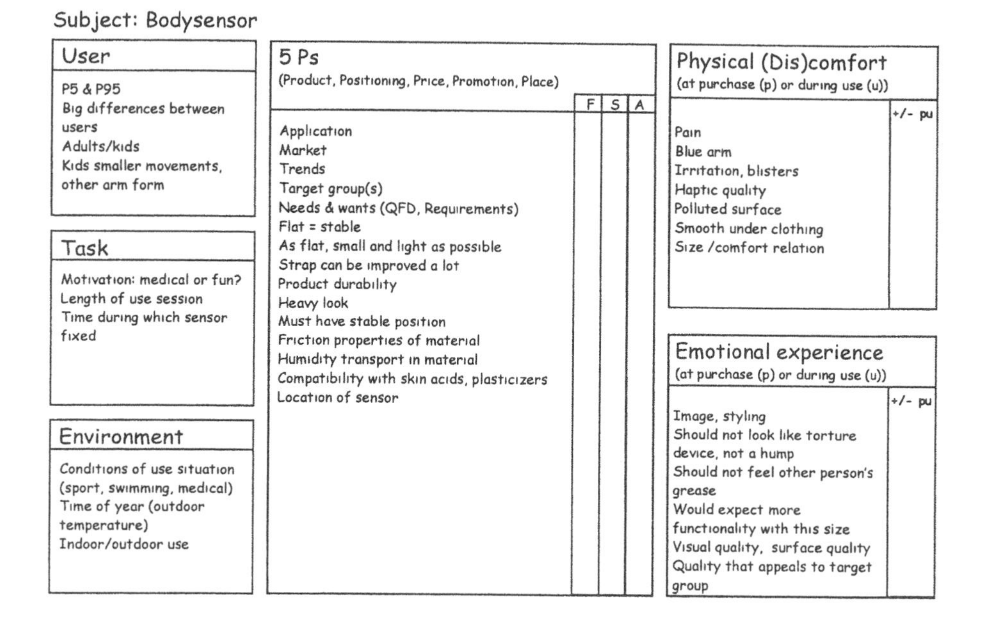

Figure 2 DfCCS session template

DfCCS session template

The session template for DfCCS sessions (Figure 2), derived from the cognitive model, helps to explore the design assignment by charting which aspects need to be covered for proper DfCCS. It is deliberately tolerant in its format to accommodate a brainstorm like session, including its multileveled discussion and text connotations.

The design team jointly fills in the template. For each topic, suggestions and key questions are given to help the team. After filling in the template, priorities are highlighted based on a special dedicated DfCCS QFD or intuitive priority setting.

We performed a protocol analysis of a DfCCS session with and without the template. The template appeared to help handle one aspect at a time instead of aspects simultaneously, thus making the DfCCS process explicit (see Figure 3).

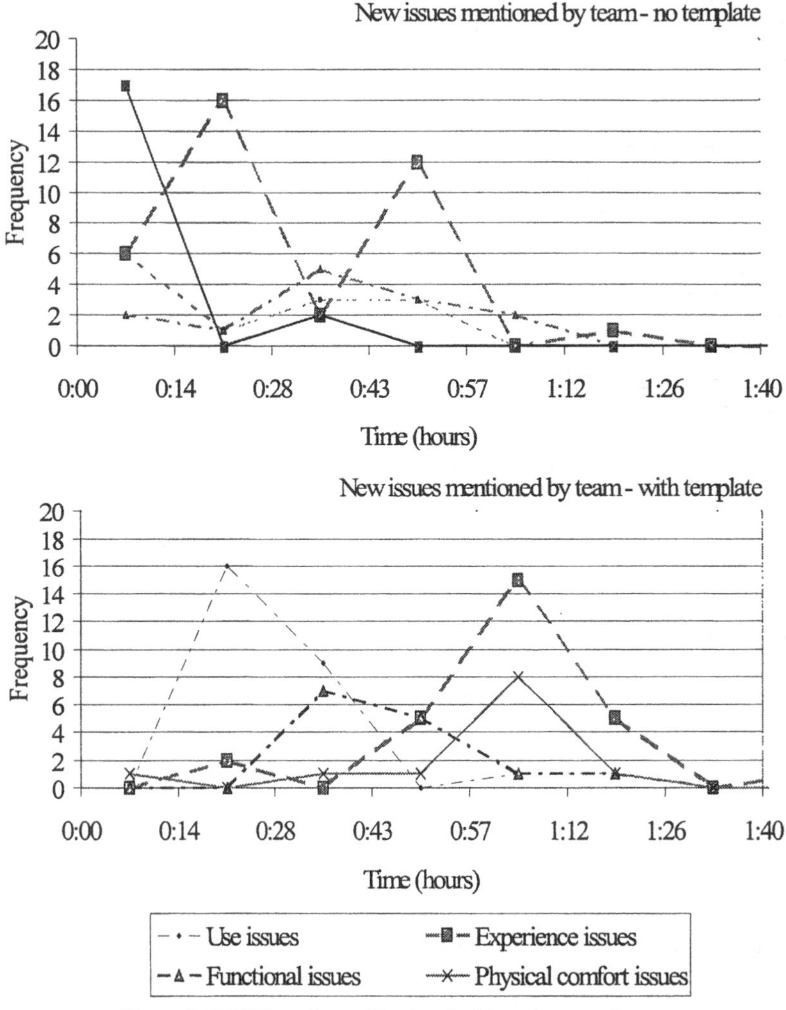

Figure 3 DfCCS sessions without and with session template

CONCLUSIONS

To make comfort and satisfaction aspects of product design more manageable, the DfCCS design process needs to become more explicit and the underlying aspects determining comfort should be clear. To ensure the stability and reliability of such an analysis, the underlying models should be scientifically validated.

At the same time, the resulting work approach should be compatible with existing product development processes. Product developers treat comfort as more than just a sensory aspect and include also experience aspects ("It looks heavy").

The DfCCS tool should not present Comfort and Satisfaction in isolation, but always take the integrative approach, including product functionality. Designers feel a need for scientific backing in the field of physical comfort rather than the field of emotional experiences.

A first evaluation suggests that the DfCCS template helps to structure DfCCS sessions.

REFERENCES

Helander, J.G. and Zhang, L., 1997, Field studies of comfort and discomfort in sitting. In *Ergonomics*, **40** (9), pp. 895-915.

Ter Hark, T., 2002, Comfort is in the buttocks of the beholder - Comfort in product development. In *Proceedings of Aircraft Interior Expo 2002,* (UK & International Press: Surrey), http://www.ukintpress-conferences.com.

Prospective design oriented towards customer pleasure

Andre Cayol
Compiègne University of Technology, France

Patrick Bonhoure
Compiègne University of Technology, France

INTRODUCTION

Nowadays in western countries, a new product rarely constitutes a ground-breaking innovation. When a new product is launched onto the market, it often displays new functions that need to be compared with those of its competitors. Currently, feedback of this type of information is slow; it is provided by specialised consumer journals. This situation is likely to change with the advent of on-line benchmarking, enabling access to up-to-date information. For companies, the challenge is now to improve overall design quality, bearing in mind differentiation. The choice for the consumer in the future will be complicated, as the number of alternatives will continue to increase. The pertinence of all the factors which contribute to the success of an innovation will (therefore) be determinant.

In this article, we are concerned with design quality, and in particular, the effect on products of introducing elements that generates pleasure for the user into the designing process. We will describe a method that has enabled the generation of prospective product concepts. Prospective Design Oriented towards Customer Pleasure (PDOCP) is a method that we have perfected via the work of CQP2 team researchers (Designers and specialists in Human Factors, Quality, methods, and value analysis). It was used for the first time in 1998 and since, we have applied it to studies involving the products of a number of French companies, including EDF, FACOM and PSA Peugeot Citroën.

THE VISIONS OF DESIGNERS AND ENGINEERS

Some aspects of design, notably functional aspects, have led designers to utilise methods, involving measurable data. The remaining types of data have been classified into subjective parameters, and have only been the object of serious evaluation in the last few years.

Design philosophy is not monolithic and has varied throughout the history of the profession. Product designers have moved from producing crude designs at the beginning of the industrial era, to a stage where production was no longer able to satisfy demand, and then to a design in which pure unornamented form, symbolises the product function. The miniaturisation of technology has brought along with it some very contradictory trends. We see now dematerialised design (where only the interface subsists and explains the functions) currently coexisting with decorated design (Memphis group, "ethnic" design and Hundertwasser movement) and all this happily sitting alongside another category: neofunctionalist design and ecological design (Puyoux and Kazazian, 1999).

How have we developed our method of Prospective Design Oriented towards Customer Pleasure (PDOCP)?

We started from our representation/model of the design process and thought that we must function like ergonomists carrying out experiments to determine anthropometrical factors to be introduced, at the very start of the design process. Ergonomists have clarified the point in time at which they have to contribute recommendations, which may lead to design constraints.

An innovation is always situated at a crossroads of a new functionality made possible by emerging technology and users who adopt it. This notion of the customer enabling the success of an innovation is very important. It highlights the notion of product attractivity. Where the mentality of the future customer has to be anticipated. As our study was destined to generate product concepts for the year 2020, we had to foresee how future customers would evolve. Good practice in ergonomic research and methods which enable design to evolve, involves ergonomists not providing standard and universal solutions, but rather recommendations adapted to each type of user, for each type of physical, social and cultural environment. This involves undertaking a sociological study of future customers to set up a clear vision of the future (Marzano, 1998).

We also based our work on a philosophical definition of the notion of pleasure. Pleasure and pain have often been considered as opposing notions. According to Plato, there cannot be pleasure without pain; one is not possible without the other. If transposed to products and the consumption of products this indicates that desire for a product engenders a desire, which is the motivation for buying leading to the pleasure of possession. In our society of abundance, this expectation is more expressed in the case of expensive luxury products. Pleasure and pain can also be found in computer games: a difficult game presents a challenge which gives a lot of pleasure once the difficulties are overcome. Nevertheless, as Aristotle remarked: simple pleasure can also be found in free and wide variety of activities.

Our research on user pleasure is based more closely on the definition of Aristotle than on that of Plato. We have sought the notion of direct pleasure linked to the satisfaction of desire and tastes in each of the phases of everyday life.

METHODOLOGY

Bibliographic Studies

In the definition of innovation, we must study the sociology of populations affected by future products. For our prospective research, we had to anticipate the users' future preoccupations. Initially we referred to studies carried out by sociologists who described the evolution of French people for the year 2025. The main directions described confirm the trends emerging in the last few years, for example: nomadism, longer life expectancy, fragmented and reconstructed families.

We also studied emerging technologies from laboratories which seemed to have an important role to play in relation to future comfort and human assistance, i.e. the convergence of transmission technologies for sound, images and digital information, which will enable possibilities for compression and treatment which were not thought of until very recently. The human/machine interaction will progress due to software developments.

We questioned the validity of different approaches that take into account pleasure to enhance the overall design quality (Hofmeester *et al.*, 1996; Marzano, 1998; Jordan, 1999; Jordan and Green, 2002). We highlighted some limitations of these approaches and this review led to the proposal of our methodology.

Family Monograph

Before starting our creative workshops, we identified the findings of sociological studies describing a typical family in 2020, which corresponded to these new trends in society. Our representative family is composed of two parents, two children aged 8 and 10 years old, one of the parent's sister who is a single mother looking for a job, living at home with her 2 year old child and grandparents who are often there for prolonged periods. We have described life sequences for each of these persons by imagining their activities and by writing the sequences.

Description of Life Sequences

We were working on creativity sessions by describing the sequence and by imagining for each person what the pleasure experiences could be and how to express them activity functions. We grouped the participants together for each of the specialities represented in the CQP2 team. The life of each person in the family was considered stage by stage. It was the tastes of the participants themselves and their pleasures in the circumstances evoked, which were used as starting points for reflections.

As an example, we took sleeping and waking. Each member of the design team explained how he or she liked to sleep and be woken up. For waking, some people explained that they preferred to listen to the radio, others liked to sense the smell of coffee, others wanted to be woken up at the last minute while others preferred to take their time in the bathroom etc.

Technical Creativity

These sessions enabled us to start the description of procedures, systems and objects which made life easier and which increased the pleasure of each of the users. Our technical documentary study enabled us to find a technological answer for most of the desires, which had been formulated. We were able to decide what the technology would be which could enable everybody's' wishes to be responded to. We worked in a group of ten persons coming from different disciplinary background. Some of them had considerable knowledge of computer sciences, new technologies, and the Internet. We started our session by discussing the different types of pleasures linked to the family's activities and we imagine what technology could be used to provide such features. We gave a description of the technical concept and of the technology that could be used.

Screening

Our principal goal was to enable the company to articulate future product concepts and after this description phase which produced many rich and exciting concepts, we decided to rule out all the innovations, which risked provoking, concerns thus posing obstacles to their adoption.

We used an approach inspired by the FAcT Mirror Method (method for complex problem resolution developed in our university by Gilles Le Cardinal) to eliminate negative proposals. We asked a group of young users trained in new technologies to declare the fears that the description of innovations inspired. We noted these fears without applying any censorship. After this listing, each participant gave a mark (between 1 and 10) at all the fears, and we calculated the mean of these marks. The most "frightening" innovations, judged according to the mean of the notes obtained, were eliminated.

DISCUSSION

For an operational product planner, the creation of a technological response be at a realistic priced in the market. In our study case, the cost was not important because the products did not have to be created straightaway, though we did not ignore that the cost of technologies diminishes progressively with their diffusion.

The results of our research were described in a user scenario using sketches to explain the function of the design proposals. Because detailed drawings showing shapes and colours are linked to trend associations, we simply explained functions without assigning to them any form, thus avoiding prejudice about the new functions.

CONCLUSION

When all products are capable of delivering adequate performance, we adopt the product, which attracts our attention, in other words, the one which will has seduced and captivated us.

Our methodology has performed very well in enabling us to identify new product concepts. These are technically possible due to the convergence of transmission technologies, sound, images and digital technologies. During the next few years, these technologies will no doubt be too expensive to be adopted by average users, but the majority of useful technologies will fall in price (as was the case with mobile phones) and these technologies will soon be widely available.

The design method, which integrates affective factors, was used experimentally in the design of prospective concepts, which had to be launched by the year 2000. Currently EDF (French Electricity Company) is investigating product concepts drawn from our research. We think that the majority of our "inventions" will be soon launched on the market, though we may only be able to judge their true merit in 20 years time!

REFERENCES

Hofmeester, G. H., Kemp, J. A. M. and Blankendall, A. C. M. (1996). Sensuality in Product Design: a structured approach. CHI 96, Vancouver, British Columbia, Canada, ACM Press New York, NY, USA.

Jordan, P. and Green, W. S. (2002). Pleasure with the use of products, Taylor and Francis.

Jordan, P. W. (1999). Pleasure with products: Human factors for body, mind and soul. Human Factors in Product Design : Current Practice and Future Trends. P. W. Jordan and W. S. Green. 11 New Fetter Lane, London EC4P 4EE UK, Taylor and Francis: 206-217.

Marzano, S. (1998). Creating value by design: Thoughts. Blaricum, The Netherlands, V+K publishing.

Puyoux, J.-B. and Kazazian, T. (1999). Conception de produits et environnement: 90 exemples d'éco-conception. Paris, France, ADEME.

Communicating product experience

Lilian Henze
P5 consultants, The Netherlands

Roel Kahmann
P5 consultants, The Netherlands

INTRODUCTION

In their book, *The Experience Economy* Pine and Gilmore (1999) state that manufacturers must experientialise their products. They should '*ing* the thing', as they say, and embed their products in an experiential brand. This means they should surround their products with experiences that make using the product more memorable. For example, the appliance manufacturer should make goods that enhance the washing experience or cooking experience.

Our clients took up that challenge and more and more they want to get a grip on the experiences evoked by, or associated with, their products. To answer their questions, usability research is done to gain insight into users, user-product-interactions, and user-product-relations. Besides the more traditional usability issues, it is important to focus on emotions and values associated with products.

As usability experts, we see ourselves as the consumers' advocates. We are anticipating the clients' need for objective information on user-product-relations by developing and exploring methods and tools in order to measure emotions in product use. Our quest to find these methods and the development of the P5 Usescan® and P5 Affectionscan® is described in an earlier publication (Kahmann and Henze, 2002). But finding and applying appropriate methods is just the first step to provide design teams with relevant information. The communication of the findings is the next and as important a step.

In our vision, it is extremely important to communicate the results of the measurements in a proper way to the designers. P5 intends to communicate its findings in a way that designers will be inspired and their knowledge increased. We would like to be the designer's facilitator and provide them with operational tools to design products that lead to intended interactions and relations. In order to accomplish this the *script methodology* is developed. In this paper we will explain this script method in more detail. Before this explanation, we define emotions and relations in the context of our usability consultancy practice.

EMOTIONS AND RELATIONS

When we look at an individual consumer, we see a transformation in product experience from buying to using to owning. Usability focuses on the using part, i.e. the interactions. Buying and owning a product focuses on a relationship (Figure 1).

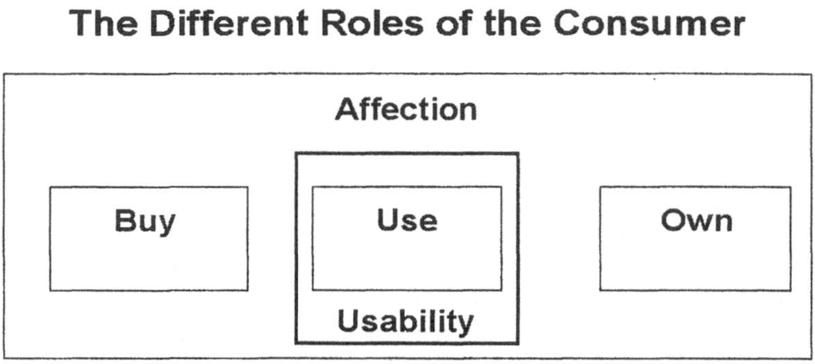

The Different Roles of the Consumer

Affection

Buy

Use

Own

Usability

Figure 1: Different roles of the user

Buying the product. In this phase the consumer has to be attracted to the product. Usually this is the area where the market researchers are active. They explore the needs of the consumer in choosing - a product. The consumers' needs result in a choice based on expected advantages (functional and hedonic) and that will subsequently result in buying the product (Goossens, 2001). Insight in to expected functional and hedonic advantages is used in design and marketing.

Using the product. When the decision is made and the product is bought the second phase starts. The product will be used for the first time. This is where Human Factors come into play. Basically the following can occur. Using the product can evoke emotions like "yes I can handle it", which lead to satisfaction and a positive (pleasant) relationship with the product. If the user does not succeed in handling the product, the user will be disappointed but the product still has a second chance. If the product fails again to fulfil the needs of the user, it will turn out in a negative or no relation to the product.

Owning the product. In time the user-product-relationship will develop based on experience and sentiments. This will not always mean one is still using the product. Many people have products in their homes they do not use anymore but still have a positive relationship with. People who still have their very first 8 mm film projector they cannot use anymore keep this product but have no problem in exchanging their latest video recorder for a new one.

We distinguished positive and negative emotions in user-product-interactions and behaviour that lead to a positive or negative relation with the product.

COMMUNICATION: DOCUMENTARY OF HUMAN PRODUCT INTERACTIONS AND RELATIONS

The P5 script method is developed to make a clear and detailed visualisation of the user, interactions and relations. The script can be seen as the screenwriting for a documentary film; - Human Product Interactions and Relations. Shooting the film, and showing it to the audience (the design team) would be the ultimate challenge. In the practice of product development the budget and time restrictions do not allow such an extra project. Nevertheless, an alternative and powerful tool is at our disposal: the script.

Script for documentaries

Professional documentary screenwriters go through the following steps (Suèr, 1996):

- Research report: summary of facts collected through research;
- Synopsis: summary of contents and target group (audience);
- Script design: Description of cast and interactions;
- Script: Detailed descriptions of scenes;
- Story board: Shooting-script, including camera positions, lighting, sound etc.

In the P5 script method, we more or less go through the same steps. Starting point is the result of desk research and usability research. Besides the written test reports, interviews, and observations are captured on video tape.

Synopsis: Contents and target group

In the first briefing with the client, contents and target group will be discussed. The client already has a script in mind (who are the consumers and what will be the outcome of product development?). The P5 script is based on objective information and will help the client to get the right script in mind.

The scripts can be used in two different ways in the product development process. The first application is as a source for inspiration for the design team. The second application is using the scripts as reference in the decision making process. Therefore it is important that the team adapts the script, they have to empathise with the actors (identify them with themselves) and recognise the situations.

Script design: Cast and context

Cast: The cast consists of the archetypes of the users. The main actor will be the protagonist, the ideal consumer in the client's mind. There is also an important role for the antagonist, the consumer that is the nightmare of all design teams but cannot be neglected. Actors have primary objectives (concrete and abstract values), sub goals and strategies and plans. Their characters are described in terms of motivation (personal goals, practical goals and drives), character (willpower, impatient, perfectionism, sceptic), decisions (how to achieve objectives, what is the reference, experience), codes of behaviour (guidance of their actions; rude, careful, clumsy) and physical characteristics (poor sighted, left handed etc.).

Context: The context of use is not only based on the physical environment but also the social environment? How do products fit into the daily life and how do they influence behaviour? For instance how does product design influence behaviour leading to reduction in energy use? In the context table for each actor, all critical interactions and relations are described, the starting point of describing the scenes.

SCRIPT: DETAILED DESCRIPTION OF SCENES

In the scenes all interactions, events and dialogues are described and visualised in a way that the designer can imagine what the impact of the design will be. The traditional segments describing a scene in film theory (exposition, crisis, confrontation, climax and resolution) are used where possible.

Story board: emphasis on relations and emotions in illustrated scenes

The medium film has an enormous potential in visualising emotions by the combination of narrative, cinematography, and sound. In illustrated scenes it is possible to emphasis emotions by using the language of the shot; how the image is photographed and framed (e.g. long shot, close up and Point of View: front, profile, bird's-eye/high angle). The impact of illustrations (pictures, video and sound) can be enormous; application has to be done in a conscious way. The pitfall is communicating objective information in a subjective way.

P5 SCRIPTS IN PRACTICE

Since 2001, we have applied the script method in our consultancy practice. In the course of 2002, after release of some of the products, the application could be illustrated in more detail. For now we illustrate the method with an anonymous case, based on our practice in the development of a complete new product. . The product development concerned a - new combination of products that should lead to a new experience for the consumers. Market research lead to the definition of the consumers' needs.

In the usability research, we gained knowledge on all interactions of the users with the product and the emotions concerned. The user-product-relation appeared to be influenced by having control in handling the product, the ease of the interactions and the sound produced by the product.

Based on the findings the client made a first checklist on consumer demands, the script they had in mind. We rewrote this into a script that showed the real practice. The cast consisted of L (the protagonist, the ideal consumer), D (the antagonist, the real consumer not to be neglected) and M (another antagonist, the nightmare of the design team). During the development process the script became increasingly detailed (more scenes could be described). In decision-making the script was used (what can go wrong) and L, D and M were used as consumer archetypes.

Now the project is in the phase of field testing. Before releasing the product, the products, the logistics and communication (user manual, advertising, call-centre, website etc.) are subject to a final check. All are based on the P5 script and the technical specifications of the designers and engineers.

CONCLUSIONS

Product developers have to focus more and more on emotional aspects. They not only need to create a product but a context for experience as well. Experiencing the product and the access to its functions needs to be as satisfying as possible.

The growing interest in user centred design means that focus on the user, in different roles, will become more significant. Usability professionals can give insight in the interactions and relations with the product. Communicating this insight can facilitate and support the creativity of designers. In our practice, the script has proven to be a strong tool in this communicating process.

Besides inspiration, the script is used in the decision making process in all phases of the product development; from the initial concepts to the development of the brand experience. Knowledge gained in market research and usability research are combined as a basis for the script. Using such a powerful tool we take our role as consumers' advocates seriously; this can lead to the ethical discussion where the borderline between consumers' needs and manufacturers drives is drawn. Our next step is to make a more interactive script. Is it possible to make a toolbox designers can use to evaluate their new ideas and concepts with the script?

REFERENCES

Kahmann, R. and Henze, L., 2002, Mapping the user-product relationship (in product design). In *Pleasure With Products: Beyond Usabilit,y,* edited by W. S. Green and P. W. Jordan, (London: Taylor & Francis), pp. 297-306.

Pine, B.J. and Gilmore, J.H., 1999, *The Experience Economy.* (Boston: Harvard Business School Press).

Goossens, C., 2001, Marktonderzoek naar hedonistische consumptie en productervaringen. In *Ontwikkelingen in het marktonderzoek,* , edited by A.E. Bonner et.al., (Haarlen, De Vrieseborch), pp. 31-239 (in Dutch).

Bordwell, D. and Thompson, K., 2001, *Film Art: an Introduction*, 6th edition, (New York: McGraw-Hill.).

Suèr, H., 1996, *Scenario schrijven voor documentaires.* (Abcoude, Uniepers), (in Dutch).

The use of images to elicit user needs for the design of playground equipment

Anne Bruseberg
University of Bath, UK

Deana McDonagh
Loughborough University, UK

Paul Wormald
Loughborough University, UK

INTRODUCTION

Images are a powerful resource to convey meanings, particularly emotional values and experiences. Their application can serve as an important tool to communicate values that cannot be expressed easily through words (Garner and McDonagh-Philp, 2000). Eliciting user needs and aspirations often presents a challenge to design researchers. This is partly because users find it difficult to conceptualise and express their ideals and wishes. Design researchers deal with the difficulties of making user (verbal and visual) data comprehensible to product developers.

Designers often find quantitative data formats, and verbatim reports, inaccessible as they are not sufficiently inspirational and can be perceived to dampen the creative process. This may be due to the time constraints that designers have typically to deal with, but also because such data formats may not convey user information at an emotional level. Empathy with users cannot only be achieved through knowledge of facts, but also requires an awareness of the user context, experience, dreams and aspirations to support a deeper understanding of underlying motivations that drive user behaviour and values (McDonagh-Philp and Denton, 1999).

Users need to be supported in expressing their needs and aspirations. Product functionality is closely interwoven with social and cultural values. Users are not always conscious of their needs or may not regard particular pieces of information as useful. Users are not always able to verbalise their emotions and reflections. It is important to give users different channels through which they can express their requirements and ideals.

Images are able to convey less tangible aspects of the users' experience like feeling and mood, giving users a medium to express themselves. Designers often prefer visual information due to its accessibility. Images can offer the designer and user a shared language aiding communication.

Designers can use moodboards – a technique producing collages of abstract images – to immerse themselves into a particular state of emotions associated with a task or product (Garner and McDonagh-Philp, 2001). This technique can be adapted for application by users to aid in communicating their ideals, aspirations, dreams, and wishes to designers.

Where adults have difficulty in conceptualising and verbalising deeper values, it may be more problematic for children. When attempting to elicit user needs for the design of playground equipment, the use of techniques that make use of images and creative activities becomes imperative.

This paper describes how mood boards and associated techniques have been applied within focus group sessions to retrieve user needs and aspirations for the design of playground equipment. It demonstrates how the techniques can be applied with both children and adults, shows what type of information can be retrieved, and illustrates the usefulness of the techniques through examples.

METHOD

Focus group sessions were used as a vehicle for the user research activities to assist in collecting various types of data. In addition to group discussions, participants carried out a range of different exercises including questionnaires, mood boards, and creative activities. Two focus group sessions were conducted on the same day, each lasting three hours. The research aimed at retrieving perspectives from both children and their parents. Two sessions were conducted, one involving parents only, and the other one focusing on children.

Children's session: The session involved six children (five boys and one girl). They were 7 to 9 years old. Table 1 shows a moderator's guide for the session conducted with the children, outlining the activities. Their parents were present in close proximity. They did not observe the children's activities, but carried out their own activities using a booklet that gave instructions for filling in questionnaires and non-verbal creative activities. These included exercises such as childhood memories of play; what does play mean (expressed through mood boards); evaluating existing playground equipment; imagining the future playground.

Parents' session: The session involved five participants, all female. During the first part, participants were asked to imagine themselves back in the role of seven year old children. Amongst other activities, parents were asked to remember impressions such as smells and sounds. By creating a mood board, they expressed visually their memories and ideals of play. For the second part, they were asked to come back to their role as a parent. This involved activities such as discussion of criteria for playground equipment and rating different types of equipment through questionnaires.

FINDINGS

User research activities involving children have very different requirements to those for adults. Adults are often initially reluctant to conduct creative activities such as brainstorming, drawing and the creation of mood boards – whereas children are more familiar with such activities.

The focus group setting was important to provide a structure, but the discussion itself proved problematic for the child group, since they tried to please, rather than voice their opinions. It had a tendency to turn into a question-and-answer session, since the children behaved as in a classroom environment. Open discussion requires social skills that children of this age have not fully developed. The children varied considerably in their ability to articulate of abstract thoughts and in the consideration of others. Confidence and maturity can vary significantly within the same age group.

Children have a much shorter attention span than adults. It is important to constantly find ways of maintaining their attention through new tasks. Hence, employment of a range of different activities is vital. Children are easy to motivate through activity and interaction changes. Moreover, most activities took much less time than expected – especially creative activities since children found it much easier to approach them. Hence, it is very important to prepare reserve exercises and alternative activities. Moreover, the moderator has to be very responsive. Dominant children tend to be especially challenging to control – since they feel easily rejected and may withdraw from the interaction. It is vital to convey the impression that moderators enjoy the activities themselves. Exercises such as rating of equipment have to be presented as a game (e.g. Task 4 in Table 1). Simple aspects such as the provision of refreshments can be used as motivators, whilst making sure they are assisting and not distracting the session (see Figure 1).

Figure 1 Children constructing their individual mood boards

Figure 2 Example of a child's mood board with descriptive terms

Both adults and children generated their own mood boards in response to the question "What does 'play' mean?" For the parents, creating mood boards helped to bring back happy and positive memories of play from their childhood. The exercise of considering what play means to them, based on their childhood experience, was perceived with much enjoyment and gave vital insight into the range of different perspectives that people have on the contents of play.

An important aspect of using the creation of mood boards to understand user needs and aspirations is to retrieve explanations from the choices of images. This can be done through labelling the pictures (see Figure 2), or through a subsequent verbal explanation of the experiences and feelings expressed. Abstract expressions for 'play' are shown in Table 2. Comments and pictures chosen for the mood boards showed that the majority of the adults, as children, like to play outside and to be close to nature. The children responded literally and tended to list activities, whilst adults perceived the question more abstractly.

Table 1 Moderator's guide for child's session.

Date: **21.4.01**	**Focus Group: Play Areas and Parks** (afternoon session)	(total: 3 hours)	
	Description	*Aids*	*Start*
	People arrive, get drinks and snacks		1.00
TASK 1 Introduction and warm-up	Game for both the adults and children to help the children settle in, match 2 picture cards. Adults sent to different location to fill in a pre-prepared booklet separately.	booklet for adults; picture cards	1.15
TASK 2 What is play? What isn't play? *(discussion/ drawing)*	If I was an alien from outer space and had never been to this place before, tell me what I could do to play? What do you do to have a good time and enjoy yourself? Explain in words what is good play/bad play. Explain with a drawing what play means to you. (Prompt with describing favourite toy).	large paper for drawing play ideas	1.30
TASK 3 *Mood board exercise* What is play?	Ask children to pick three pictures which remind them of play from images on table; Task: think of something which you enjoy doing, is fun and exciting; then children explain their choices (e.g. what is the best thing about playing?)	mood board	1.50
TASK 4 Rate and name equipment *(rating and discussion)*	Giving a name to pieces of play equipment. Write names on sticky labels, and stick onto pictures as a group activity (tick a yellow star on your favourite, exciting and think about why; stick a red triangle by the one which is your least favourite and say why (least fun); green oblong sticker by the one which you think is most fun) What games do you think they would play? Ask what do you think they eat; what do they watch on TV; what music do they listen to; what clothes do you think they would wear?	pictures of equipment	2.05
BREAK			2.20
TASK 5 Magnet play area What would you like in a play area? *(choosing pictures)*	Use picture of unattractive urban environment. What would you like to be there? Provide a number of items printed onto sheets to cut and stick (e.g. trees, animals, play equipment, water, sand, seats, benches, textures – ask children to pick out good and bad items before they start sticking. Choose most important item first (6 items so no fighting). Stick magnetic strip on chosen pictures (with help) to stick them on their section in the play area marked on the board.	images, magnets, metal surface, paper sheet	2.30
TASK 6 *Draw* your dream play Area	Ask children to draw a dream place to play, using a variety of pens, pencils, crayons, stickers etc.; when they have had enough time ask children to explain their pictures. If you could choose anywhere to play, what would it be like? Make clear that they can draw anything they want.	paper, pens etc.	3.00
End session	Ask questions to find out what they enjoyed and didn't enjoy. Distribute goodie bags and disposable cameras to record their play 'space' – what they consider to be play (to return it in the envelope provided within a week, include SAE)	evaluation sheet	3.30 4.00

Table 2 Terms expressing 'play'

Adults	Children
sun; playing with friends; playing alone; happiness; water; enjoyment; imaginative play; freedom	fun; playing games with friends; going to the park; swimming; going to the cinema; playing on the computer

Parents were reluctant to return to the parental role afterwards. Although the session assembled a culturally diverse group (e.g. English, Eastern Europe, Africa), they shared the view that the essence of play was as a social activity.

The children clearly conveyed the importance of movement for the design of playground equipment. The favourites were climbing trees or climbing frames. The children also considered the role of the parents. They wanted the adults to be within sight, but far enough away to dissociate from play. They were concerned about the comfort of the parents – but mainly from the perspective of being able to stay longer at the playground!

CONCLUSIONS

User research techniques based on group sessions can be used as a mediation tool between users and designers. It is vital to enable effective communication. Activity sessions provide opportunities for designers to integrate a range of activities and tasks that support the elicitation of user information, experience and dreams in diverse formats, whatever the age of the user. Group work can exploit the synergetic effect created through discussion and activities conducted within a shared environment.

Designers benefit from the breadth of different activities that can be employed, since the choice of exercises can be adapted flexibly. In comparison to written reports, visual data can provide a more immediate format, which is more easily accessible and understandable by designers. Designers gain particularly from the combination of visual material with spoken words.

Images help create a shared language between the user and the designer. User needs can be expressed in non-verbal ways, through the use of images. Research activities presented as play can provide rich data. The closer designers are involved in research, the deeper they can immerse themselves in the user's world. The availability of inspirational material not only helps understanding, but exposes one of the most important resources of designing: learning through play.

REFERENCES

Garner, S. and McDonagh-Philp, D., 2001. Problem interpretation and resolution via visual stimuli: The use of 'mood boards' in Design Education. *The Journal of Art and Design Education*, **20** (1) pp. 57-64.

McDonagh-Philp, D. and Denton, H., 1999. Using focus groups to support the designer in the evaluation of existing products: A case study. *The Design Journal* **2** (2) pp. 20-31.

EVALUATIVE TOOLS

Researching users' understanding of products: an on-line tool

Mirja Kälviäinen
The Kuopio Academy of Design, Finland

Hugh Miller
The Nottingham Trent University, UK

INTRODUCTION

Over the years, there have been various attempts to identify the nature, character, social or personal identity of designed objects (Janlert and Stolterman, 1997, Shackleton, 1996, Chuang and Ma 2001, Veryzer, 1997, Nagamachi, 2001). These aspects are part of what makes objects appropriate, pleasurable, or marketable, and understanding how they are perceived by users should be helpful in the design process for producing enjoyable objects.

We have been developing a method, based on free-sorting and multi-dimensional scaling, for describing the social and design dimensions of objects, as seen by users (Miller and Kälviäinen, 2001). The method makes it possible to produce a well-categorised pool of examples, which might be useful to designers, from the responses of users who might not have ready-made schemes for categorising or describing objects.

This paper describes further research aimed at developing the technique into a tool which can be used to support design work. We describe interviews with designers about what information they might need from such a tool, and where they might use it in the design process.

BASIC METHOD

We gave respondents thirty six photos of living room chairs, representing a wide range of styles and constructions, mounted on 11x13cm cards, and we invited them to sort the cards into piles of chairs that were similar in social ('how they fit into people's lives') or design ('how they look') terms. They were told that they could use as many or as few piles as they liked, and could move cards around between piles until they were satisfied with their sort. Once they had completed a sort, they were asked to describe each pile. We carried out the study in Finland and in Britain, with 30 respondents in each country.

The stimuli we used were small two dimensional representations of pieces of furniture – which are really large three dimensional objects meant for sitting on and touching. This raises the question of how much this can tell us about people's preferences for the real objects. When people make decisions about purchase or use of furniture, then seeing and sitting on the real thing is important.

Our view is that rather than being concerned with someone's choice of a chair for themselves, we are interested in finding out about people's *ideas* and *impressions* about a range of chairs and their understanding of the meanings of chair design for a range of people. We want to find out how people look at these things (metaphorically), so looking at pictures (literally) seems appropriate. All the same, people are accustomed to mapping from pictures to objects, and our respondents were quite ready to give judgements about aspects like comfort or feel of the fabric just from the pictures. If this became a design tool, it would be a conceptual and exploratory tool, rather than a guide for the final, material version.

Information on the groupings for all the respondents was pooled separately for the social and design sorts to produce two confusion matrices which showed how often each example was grouped with each other example. This matrix was then subjected to multidimensional scaling analysis. We found that a three-dimensional analysis adequately represented our results, with most of the information accounted for in the first two dimensions. This analysis was used to produce visual plots of our examples placed in a space defined by the major dimensions. Each dimension was labelled by reviewing the descriptions applied to those examples which were most strongly differentiated by that dimension. Fig 1 is an example of such a plot.

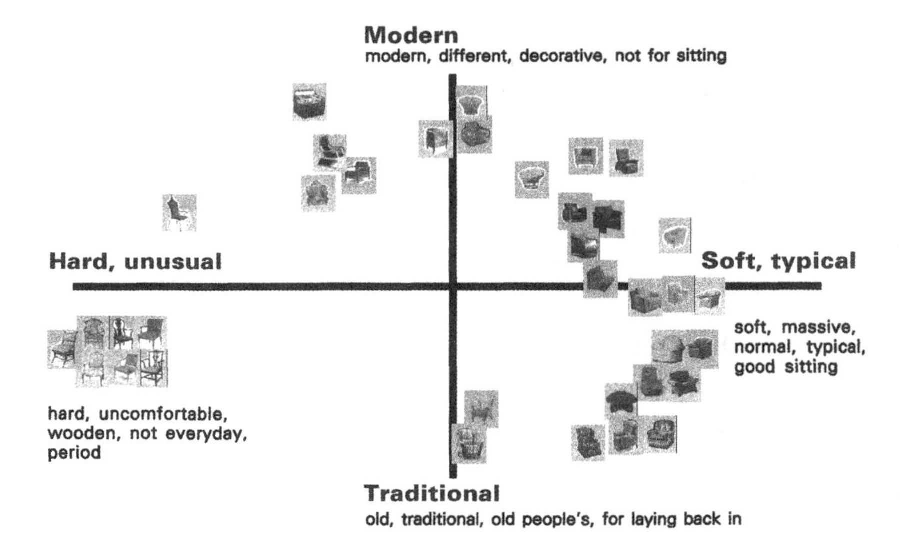

Figure 1 Two dimensional plot from the Finnish 'design' sort

Descriptions which were commonly used in the social sorts included 'comfortable-not', and 'modern-traditional', but sortings based on the kinds of people who might own such chairs, the kinds of room in which they might appear, or the kinds of activity they might support, were also apparent. Although descriptions of groupings in the 'design' sort were largely visual, we got the impression that they were closely tied to social aspects, and the construction 'looks like [some social categorisation]' was often used. The grouping of 'hard' and 'unusual' may be artefactual. There were several 19th century chairs which were commonly grouped together, and seen as 'unusual' or 'not for ordinary places' (or ordinary people) - and these chairs were also obviously wooden and less padded than many of the other examples, hence also 'hard'. It is also possible that the 'usual' kind of chair for a living room is 'soft' for 'good sitting', so the dimension may represent a design judgement on what is most appropriate for the setting. 'Comfortable' occurred in both areas, but was perhaps more literal and ergonomic in the design sorts, and metaphorical and personality-related in the social sorts. There were considerable similarities, both in categories used and in structure, between the social and design sorts.

We wanted to avoid anticipating the ways in which users might describe objects and the conceptual structure within which they might organise those descriptions. This might be a good way of understanding users' understanding of the objects but it isn't an approach which seems likely to produce explicit material design guidelines. In fact the study showed how non-designers did not see the world in the object-centred way that designers use, based on material and form. Their views about objects were social and private life centred. It was more important how a chair fitted into everyday life than how it was constructed. So, in the 'design' sorts, people used few descriptions which related to any specific design or constructional aspects of the objects, and concentrated on rather general functional, emotional, or social labels. This gives rise to another kind of design guidelines, which we feel would be useful for designers.

RESEARCH WITH DESIGNERS

In the course of doing the research, we showed the materials and the technique to furniture designers, and several commented that they would like to see examples of their own work included, to 'see how they came out' in relation to the overall pattern. This raised the question of how the method could be systematically used in the design process. Another reason for investigating how this approach might be used as a practical tool is that we are working on an electronic implementation of the methodology This would make it possible to use it very flexibly, with variation of the examples used, immediate analysis and feedback, and the potential to get information from many people, quickly and easily (the present method, though fairly quick and easy for the respondents, is slow and labour-intensive for the researcher). So we need to consider how and why designers might find the method useful, and just what kinds of input and output would be most useful for them.

We interviewed a small number of designers and design project managers in Finland to explore the usefulness and possible development of this method. The respondents were given a description of the sorting task and shown the cards and the plots that resulted from the analysis. They easily grasped the point of the research and found the output understandable, interesting and useful. We discussed possible software development and then asked: "As a designer, how do you see you could use this method as computer based software tool?"

Our respondents thought this tool might be especially suitable at the beginning of the product development process, where it is necessary to scan for the right direction for the design. For this scanning a wide variety of picture material should be acquired or produced, even more than it would be usual for a designer to acquire in the first phases of visualisation. This tool is naturally applicable several times during product development. The nature and purpose of the application changes according to the needs of each stage in the product development.

The designers felt very comfortable with the kind of pictures the research produced. An analysis which gives them groups of pictures with different explanations of how the users saw them was useful from their point of view. They felt capable of using this kind of result because that is the way they would work anyway: it is just that these visuals might come from a more informed basis.

The method could also be used to aid discussion between designers, managers, and clients during the product development process. Designers emphasised the importance of discussion with managers, because they felt that speaking the same visual language within the company was a constant problem

It is important that an electronic version should maintain the direct involvement with the images and the relationships between them which is possible when physically sorting cards into piles. It should be possible to see most of the collection of examples at any one time, and assignment of examples to groups or movement of examples between groups should be done simply and directly. It could be used with commercial customers or final users, and an Internet compatible version would offer international possibilities.

Figure 2 Examples of 'pleasurable' (left: "soft, good to sit in, resting place, homely, brisk, happy, stylish, splendid, clear") and 'displeasing' (right: "upsetting, heavy, gloomy, horrible, irritating") chairs

EXTENDING THE METHODOLOGY

In our pilot, nondesigners were not very able to categorise the products by the visual appearance/design. They went easily into other attributes, e.g. product affordances ("comfortable") or emotional descriptions ("distressing") and did not describe concrete product features. However, a design analysis would be important for the designers and companies in deciding their visual directions. This is outside the scope of the method described here, but it could be carried out by a visual specialist after the software has produced suitable groupings of products.

The Finnish respondents were asked to sort the chairs into 'pleasurable' and 'displeasing' piles. Our design respondents were particularly interested in profiles derived from this sort. There was strong discrimination across the sample, and agreement on descriptors like soft, splendid, good to sit in, brisk, happy, stylish, clear for the liked chairs and horrible, ugly, heavy, gloomy, distressing, strange for the disliked ones. As before, many descriptors are emotional rather than design terms. Examples of the most pleasurable and displeasing chairs are in figure 2.

It was also suggested that there was a need for a picture archive with 'standard' examples, and examples which could be associated with particular market segments and personas. This might be developed into standard sorts against which developing products could be quickly evaluated.

CONCLUSIONS

In product development it is necessary to understand the translation of concrete product features into the images and meanings attached to the product. This picture research tool helps to show how the holistic image of the product and its exterior features attach the product to certain experiential mental images: the purpose of use, user group or product personality. The tool could aid the development and definition of brand image and produce information about product visualisation, acceptable product features and product image. The same picture research methods might also be used in more traditional usability research by using pictures that represent various situations in which the product might be used.

The designers and design companies we approached could see the usefulness of this approach, and so we are encouraged in further development of the method. We intend to conduct further piloting with the companies, following through their design process to see what they really need, how this method can be applied, and how the software can best support the research and analysis process. We will also work with the Lahti Design Research Centre to review existing design tools in this area.

REFERENCES

Chuang, M-C. and Ma, Y-C., 2001, Expressing the expected product images in product design of micro-electronic products. *International Journal of Industrial Ergonomics*, **27**, pp. 233-245.

Janlert, L-E. and Stolterman, E, 1997, The Character of Things. *Design Studies*, **18**, pp. 297-314.

Miller, H. and Kälviäinen, M., 2001, Objects for an Enjoyable Life: Social and Design Aspects. In *Proceedings of the International Conference on Affective Human Factors Design (Singapore),* edited by Helander, M., Khalid, H. M. and Ming Po, T. (London: Asean Academic Press), pp. 487-494.

Nagamachi, M., 2001, *Research on Kansei Engineering: Selected Papers on Kansei Engineering* (Tokyo: Nakamoto Printing).

Shackleton, J. P., 1996, *The Application of a 'Prototype Theory' Framework to the Modelling of Product perception and the Emergence of New Product Groups.* Unpublished doctoral thesis, Chiba University, Japan

Veryzer, R. W., 1997, Measuring Consumer perceptions in the Product Development Process. *Design Management Journal*, **8**, pp. 66-71.

Getting what you want, or wanting what you get? Beyond user centred design

Wendy Olphert
Loughborough University, UK

Leela Damodaran
Loughborough University, UK

INTRODUCTION

New technology offers the potential for tremendous benefits to people and society. Sackman, in his influential book published in 1967, discussed ideas for augmentation of human capability through the concept of 'human-computer symbiosis', and anticipated a world in which people would be freed from the drudgery of routine tasks and empowered by computer technology to expand their horizons and creativity. But, as computer technology proliferates and becomes ever more sophisticated, how close are we to realising this inspiring vision? Although there are undoubtedly many benefits from the advent of new technology, it could also be argued that, in many cases, the technology has simply brought new kinds of drudgery and different kinds of routine tasks. However these consequences do not usually arise from the technology itself, but from the way it is designed, implemented and used. Williams and Edge (1996) note that there are choices (though not necessarily conscious choices) inherent in both the design of individual artefacts and systems, and in the direction or trajectory of innovation programmes, and that these choices may have differing implications for society and for particular social groups. If this is the case, then technology can be seen as negotiable, with scope for particular groups and forces to shape technologies to their ends, and the possibility of different kinds of technological and social outcome. This then poses two kinds of problems: what are the ends which people want technologies to serve, and how can people best influence the design and development of technologies to ensure that their objectives are achieved?

DISCOVERING REQUIREMENTS

The first step in the process is to establish what people want to achieve, and how technology can help. It is now widely recognised that successful systems and products begin with an understanding of the needs or requirements of the users, in other words a user-centred approach to design. As instantiated in ISO 13407, user-centred design begins with a thorough analysis of the characteristics of users, their tasks and the environment or context of use. This analysis leads to the formulation of a set of requirements which are then used to produce design ideas and concepts that are prototyped and evaluated by users. This process, and the active involvement of users in reviewing and testing design options, should certainly ensure that when the end result is delivered, the users will find it acceptable. But whilst this approach might lead to users wanting what they ultimately get, that is not the same thing as helping them to get what they want.

Although the form and direction of future technologies may be negotiable, there are many reasons why we may not exercise real freedom of choice. At the design level, earlier technological choices tend to pattern subsequent development, and options initially selected from a range of possible alternatives may become entrenched. New technologies tend to develop cumulatively, erected upon the knowledge base and social and technical infrastructure of existing technologies, and where increasing returns are sought for investment, this can result in 'lock-in' to established solutions (Williams and Edge, 1996). Design methods and approaches can also act to restrict choice. Conventional approaches to systems design have tended to be aimed at eliciting, and satisfying, functional (task related requirements) and have ignored other kinds of user requirement. However the non-functional requirements of users may actually be equally or even more important for the effectiveness as well as the desirability of the system.

At the level of the individual user, our experience of existing technologies and products will influence and constrain our expectations about the "shape" of future technologies and products. In the workplace, we are not used to systems which are rewarding and fulfilling to use. This is not entirely surprising, since very often it is the physical characteristics of the users which is taken as the starting point (the 'ergonomics') for the design, rather than the psychological characteristics and, more importantly, the 'higher order' (Maslow, 1954) needs or wishes, of the users. Herzberg (1959) draws a distinction between 'hygiene' factors, which cause dissatisfaction at work, and 'motivator' factors – which produce satisfaction. To motivate and fulfil users of systems in the workplace, it is not enough to meet their 'hygiene' needs – systems should help users to achieve their own personal goals, and be consistent with their values and aspirations. Maslow (1954) suggested that humans are ultimately motivated towards self-actualisation, the development of our full potential, "to become everything that one is capable of becoming". There are now many applications of technology, in leisure and education for instance, where interaction with computers is highly enjoyable and rewarding for the individual. Particularly perhaps as the work/non-work boundary becomes increasingly blurred, users will expect systems in the workplace which minimally deliver the same qualities of interaction (joy of use, not just ease of use), and maybe even to help them to achieve their own particular peaks of human achievement.

One of the major problems with requirements such as these is that they are hard to elicit. In many cases they cannot be directly observed or easily articulated; they may, for example be embedded in organisational structure and policies, or they may not even reside in the users' consciousness. Robertson (2001) distinguishes between three classes of requirements: conscious requirements, unconscious requirements, and undreamed of requirements. Conscious requirements are those that stakeholders are particularly aware of. U nconscious requirements are those that stakeholders do not realise that they have. Reasons for having unconscious requirements might be that stakeholders are so used to having a requirement fulfilled that it does not occur to them to express it. Undreamed-of requirements are those which do not occur to stakeholders because they cannot imagine what it might be like to have access to a new kind of technology or product. Unless we can elicit or bring these requirements to the surface at the design stage, there is no prospect of designing products and systems which deliberately aim to meet them.

Even if we find ways of eliciting these goals, values and aspirations, it is often difficult to link these to the features of a particular product or system designs. Prototypes and simulations can be a good way of finding out from users whether the features of a design meet their needs, and the more realistic the representation, the easier it is to evaluate – but research also shows that as design representations become more 'concrete', evaluations become more focussed, i.e. the scope for imagination of alternatives becomes much more limited. Through successive iterations of prototyping and evaluation it may be possible to ensure that the qualities and features that the evaluators do not like are designed out, but it is harder to ensure that qualities and features that they may have liked are 'designed in'. To do this requires a different strategy from the conventional systems analysis approach: one which begins not with analysis but with imagination, and which encourages the widest exploration of opportunities and possibilities before commitment.

A further problem is in deciding who is the 'owner' of the requirements. In addition to the potential users of a system or product, there are numerous other stakeholders who will have a legitimate 'stake', or interest, in a system. Most of the stakeholders are likely to have requirements of the system, and the requirements of some are likely to conflict with the requirements of others. Satisfaction of different sets of requirements will lead to different consequences for the various stakeholder groups, and those with different kinds of 'stake' in a project should be able participate in exploring the implications and shaping the eventual decisions.

TOOLS FOR SHAPING TECHNOLOGY

In order to overcome the difficulties raised by determining requirements, designers need tools which help people to imagine or 'surface' their requirements, and which help people to explore and evaluate design options and their implications. Although most conventional approaches to system design do not make use of them, there are many potential candidates, and some of the most promising are described below. The two objectives for the tools are not mutually exclusive – many tools will have benefits for both parts of the requirements determination process.

Imagining and surfacing requirements

The most important features of successful techniques for supporting imagination and surfacing requirements is that they should enable people (not just stakeholders, but also designers) to break away from the constraints which tend to arise from the existing situation or current ways of doing things ("out of the box thinking"). Brainstorming is a well-known technique which encourages such thinking, but there are other techniques which may help to 'break out' from current situations or ways of thinking, such as 'future search' techniques which are widely used in strategic planning. Search conferencing for example (Emery,1993) starts with the question "Where do you want to be 10 years from now? and, once participants have arrived at a defined goal, then seeks to establish an agreed process by which it can be achieved. Focus Groups, widely used in marketing, have also been used by designers both as a way of eliciting requirements, and as a way for designers to gain additional insight into the needs and aspirations of users (Bruseberg and McDonagh-Philp, 2001). The use of 'rich pictures' as in Checkland's Soft Systems methodology (1981) can help stakeholders to map, explore and understand a complex 'problem space', and thereby help to surface obscure requirements. Robertson (2001) lists a variety of techniques to help with what she calls 'trawling' of requirements, above and beyond the conventional step of interviewing stakeholders. Although some of these techniques may already be familiar to systems designers (such as prototyping and use case workshops), there will be many which are not, but Robertson describes how these have been successfully used in the requirements discovery process.

Envisioning and evaluating design options

The requirement here is for techniques which not only allow stakeholders to visualise the form of design solutions, but which provide sufficient detail for them to be able to assess the 'impact' of the design at different levels – at the individual level, for ease of use, joy of use, etc. and at the collective level for organisational and/or social impact. Sketches, prototypes and mock-ups have long been a part of the designer's repertoire of tools, and each of these forms of representation offers stakeholders the opportunity to explore and evaluate aspects of design. Lower-fidelity representations such as sketches or diagrams may be better for testing out concepts because they are seen as more 'disposable' and less costly, but they are only capable of conveying superficial or limited aspects of a design, and may be capable of being interpreted in widely different ways by different stakeholders. Scenarios and use-cases have become popular in the requirements engineering community as a way of exploring and defining requirements, but these too can lead to different interpretations and resulting requirements. In order to explore the implications of designs more fully, more realistic representations are needed, and VR technology is increasingly being used as a tool for this purpose. Davies (1999) for example describes the 'Envisionment Foundry' – a virtual reality tool which has been developed to support the participative design of work environments. Other experiments, such as Nokia's Studiolab (Kaasinen, 2002) are currently exploring the use of VR to support design evaluations. Simulation environments such as this may be relatively costly to set up initially, but the representations which they enable, have the twin benefits of being highly realistic and yet relatively easy and cheap to change.

CONCLUSIONS

If users are to get what they want, instead of simply settling for 'wanting what they get', then this paper argues that the design of systems needs to look beyond the current repertoire of techniques for user-centred design for techniques that stimulate imagination, encourage exploration, promote participation, and enable exploration and debate of the implications of technology and design.

REFERENCES

Bruseberg, A. and McDonagh-Philp, D., 2001, New product development by eliciting users' experience and aspirations. *International Journal of Human-Computer Studies*, **55** (4), pp. 435-453.

Checkland, P., 1981, *Systems Thinking, Systems Practice*, (London: Wiley).

Davies, R., 1999, *The Envisionment Foundry – a Virtual Reality tool for the Participatory Design of Work Environments*. Proceedings of the Workshop on User Centred Design and Implementation of Virtual Environments 1999, Kings Manor, University of York, pp. 23-43.

Emery, M., 1993, *Participative Design for Participative Democracy*, (Canberra: Australian National University: Center for Continuing Education).

Herzberg, F., Mausner, B. and Snyderman, B., 1959, *The Motivation to Work*, (London: Granada).

ISO DIS 13407, 1997, *User-Centred Design, ISO Draft Standard*, International Standards Organisation, 13407.

Kaasinen, E., 2002, *Personal Communication*.

Maslow A., 1954, *Motivation and Personality*, (New York: Harper & Row).

Robertson, S., 2001, Requirements Trawling: techniques for discovering requirements. *International Journal of Human-Computer Studies*, **55** (4), pp. 405-423.

Sackman, H., 1967, *Computers, System Science and Evolving Society*, (New York: Wiley).

Williams, R. and Edge, D., 1996, *The Social Shaping of Technology, Research Policy*, **25**, pp. 856-899.

Making sense: using an experimental tool to explore the communication of jewellery

Jennifer Downs
Sheffield Hallam University, UK

Jayne Wallace
Sheffield Hallam University, UK

The researchers, who are both jewellery designer/makers aim to communicate through their work by creating inalienable objects, which encourage personal, social and physical interaction. In order to understand and relate to the users there is a need to explore viable methods for user research that enable evaluation. As this is not a perfect problem this pilot study does not produce a benchmark, but suggests new ways to analyse design and craft practice. The authors have undertaken an exploratory exercise into the use of 'Kelly's Repertory Grid Technique' to analyse perceptions of twelve jewellery objects. This pilot study was aiming primarily to discover how others perceived the work of the makers and also to open up the possibilities of applying a scientific tool to design. George Kelly devised the repertory grid technique in the 1950's as a clinical psychology method (Kelly, 1955), initially used to examine and measure changes in attitudes of individuals. Here it is used to elicit information regarding individual perceptions of jewellery. Findings created viable ways to explore personal constructs that influence peoples' relationships to objects.

TECHNIQUE PROCEDURE

Six individuals were selected from a range of age groups, gender and background. It was important that none of the participants came from an art and design background thus avoiding people who regularly engage with, and discuss objects. The researchers also tested themselves in order to experience the procedure, considering how they view the pieces through this framework.

The participants were presented with a board on which were placed twelve items of jewellery and supporting images showing the pieces when worn (as illustrated in Figure 1). The images were important to contextualise the objects and present them as intended by the maker or owner. The six trials were video recorded in order to gather qualitative data. A second grid is used to record the elicited responses and the attributed numerical values (as illustrated in Figure 2).

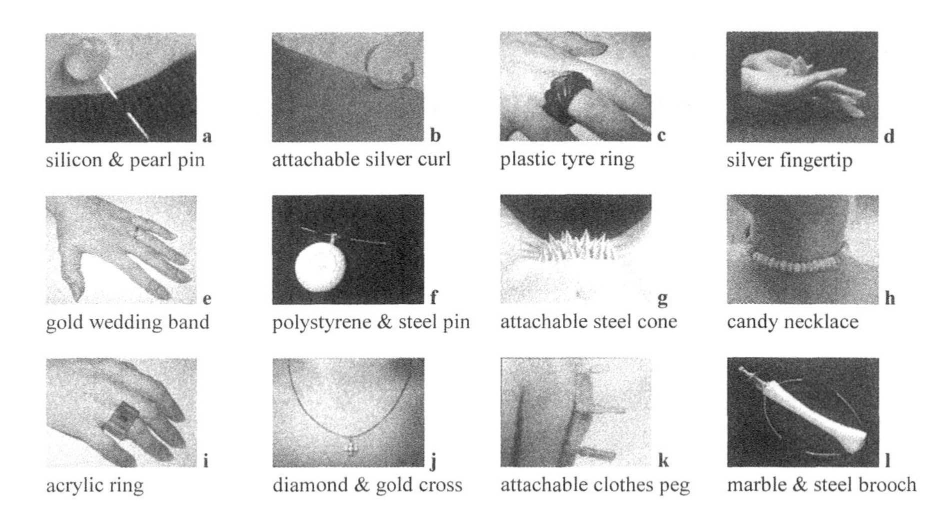

Figure 1 Supporting images of the twelve elements as they were presented on the board, plus text.

1	a	b	c	d	e	f	g	h	i	j	k	l	5
delicate	2	**1**	4	3	3	**1**	2	3	3	4	5	**5**	**brutal**
emotional													rational
accessible													inaccessible
permanent													non permanent
controlled													uncontrolled

Figure 2 Grid used by participants during this pilot study showing the four sets of given constructs provided by the researchers and (in bold) illustrating an example of a participant's chosen constructs and attributed values. The letters at the top of the grid were shown with images that represent each element.

Step 1 Familiarise yourself with the objects. Select three of the objects, two of which are similar to each other and one of which is in some way different.

Step 2 Use words to describe this similarity and difference.

Step 3 Place these terms/constructs on the grid. The constructs and the objects described as 'similar' are always given a score of one i.e. in Fig 2 the participant saw **(b)** and **(f)** as being 'similar' and chose the construct **'delicate'** to describe this similarity. In the same way the constructs and the objects seen as 'different' are given a score of five, seen here with element (l) chosen by the participant as being 'different' to the other two and described as being **'brutal'**.

Step 4 The remaining elements are then all rated by the participant on a scale of one to five i.e. in Fig 2 a rating of one would indicate that the element was perceived as 'delicate' as opposed to a rating of five which would indicate the element was perceived as 'brutal'.

Step 5 This process was repeated nine times, ten times in total. Six times with constructs elicited by the participants and four times using constructs provided by the researchers.

PROVIDED CONSTRUCTS

Four sets of constructs were provided for all the participants: emotional/rational, accessible/inaccessible, permanent/non permanent, and controlled/uncontrolled. This provided common elements, which could be considered by all six participants. This guaranteed a consideration of the elements regarding emotional content. It is arguable that each person brings with them their interpretation of the words, however the terms can be further cross referenced through the grid technique, which then develops individual interpretation of each participants' response. The researchers had their own perceptions of their work: the objects made were meant to be transmitters of meanings and ideas; meanings that they wished to be perceived by others. The motivation, for the researchers, in making jewellery is to communicate. The two pieces made by the researchers were the curl (artefact b) and the fingertip (artefact d). The maker's aim through the curl was to express notions of existence, by questioning permanency and through fragility, encouraging the user to handle the object sensitively and with consideration for its content. The fingertip covers the fingerprint of the wearer, shielding an element of identity when worn and leaving a visual memory of the person when not. The maker was considering the marks people leave on each others lives, accepting loss, yet allowing their presence to remain. These concerns led the researchers to their choice of given constructs.

ANALYSIS OF DATA AND FINDINGS

Qualitative analysis drew out questions that could be answered through quantitative analysis and vice versa. There was evidence to suggest that there is an overriding response to jewellery concerning how and what one should wear in a social context. There are concerns relating to appropriateness, tradition, taste, value, familiarity, material, comfort and gender. This is worthy of further analysis and testing, to question and establish jewellery as a social indicator. (Dant, 1999) There are further indicators suggesting a relationship between gender, occupation, social status and how one perceives jewellery. (Bourdieu, 1984) The terms religion/man, ancient/modern, childhood/death were elicited from a male with a background in philosophy and politics. The terms erotic/non erotic, delicate/brutal, medical/non medical were elicited from a female dental hygienist. The terms non-formal/formal, careful/careless, traditional/non traditional were elicited from a female information technology researcher.

Analysis has begun with looking at the data gathered relating specifically to the researchers' own work i.e. the curl (b) and the fingertip (d). The data from all six participants has produced an aggregate exposing a significant positive correlation between the curl and the fingertip. However the curl correlated positively, even 'more significantly', with the cone (g), unlike the relationship between the fingertip and the cone which was seen negatively. It is difficult to deduce why this is, initial inferences question possible links to material, colour, image, style and femininity. Material and colour could be discounted as all three elements were metallic and silver in colour and should therefore have equally correlated on those counts, but did not. The notions of image style and femininity may be the common bonds. One female participant saw all three elements as both erotic and delicate as opposed to non erotic and brutal which could suggest a connection to gender, sexuality and in this case the female form. This can be explored through additional analysis of data or further testing. On closer interrogation, through looking at the objects, possible explanations became apparent. The cone and the curl both have strong semantics; both could be seen as close to 'pure forms'. From the twelve elements used nothing is as visually and as singularly physically sharp as the cone and nothing is physically lighter than the curl. Both represent extreme physical forms as opposed to the other objects, which are more ambiguous.

The relationship between the curl (b) and the fingertip (d), the polystyrene pin (f) and the gold wedding band (e) is another interesting example (as illustrated in Figure 3). The three elements that correlated most highly with the curl are the cone (g), fingertip (d) and polystyrene pin (f) and the three elements that correlated most highly with the fingertip (d) are the curl (b), cone (g) and gold wedding band (e). The elements of difference between the curl and fingertip are therefore the polystyrene pin and the gold wedding band.

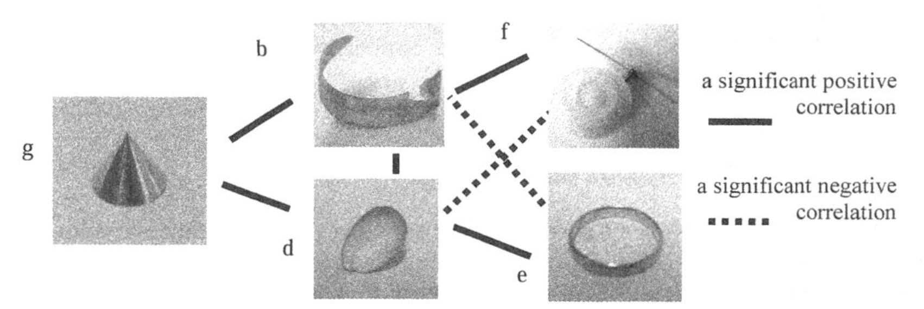

Figure 3 diagram of significantly correlating elements to the curl and fingertip.

A possible explanation could be that the curl and polystyrene pin are perceived in a form and material sense and the fingertip and gold wedding band are seen to represent culturally and socially established concepts. For example the notion of a fingerprint representing identity and a gold wedding band representing marriage are both commonly shared concepts within western culture.

When analysing the constructs given by the participants it became apparent that they were descriptive of form and function rather than emotional content. This leads to question what factors determine the level at which one engages with objects? Possible suggestions are that: 1) the test itself was an alienating process because: i) the researchers were not able to assist the participants with their responses therefore creating an artificial environment. ii) The participant felt suspicious of the situation and feared unwittingly revealing aspects of their personality or providing 'wrong' answers. iii) The individuals were being video taped and possibly felt intimidated and inhibited. 2) It is possible that these individuals had no interest in the meaning jewellery may have. Other questions relate to how much and what kind of information needs to be presented with an object in order for specific meanings to be read.

It would be worthwhile testing the same participants again with the same objects, providing narratives and images, which differ in each round of testing, in order to compare results. It is often debated whether a craft object is 'successful' if it relies on a narrative to accompany it, (Dormer, 1997). Further exploration will be made into this, using the grid method. There were similarities between the participants' choice of correlating elements and the researchers' correlations. So does narrative have an impact on perception? The researchers knew the intended narrative of the pieces, but the participants did not. Further exploration is required. Do makers who are familiar with analysing the form and function of objects, engage and respond to objects on a differing level to non-makers (Carey, 1992)?

CONCLUSIONS

The value of this method of eliciting perceived meanings, is not as a prescriptive tool for designing and making, but as a post design tool. It provides an individual's complex and specific understanding of an object, its context and how it is communicating. This information supplies a designer or maker with a more thorough understanding of the environment in which their work exists. Surely an understanding of one's audience and stage can only lead to a better performance? The method allows the practitioner to understand the decisions they will have made during the design/making process. The resulting transparency can demystify the sometimes intimidating task of creating, leading to more self aware, 'emotionally intelligent' professionals. (Goleman, 1996) There is a worry that using a scientific method may limit creativity, however a freedom can result from a closer understanding of the meanings and emotions in play. The researchers found the method to be inspiring, creating a clearer, broader view of their own practices. Meaning does not come from an object, but from the experiences brought to it by the user and as such objects are perceived differently by different people. (Kelly, 1995) Within jewellery practice there is a constant search for new methods by which to advance and further understand what is happening and how work communicates on a social and physical level. The use of this method raised many questions, primarily: how can makers express an idea most effectively, what are the future possibilities of applying this social scientific method to the creative process, and finally, how can jewellery make sense?

REFERENCES

Bourdieu, P., 1984, *Distinction* (London: Routledge)

Carey, J., 1992, *The Intellectuals and the masses: Pride and prejudice among the literary intelligentsia 1880 –1939,* (London: Faber and Faber).

Dant, T. 1999, *Material culture in the social world.* (Buckingham: Open University Press).

Dormer, P. 1997, *The culture of craft: Status and future*, (Manchester: Manchester University Press).

Goleman, D. 1996, *Emotional intelligence*. (London: Bloomsbury).

Kelly, G. 1955, *The psychology of personal constructs Vol. 1, A theory of Personality*, (New York: Norton).

Semiotic product analysis

Anders Opperud
Volvo Technology Corporation, Sweden

INTRODUCTION

Humans have through all times struggled to understand the reality they are surrounded with. In our everyday life we search for usable hints and signs, which can guide our interpretation of reality. When we first perceive a product, the attention will be drawn to signs that can help to identify and categorise the product. This mental identification process, which usually takes milliseconds, creates our opinion of what kind of product it is, what qualities it has and even who the user might be. What is it that really happens in this process?

This paper gives an introduction to a method that is able to map this mental process, and also gives a brief presentation of three of the theories that this method is based on. The method, called Semiotic Product Analysis (SPA), was developed within a master thesis called "Is Beauty in the Eye of the Beholder? – A semiotic study of cellular phone design" (Opperud, 2001), and is today utilised at Volvo Technological Development Corporation.

THE SOCIAL CONSTRUCTION THEORY

The social construction theory, expounded by George Herbert Mead (Dittmar, 1990), focuses on the fact that physical objects function as communicators of social meaning between people. The physical object becomes a sign with symbolic qualities that people can interpret and thereby retrieve information about both personal qualities of the user of the product, as well as qualities of the product itself. This information could be the users personality, group belonging, political values, leisure interests or socio cultural status.

The theory is based on the fact that people interact with the world and participate in social activities, leading to a continuous construction and reshaping of common cultural and social meaning systems. In other words, there exist a social meaning system as a context for mediating social and cultural knowledge, which we use to construct our subjective reality. Mead was especially interested in the unique role that symbols and the common social meaning systems have for the development of an individual identity. A mother, who shows a picture in a book to her child and says that the one who lives in this great house is a clever and successful person, introduces this meaning system to her child. This meaning system later helps the child to navigate in the social and cultural context where material products are increasingly more important as communicators of values. In this reality, it is the designer's job to decode the common values and opinions that exist in the culture, and reproduce them into forms that embody of the appropriate symbolic meaning.

COGNITIVE PROTOTYPE THEORY

The cognitive prototype theory (Rosch, 1978) explains how we tend to build our understanding about new products by attributing prior experiences into our perception of the product. According to the theory, we structure our experience in mental schemata, containing mental prototypes that represent the attributes that a particular product should display to belong to a certain mental category. A product is considered to belong to a category, if its attributes match those of an archetype of that category over a certain level. Let us say that our mental archetype for a chair displays these attributes: Four legs, seat and backrest. If we see an item of furniture with four legs and a seat, but without a backrest, we probably would not say it were a chair, but a stool. In this way, we usually have an opinion about how things should look like. In the same way, experience has taught us that it is not always helpful to have rigid boundaries between categories (is tomato a fruit or a vegetable?). This approach is also one that designers utilise. A product in one category can be loaded with attributes from a product from another category, and thereby may be perceived as having the same qualities as the products of that category. This was especially significant with the Sony Sports Walkman, which had certain qualities like water and shock resistance. Products from other companies with the same yellow colour, attracted complains that their products did not resist water and dust, even when the products never intended to possess these qualities.

THE SEMIOTIC THEORY

The SPA method uses Charles S. Peirce's semiotic theory (Nöth, 1990) as the main theoretical foundation. This theory involves a semiotic model which describes how people interpret the signs they interact with in their everyday life. The definition of a sign used here includes all the possible product attributes that may function as a trigger for the impressions of a product. The nature of a sign is that it stands for or represents something other than itself. The reason why a specific product design evokes certain thoughts, emotions, impressions or associations is because it displays signs that consciously or unconsciously trigger these reactions. The mental archetype schema and the common social meaning structures described earlier in this paper, can help in explaining the particular reactions.

According to Peirce's semiotic model, the mental interpretation process is separated into three different stages or categories. In the original semiotic theory (Peirce, 1991), the categories are described in a highly abstract way and the definitions are modified according to the SPA method to be usable for practical design work:

- The *representamen* is here defined as the concrete physical aspects and attributes of the design (shape, colour or material).
- The *object* is here defined as the spontaneous impression that the product evokes in the user (e.g. an association, a metaphor, or an analogy).
- The *interpretant* is here defined as the subjective meaning or experience of the product, which the person constructs when mentally connecting the representamen and the object in a context.

One might say that the representamen represent the stimuli, the object represents the reaction to the sign, and the interpretant represents the subjective understanding of the stimuli and the reaction in its context. These three categories interact with each other in a circular process called semiosis (see illustration 2), where one semiotic circle makes the foundation for another following semiotic circle. These circles reflects in a simplified way how humans tend to relate one thought to another

THE DESIGN OF THE SEMIOTIC STUDY

The study in Opperud (2001) was performed according to the phenomenological scientific approach. The test sample was a group of 6 students from psychology courses and 6 students from design courses, equally balanced according to sex and with ages from 21 to 29 years. They were interviewed about the thoughts, feelings and associations they experienced when looking at three different mobile phones: Ericsson R310s, Ericsson A2618 and Nokia 8210. Every aspect of both their responses and their reactions to the attributes of the products were carefully followed up in the interviews. Their responses were coded and categorised according to the semiotic theory.

Figure 1 Nokia 8210

THE RESULTS OF THE SEMIOTIC STUDY

The results presented in this paper are an example of one the many mapped train of thoughts that the study captured.

The first thing that one of the persons from the sample said when looking at the Nokia 8210 (as shown in Figure 1) was "-this is a cellular for people living an urban lifestyle". This was the impression she got from the design. Further investigation revealed that there were many things that affected her and led to what we call the final semiotic interpretant. By following her associations backwards, from the interpretant via the associated object to the representamen that triggered it, we discovered that the impression could be explained by certain elements of the design. The following illustration shows the principle of semiosis when an interpretant in the semiotic circle becomes a representamen for the following circle, and thereby functions as a trigger for further mental reasoning.

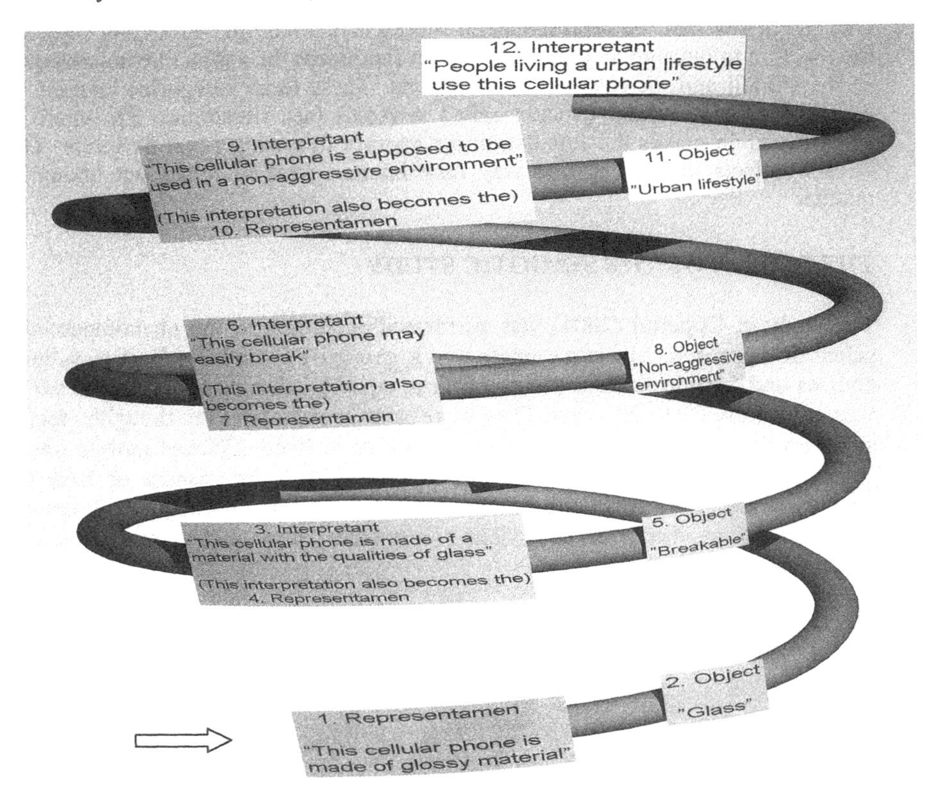

Figure 2 The semiosis process

CONCLUSIONS

The illustrated semiotic spiral (Figure 2) is an example of how this method revealed information that could be structured according to the semiotic theory. Using this technique made it clear that it was the glossy material on the cellular phone that made the test person associate to glass. According to the cognitive prototype theory, the test person concluded that this material belonged to the glass category, and therefore should have all the quality of glass. Together with experience from other cellular phones and her socio-cultural information extracted from common social meaning systems, she could also identify who the potential user of this product might be. The test person had not seen this cellular phone before. If she had seen it before or seen advertising for it, her social meaning system could have been affected and this experience should be possible to observe in her reasoning process.

The results show how the SPA method creates knowledge about how specific design elements and attributes may affect a potential customer. This should be a valuable tool for a designer who wants to design products that express the attractive qualities and characteristics for the intended customer group.

The SPA method may be used at several stages in the product development process. It can be used as a tool at the beginning of the process when it is desirable to distinguish the elements which function as attractive product attributes for a particular target group. On the other hand, the SPA method may be used for evaluating how a potential customer experiences and reacts to a prototype or for comparing and benchmarking the design of competing products[1].

REFERENCES

Dittmar, K. 1992, *The Social Psychology of Human Possessions*, (Hertfordshire & New York: Harvester Wheatsheaf and St Martin's Press).

Nöth, R. 1990, *Handbook of Semiotics*, (Bloomington & Indianapolis: Indiana University Press).

Opperud, A. 2001, *Is Beauty in the Eye of the Beholder? En semiotisk studie av mobiltelefondesign.* [Master thesis]. (Göteborg: University of Göteborg).

Rosch, E. 1978, *Cognition and Categorization*. (Hillsdale, N.J. : Erlbaum)

Peirce, C. S. 1991, *Peirce on Signs: Writings on Semiotic*. (Chapel Hill: University of North Carolina Press).

Wikström, L. 1997, Methods for evaluation of products' semantic functions. Paper presented at The Second European Academy of Design Conference (EAD) in Stockholm 1997.

[1] The results of the SPA method can in a fast and easy way be quantitative secured by complementing a semantic scale test like the PSA method developed by Wikström (1997). This methodology is based on statistical analysis of how a group of people evaluate certain semantic meanings generated by a product.

Experiential design in a virtual character system for exploring mood dynamics and affective disorders

Nicholas Woolridge
University of Toronto, Canada

David Kreindler
University of Toronto, Canada

Charles J Lumsden
University of Toronto, Canada

INTRODUCTION

The affective disorders (ADs hereafter: depression and bipolar disorder, plus variants) are a global health threat (e.g. Frank and Thase, 1999; Solomon, 2001): ADs have a devastating impact on their victim's lives and on society because they are widespread, long lasting, often untreated, and still poorly understood despite improvements in treatment. In North America alone some 18% of the population (about 20 million people) endures chronic depression, while bipolar disorder afflicts an additional 3-4 million. Bipolar disorders ("manic depression") now ranks second as the cause of death in young women, third for young males.

The worldwide effort to develop improved means of understanding and treating affective illness has recently benefited from several striking discoveries about how mood changes over time (Gottschalk *et al.*, 1995). It is now understood that, even in the most profound depressive state, mood and emotions do not simply flatline at some fixed nadir. Rather, there is a churning of affective tone that creates complex patterns of mood change, with time scales stretching across days, weeks, and months. These patterns resemble the chaotic dynamics of non-linear systems.

Unfortunately, the mathematical methods used to analyse mood change over time are notoriously opaque to non-specialists, and even among experts, there are steady controversies on the merits of specific non-linear models, procedures, and algorithms. Recently, we have become interested in the possibility that experiential design may offer new alternatives for representing the complex temporal patterns of mood change in a concise, potent, yet widely accessible manner. Our specific approach is based on the idea that, by harnessing the human brain's highly evolved capacity for face perception, these data can be used to recreate the mood experience on computer generated (CG) human faces. Animation of these faces with the mood data might then efficiently "decode" key qualitative and experiential properties of the affect dynamics. Just as time-lapse cinematography can, with startling clarity, make visible a flower's opening and closing through the day, the simulated flow of mood experiences across a CG face may render evident basic patterns of mood change normally taking days, weeks, or months to assert themselves.

This paper outlines our approach to this new phase of our mood dynamics research and summarises our progress to date. While preliminary, our results suggest that experiential design, and designers, may stimulate powerful new technologies for understanding and treating psychiatric disease.

OUR APPROACH

In order to map the dynamics of mood change with new precision, we are following a control group of normal (*N*=20) and an AD group of rapidly cycling bipolar (*N*=20) subjects for a period of 18 months using a specially designed, 19-item mood assessment questionnaire (Kreindler et al., 2002). Every item in the questionnaire is programmed as a visual analogue scale (VAS) on a PalmOS®-enabled cellular telephone and activates twice a day on the Bell Mobility (Canada) network. Each VAS is a horizontal line 4 cm long displayed on the digital screen of the cell phone. Using a small electronic stylus, the subject places a mark on the scale, at a position reflecting their current experienced intensity of a mood factor such as sadness or euphoria. The pixel position of the mark on the scale line is recorded by the software as the response datum for that VAS. After each session, the subject response data is automatically encrypted and transmitted by cellular telephone channel to a central database for processing and analysis. This on-going data collection protocol constitutes a time series record of human mood change of unprecedented detail and length.

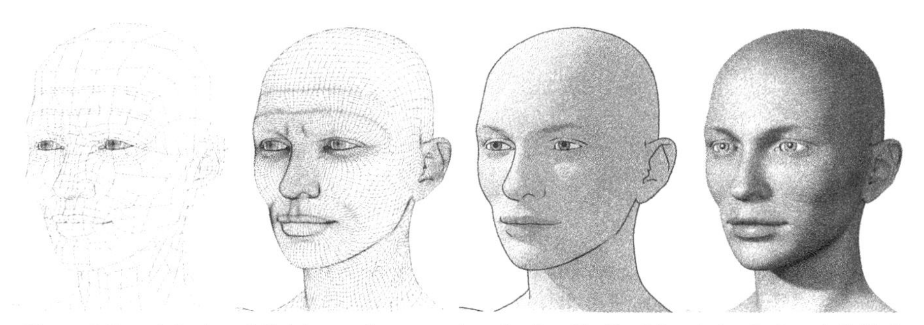

Figure 1 From left: the subdivision surface control mesh of an idealised female head; the subdivided result; a non-photorealistic rendering (used for the illustrations in this paper); and a photorealistic rendering.

Our computer-based facial animation system is being designed to help with the challenging task of representing and exploring the mood data stream for each subject. Two idealised 3D CG heads, one male and one female, have been designed and implemented (Figure 1; Woolridge, 2001) using subdivision surface modelling in the Cinema 4D XL 3D computer animation system (C4DXL hereafter; Maxon Computer GmbH, Friedrichsdorf, Germany). Keyframe poses representing emotional expressions and facial mood behaviours have been composed for each virtual character. Because of their clinical importance and their animation "readability", we have to date focused on facial expression models for four of the scales: depression, mania, overall bipolar mood, and fatigue (Figure 3). A morph target typifying a full intensity facial expression of each visual analogue scale (e.g. "Saddest ever") was created by manipulating face mesh target points (Woolridge, 2001) emulating the actions of the facial musculature controlling the eyebrows and lids, mouth, and cheeks.

The Morph plug-in environment for C4DXL (Vreel 3D Entertainment, Menden, Germany) was then applied to construct a "slider bar" control element for that mood scale in the C4DXL graphical user interface. There was one slider bar for each mood VAS with the exception of the facial animation pose defining a bipolar mood VAS datum. For the bipolar scale two sliders was used, one corresponding to degree of positive affect expression and one to negative affect expression.

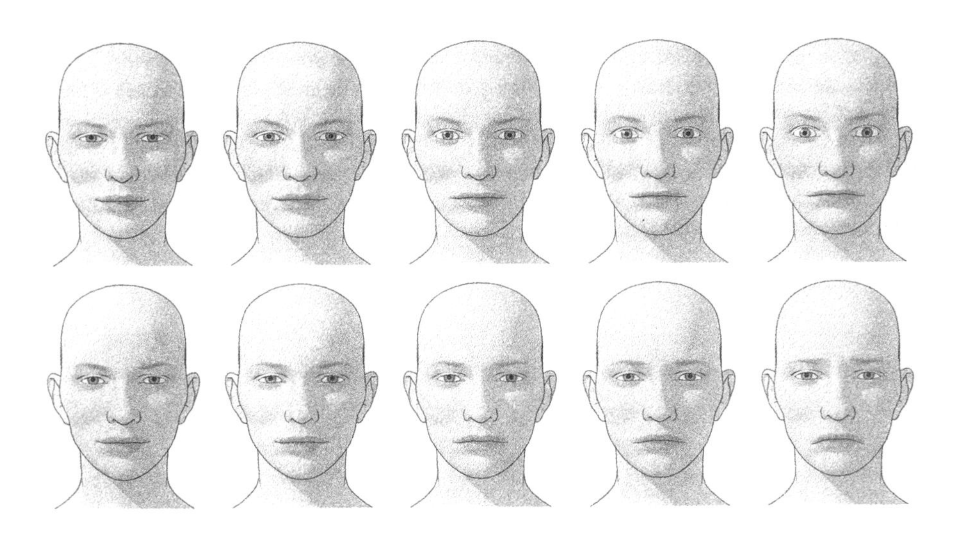

Figure 2 Expression morph series for mania (upper) and sadness (lower). Expressions are represented at 0%, 25%, 50%, 75%, and 100% from left to right.

Moving a slider control smoothly changed the CG face from the neutral or "rest" position to an expression that was software interpolated between the rest and the extreme morph limits (Figure 2). Each position of the slider control corresponded uniquely to a numerical coordinate on the interpolation trajectory of the morph targets, defining a unique facial pose. This value could then be matched to a numerical value on the corresponding visual analogue scale in our questionnaire. Each such facial pose was then established as a keyframe pose in the C4DXL animation track for the time series. Final animations were then rendered.

Mood data input to the system was later facilitated by a composing a C4DXL script that fed the mood scale time series data, in tab-delimited file format, to the morph target interpolation process, allowing us to quickly map a specified stream of mood time series data onto a pathway or trajectory through the space of keyframe poses, generating a real-time animation of the mood change pattern in the virtual character.

PROGRESS TO DATE

The long-term goal of this project is explore two hypotheses. First, we anticipate that this type of experiential design can supplement traditional quantitative methods of time series data analysis in the search for concise models of complex affective behaviour. Second, we anticipate that observers watching such animations will form reliable expectations about how the mood state will change. These experiential predictions may be an intriguing adjunct to current, numerical methods of forecasting change, all of, which have distinct limitations when applied to complex, chaotic time series data.

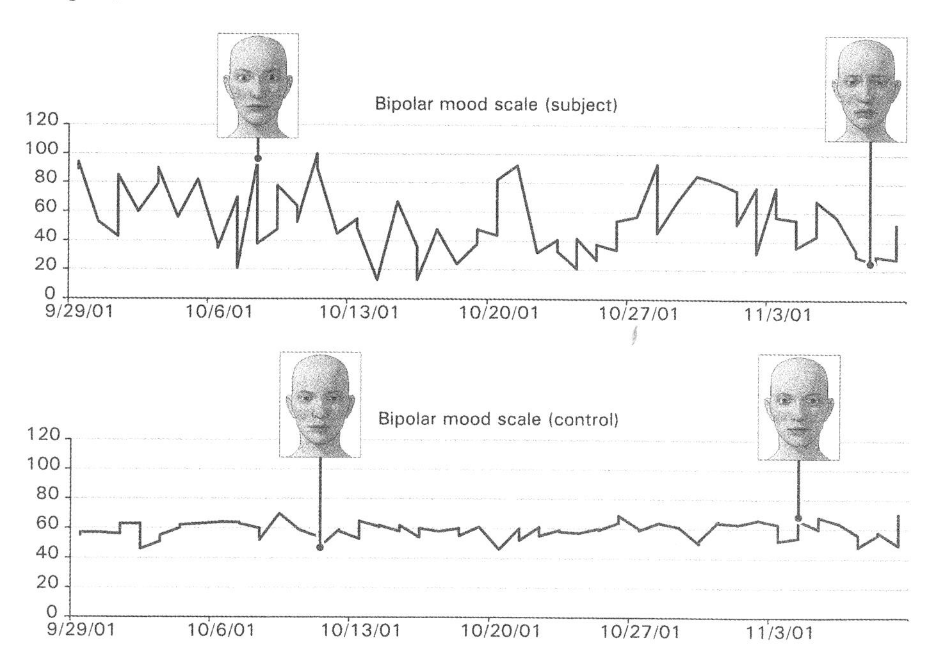

Figure 3 Mood time series: top, bipolar subject; bottom, normal.

To date we have focused on the process of CG face creation and identifying the principal challenges to effective design of the mood flow experience on such CG structures. We are first studying bilaterally symmetric CG faces, in which the same graphic mesh, mirror reflected, defines each half of the face around the vertical centreline through the mid-nose. Although such facial models have well documented limitations (for instance, blinking is allowed, but not winking, and left-right anatomical differences are excluded), their symmetry provides advantages to both the animation process and the experiential design. The symmetric mesh is relatively efficient to render into keyframe poses and into final animations; this is a significant consideration in a process where many thousands of keyframe poses, implementing the mood data streams for each subject, must be constructed, interpolated, and rendered across 40 subjects. Experientially, the symmetry of the face meshes simplifies the visual complexity of the rendered face and thus reduces the visual workload required to "read" animated changes: there are no left-right differences to track. Character "appeal" is also enhanced given the widespread viewer preference for more symmetric face designs in judging physical attractiveness. It is of course possible that important clues to understanding another's mood and emotional state are expressed through facial asymmetry. This will be explored subsequently as we advance to more computer-intensive animations based on asymmetric face meshes and textures.

The time-lapse effect, in which months of affective experience are compressed by animation into seconds or minutes of visual experience, is a striking feature of our renderings to date (Figure 3). On each mood scale examined thus far, the animated sequences convert abstract numerical data into an experiential flow in which impressions of general trends or directions in mood change emerge against a background of fast, rather random looking changes in feeling. We are therefore at the point in which statistical testing of observer ratings of mood state, and intuited directions of likely change in mood, can be compared to the predictions of numerical methods from non-linear chaotic dynamics.

DISCUSSION

It is clear that the illusion of life created by successful animation design depends not only the animator's ability to express the narrative and characterological essence of a scene (Thomas and Johnson, 1981). It also requires a strategy for the *mise en scene*. Thus, the work begun here also suggests a role for computer-generated speech and musical structures in mapping the hidden structures of mood. We are encouraged by our work to think that experiential design will be of increasing prominence in the worldwide effort to understand and treat the major affective illnesses.

Work supported in part by the Bell University Labs Program, University of Toronto.

REFERENCES

Frank, E. and Thase, M.E., 1999, Natural history and preventative treatment of recurrent mood disorders. *Annual Review of Medicine*, **50**, pp. 453-468.

Gottschalk, A., Bauer, M.S. and Whybrow. P.C., 1995, Evidence of chaotic mood variation in bipolar disorder. *Archives of General Psychiatry*, **52**, pp. 947-959.

Kreindler, D., Levitt, A., Woolridge, N. and Lumsden, C.J., 2002, A new instrument for mood self-report. *Journal of Abnormal Psychology*, submitted.

Thomas, F. and Johnson, O. 1981, *The Illusion of Life*, (New York: Hyperion).

Solomon, A., 2001, *The Noonday Demon: An Atlas of Depression*. (New York: Scribner).

Woolridge, N., 2001, Modeling a human head. In *The Cinema 4D XL Handbook*. edited by Watkins, A.(Hingam, MA: Charles River Media:). pp. 191-230.

Emotional factors in design and their influence on purchase decisions

Tore Kristensen
Copenhagen Business School, Denmark

Gorm Gabrielsen
Copenhagen Business School, Denmark

INTRODUCTION

This chapter is concerned with the possibility of design testing. The aim was to test, by quantitative methods whether we could detect emotional differences between design and whether we could identify the preferences or value of design. In two experiments, we found evidence that testing was feasible and that we could establish a value. In addition, we found that people's design competency affects their valuation of design. This issue is urgent when we speak of plagiarism, because in order to avoid it, we need people to be able to distinguish and to value the "original" compared to cheaper alternatives.

In this paper, we want to explore the emotion, content and evaluation of design. Are we able to use people's emotions and reasons to distinguish between designs? Also, are we able to establish measures for the valuations of design? Based on two experiments, we evaluate how methods can be evolved to explore the emotional and valuations aspects of design. The need for design testing of design performance will be greater as design used as a tool in business. Better testing procedures are likely to enhance manager's willingness to improve design. The issue is particularly important for designers and companies who emphasize well-designed products and who may suffer from copies and plagiarism. It has become essential to find a way to test whether people are able to identify good design and understand how they value it.

The first experiment: Do people distinguish the emotional aspects of design?

To develop a simple test procedure, we developed a standardised research instrument for testing visual designs. The test included: Collecting data from 6 different sample groups, contrasting the actual design of an existing travel agency and design proposed. The design proposed may be seen as an attempt to change the existing design to improve market position, or to capture the market by emulating an existing design. We obtained graphics from a bookstore and a travel agency, to be presented in its original form and contrasted with the design proposed. The respondents reported their emotions; attitudes and preferences according to word associations describing emotions and feelings, Eich et al. (2000) based on Richins (1997) test battery. The respondents were asked to score associative words on a 5-point scale. The data was analysed in accordance to a factor analysis. This resulted in a variety of significant factors explaining the discriminations obtained in the testing procedure. We tend to believe that people do not pay much attention to the corporate design of a company; they usually focus on the company's offer or products with design considerations being marginal. We believe that design can influence people's perception of a product or service when there is limited differentiation between competing products or there is quality uncertainty. The test design follows a standardised model using measures inspired by the standardised measures used in advertising testing, the ELAM test, Petty and Cacioppo and Schuhmann (1987). We also tested how easily the designs were identified.

Table 1: Recognisability

Recognisability	Kilroy	Subgate Travel	Samfundslitteratur	Subgate Books
Avr. Points per resp.	0.36	14.42	0.63	15.93

In Table 1, the Subgate design proposals score remarkably higher on average both as a travel agency and as a bookstore. The lowest score assigned to the real travel agency and bookstore may be ascribed to the new design being unfamiliar compared to the designs of well-known companies. The lack of recognisability for the known companies may be a result of a lack of recognisability in relation to the existing designs of these companies, as they are known in advance. In Table 1, the artificial Subgate designs score remarkably higher on average both as a travel agency and as a bookstore. The test indicated that recognisability favoured the artificial designs, probably because a new design provokes attention. On the other hand, this does not necessarily lead to preferences, because existing designs have a stronger foundation, due to the "mere exposure effect", Zajonc (1980). We found the new design lacking credibility. The conclusion is that we were able to identify the emotional, attitudinal and preferential aspects of the designs and how they were discriminated, based on emotional reactions.

The second experiment: can we assess people valuations?

The second experiment, Kristensen et al. (2002), tested people's ability to identify design quality in a product category and thus to relate to product price. In addition we asked whether the design awareness of the respondent is a significant factor. The product category is rice paper lamps. The Japanese designer Noguchi designed some of these lamps. The Noguchi product is distinct and expensive, but there are other lamps and cheaper copies on the market. Other lamps are copies or inspired by art and high-class design. The cost varies between 12 or 85 dollars. We were interested in determining whether people were able to tell which lamp was the original and if they could, how much were they willing to pay for it. The experiment took place at an exhibition in a design museum. The lamps were priced at 12, 18 and 85 dollars respectively, but the price information was not associated with particular designs. But all three prices were assigned to each lamp. People were given the information that Noguchi designed one of the lamps and asked to select which one. This information was compared to the respondents' level of deeper awareness. The data were analysed using paired comparisons with continuous ratings. In the first part of the experiment, a preference scale for the lamps was estimated. In the second part a conjoint analysis for paired comparison measurements was performed, which gives separate preference scales for lamps and prices. A total of 107 respondents participated in the experiment. In the first part of the experiment, in which only lamps were compared, the preference scale was estimated.

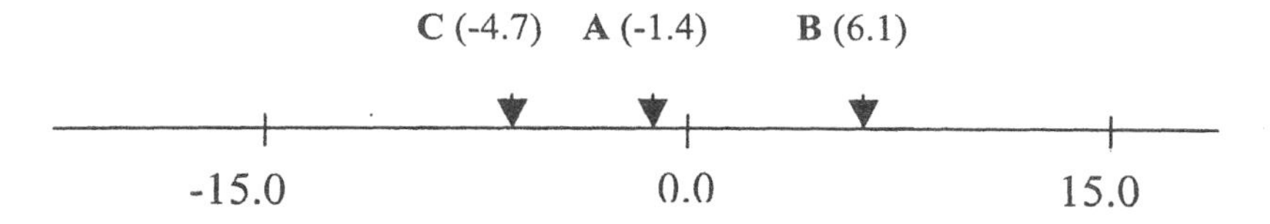

*Scale goes from a low of –25.0 to a high of 25.0

Figure 1: Preference scale of lamps

As there is no point of zero in a scale estimated from paired comparisons the preferences are represented with the sum being zero. It is seen that lamp B is the most preferred when no information is given about prices or origin of lamps. Furthermore, the difference between C and A is smaller than the difference from A to B. Although the scale goes from a low of –25.0 to a high of 25.0 the estimated common preferences only reach from –4.7 to 6.1 reflecting a high degree of heterogeneity among the participants. The conjoint analysis separates the effect of lamps and the effect of prices into two separate scales shown in Figure 2.

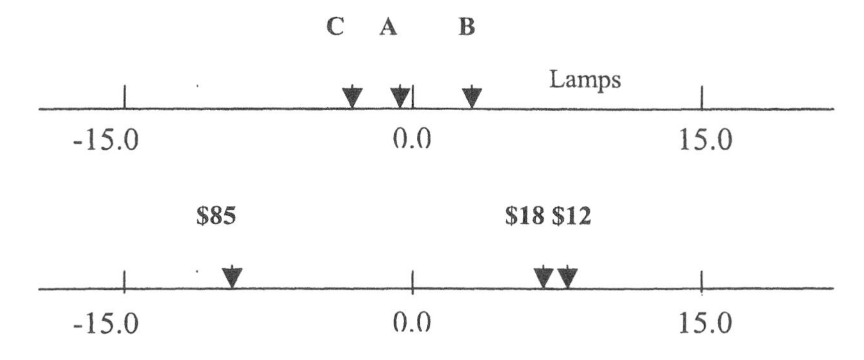

Figure 2: Estimated contributions to preferences from lamps and prices respectively

The preferences concerning the prices are as expected such that the preferences decreases as the price increases. The preferences for the prices $ 18 and $ 12 are almost the same, whereas the preference for the price $85 is very much lower. This is a simple negative ranking of prices, which shows, not surprisingly, that people in general prefer a cheaper to a more expensive price, not taking the product into consideration.

The preference scale for the lamps shows the same ordering of the lamps as in Figure 1 and also the same relative distance between the lamps, however, the preference scale for the lamps in Figure 2 is more compressed compared to Figure 1. The interpretation of this is that the difference between the preferences seems to diminish. When the prices are attached, the comparisons are expressed by the willingness to pay for each individual lamp. What we see is not the immediate ranking of product. We get the more consequential type of ranking from asking the respondents how much they are willing to sacrifice to acquire each of the lamps. This is the side of the price that concerns the perceived value in monetary terms.

The participants were asked to rate their design competence (high/low). In the analysis data was divided according to whether the persons claimed themselves as having low or high design competence (46 and 61 persons respectively. However, the preference scales for the prices were significantly different, Figure 2 and 3. The preference scale for prices in the segment with low design competence was highly compressed compared to the segment with high design competence. This means that the higher, the competence, the more weight is attached to the price. The competent respondent is more willing to pay a high price for the original and less willing to pay for a copycat.

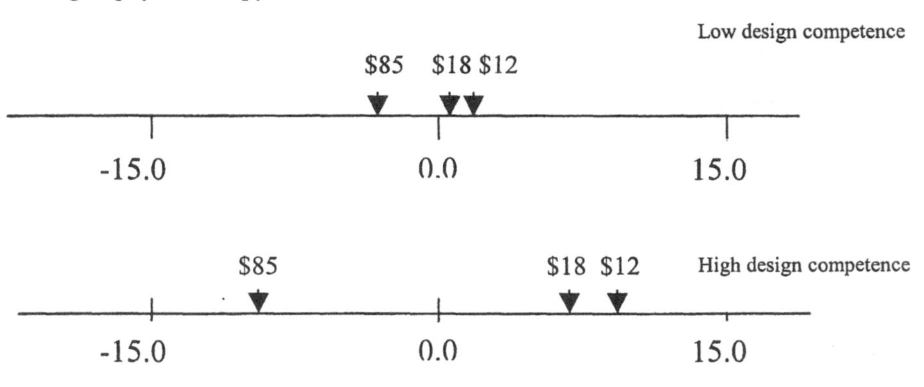

Figure 3: Estimated contribution to preferences from prices according to design competence

The addition of another variable also changes the result. The respondents were both asked whether they thought of themselves as having low or high design competence and whether they were working professionally with design. There seems to be very clear signs that increasing competency implies better discrimination. It also seems more urgent for these people not to pay a too high price for the "wrong lamp". This is rather urgent since plagiarism is a problem.

Less competent people care much less whether they have the "right lamp". They are generally less willing to pay and cannot appreciate the value of the "right lamp".

Discussion and Conclusion

The experiment is an application of a well-known method in marketing. We have established the possibility to measure the respondents' emotional, valuations and purchase motivation for different designs. We also found that the valuations depended on the design competencies of the respondents. So, there seems to be no simple answer to the value question. That may rather be found in a real market experiment with a broader distribution of competencies. Further research may take advantage of the developments in sensorics and senso-metrics. These disciplines have been dealing with recognition and characterisation of food, fragrances and drinks for many years. There is an existing body of knowledge that can lead into manageable testing procedures that we expect may be applicable for other product categories as well. While there exists a well-defined set of characterising terms for fragrances, food and drinks, such terms may have to be invented for other products. Also here, there are insights to be learned from the recent developments in vision research and cognitive science. The two experiments both give some validation of the claims, but further research is essential. In particular we need to experiment with other sensory modalities than the visual.

REFERENCES

Davcsyk, Spencer (2000) "Aesthetic Functionality in Trade dress: Post-secondary Aesthetic functionality proposed", *Commercial Law Journal* Vol. 105 pp. 309 - 330

Eich, Erich, John Kihlstrom, Gordon H. Bower and Pamela M. Niedenthal (2000) Cognition and Emotion *Counterpoints Oxford University Press, Oxford*

Gabrielsen, Gorm (2000), "Paired comparisons and designed experiments", *Food Quality and Preference,* 11, pp 55-61

Gabrielsen, Gorm, Tore Kristensen and Flemming Hansen (2000)"Testing Corporate Design" *in Corporate Communication International Journal Volume 5 Number 2, May 2000*

Kristensen, Tore, Gorm Gabrielsen, Ricky Wilke and Judy Zaichowsky (2002) Plagiarism, Trade-dress and the value of Design *Paper at the European management Conference Stockholm 7 - 9 May, 2002*

Petty, R.E., Cacioppo, J.T. and Schumann, D (1986) Central and Peripheral Rates to Advertising Effectiveness: The Moderating Role of Involvement. *Journal of Consumer Research* 10

Richins, M.L (1997) Measuring Emotions in the Consumption Experience. *Journal of Consumer Research* 24

Zajonc, R.B. (1980) Feeling and Thinking: Preferences need no inferences. *American Psychologist 35, pp. 151 - 175*

EMOTIVE EFFECTS OF VISUAL PROPERTIES

Using “visual/verbal interplay” to tap into collective memory and shared understanding

Susan M Hagan
Carnegie Mellon University, USA

INTRODUCTION

Communication design still searches for a theory of invention and composition that includes not only “looking” interests but also visual/verbal interaction. My work indicates that interaction revolves around one of three kinds of look/read meaning integration. I call the first “parallel play,” the second “sequenced play,” and the third, “visual/verbal interplay.”

Parallel play, sequenced play and interplay all serve important communication functions. In each, visual and verbal “modality selections” can tap into collective memory and shared visual understanding. However, visual/verbal interplay has the potential to create collaborations so tightly semantically and structurally linked that concept meaning would be significantly altered if the words and visuals were isolated.

PARALLEL PLAY

Parallel play connects visual to verbal information primarily by way of visual juxtapositions such as the example created by Josef Muller-Brockmann (1981; see Figure 1).

Figure 1 Parallel Play
(Grid System, Josef Mueller-Brockmann, ISBN 3721201450, Verlag Niggli AG, Switzerland)

Muller-Brockmann's work typifies elegant parallel play. We perceive the visual and verbal collaboration primarily because the two elements are close together. Visual proximity encourages groupings (Arnheim, 1974). Even though the words are too small to read, this layout still holds together. In terms of the collaboration between word and image, the text is important for its placement on the page rather than its content. The composition does not depend heavily on meaning relationships between the text and the visual nor does it rely on particular images to hold the work together. Of course connections between the two exist, but parallel play invites audiences to create loose connections between visual and verbal information in their own time and at their own pace. Muller-Brockmann's grid is one way to create these loose connections. Although it will be the only approach discussed in this paper, it is not the only method.

VISUAL/VERBAL INTERPLAY

Like parallel play visual/verbal interplay also depends on placement to hold the two modalities together. But interplay must modify as well as add to the composing practices used in parallel play in order to create compelling visual/verbal collaborations that enhance an audience's experience. In order to see how visual/verbal interplay creates an experience that differs from the parallel play shown earlier, two examples of interplay will be presented. First you will see the word separated from the visual. Then you will be able to view the visual/verbal composition. In both cases, the experience is never quite the same when the visual is separated from the verbal. In both cases, visual meaning links to the text not just because the two share the same space, but also because they are co-dependant contributors to the meaning of the message.

Example 1 The text below is separated from the visual.
The example of Visual/Verbal Interplay can be seen in Figure 2.

Breaking Out: A Black Manifesto, by Dick Gregory
George Wallace was half right when he said, "Gregory's not funny any more." The whole truth is that Dick Gregory is far from merely funny these days. His years of immersion in the caldron of our racial strife have transmuted the master comic into a wise eloquent and lacerating spokesman for black aspiration, a one-man hotline from the heart of the ghetto to the conscience of White America. Speaking at rallies and on college campuses across the country, running for President, publishing an insightful book (Write Me In!), Gregory has spread the word of black liberation with wit and passion. Now that Malcolm X and Martin Luther King are gone, Gregory's is one of the few truth-telling voices left in the land; and for all its jokes, it leaves no bitter truth untold. Turn this page to find out why it makes George Wallace squirm.

Figure 2 Interplay (Courtesy of The Herb Lubalin Study Center of Design and Typography, The Cooper Union School of Art, New York)

What compositional considerations give the visualisation of this text new meaning possibilities that imply no "breaking out" could happen? This work relies on subconscious cognitive interests (Solso, 1994) that can only take on directed meaning when under the influence of the text. In this example, text is crucial. Without it the visual emotional cue could not convey a directed experience because visual meaning is, for the most part, privately selected and interpreted (Solso, 1994). However, some visual attributes do seem to communicate across audiences. In this particular case, the work relies on our tendency to be depressed by dark colours (Solso, 1994). Lubalin weaves these cognitive tendencies with explicit word meaning in order to create co-dependant meaning interaction between the words and the visuals.

This work not only relies on the depressing effect of dark colours, it also employs a "spatial story" (Turner, 1996) that simulates physical constraint. A spatial story explains an abstract concept by translating it into an action we can more easily understand. "Shame forced him to confess" is easy to understand even though shame has no physical presence. In the same way, the words "breaking out" have no actual physical presence, but their constraint in a tight space makes it appear that no breaking out can occur. The affect is ironic. Note how that affect changes when the dark colours and typographic constraints are removed in the example Figure 3.

Figure 3 Visual constraints removed

In "Breaking Out," the arrangement of visual and verbal information creates an effect that communicates with visual immediacy and verbal direction. It would not be possible to do this using either modality alone.

The second example by Uwe Loesh communicates interplay differently.

> **Example 2** The text below is separated from the visual. The example of Visual/Verbal Interplay can be seen on the following page (Figure 4).
>
> History as Argument
> Ludwig-Maximillans University
> September 17 to 20th, 1996

In his book, *The Image and The Eye*, Gombrich (1982) uses the divisions of language conceived by Karl Buhler to explain the workings of images. Buhler categorises language functions as expression, description and arousal. Gombrich believes visual information functions best at the level of arousal of emotion. Images in isolation cannot describe or express the abstractions needed to make specific statements concerning what has been happening or what could happen. On the next page you can see how a combination of visual and verbal information begins to tap into the best of both worlds with clarity and immediacy.

History as Argument, taps into our collective memory (Radley, 1990), by combining the meaning possibilities in the visual information with explicit word meaning. Words constrain the meaning of visual cues (Hoffman, 1998). In combination with text, images encourage fewer, but sometimes more powerful, interpretations. *History as Argument* constrains the situated meaning of the utilitarian shredder from a neutral object that can protect privacy to one with a more sinister "enduring quality" (Radley, 1990).

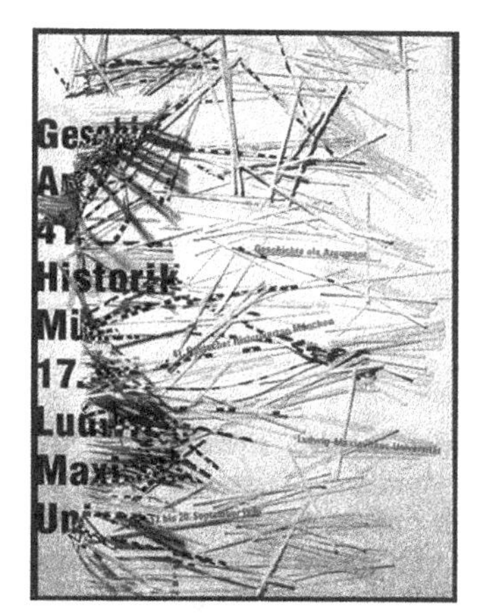

Figure 4 Visual/verbal Interplay (Courtesy of Professor Uwe Loesch)

CONCLUSION

Parallel play is often the best available means of visual/verbal collaboration. That method allows the audience to create loose connections at their own time and pace. But at times we must create the invitation to read *and* look with an eye and *ear* toward visual/verbal interplay rather than parallel play. The tools we need to create these tight collaborations must include the compositional interests used to build parallel play. But in visual/verbal interplay, we must also communicate a single concept by destabilising the dominance of the text or the visual when a concurrent interest in both modalities can communicate emotion as well as description with more immediacy and power. Many writers and designers do not yet know how to fully connect and interweave one modality to another. But soon we will.

REFERENCES

Arnheim, R., 1974, *Art and visual perception: A psychology of the creative eye*, (CA: University of California Press).

Gombrich, E. H., 1982, *The image and the eye: Further studies in the psychology of pictorial representation*, (New York: Cornell University Press).

Hoffman, D, 1998, *Visual intelligence*, (New York: Norton and Company).

Muller-Brockmann, J, 1981, *Grid systems in graphic design: A visual communication manual for graphic designers, typographers, and three dimensional designers*, (New York: Hastings House Publishers).

Radley, A, 1990, Artefacts, memory and a sense of the past, In *Collective remembering*, edited by Middleton and Edwards, (United Kingdom: Sage), pp. 46-59.

Solso, R.L., 1994, *Cognition and the visual arts*, (Cambridge: MIT Press).

Turner, M., 1996, *The literary mind*, (New York: Oxford University Press).

Lyrical visual

Frank Holmes
University of Gloucestershire, UK

A VISUAL EXPOSITION OF THE SUNG WORD

Think back to when you were a child, a teenager. Do you remember the days of sitting in your room, in the darkness, listening to your favourite records? Do you remember playing them over and over again? As you listen to the music, you begin to dream. You become part of a scene. The words of the song are sometimes those you say. The music becomes the backdrop - a vivid adventure.

I have always been intrigued by the power of music and particularly of song, to create a form of visual expression. The tone and emotion of the singers voice, plus the combination of musical instruments, melody, rhythm and production mix, all add to develop a sense of colour and form. A song can take you to a particular place and time. You can feel and see something.

Nowadays, with the introduction of MTV, Q, Kerrang and other digital video music channels, we are presented with an instant vision of song, someone else's visual interpretation. The pop-promo video has coloured our own unique way of seeing things. We are now more inclined to see things as others do. Perhaps we may identify with the words of international songwriter and record producer, Trevor Horn and his song, *Video killed the Radio star* (1979). In this recording he suggests that image has become more important than the music it represents.

Many recording artists have expressed doubt about the use of visual representation for the promotion of musical work. In Clinton Heylin's biography of Bob Dylan, *Behind the Shades* (2000), Dylan talks of his concern about the video which will accompany his song, *Neighbourhood Bully*. "We wanted to do Neighbourhood Bully, but it's difficult trying to explain to someone what you see. I haven't really found anybody that really thinks a certain way".

Lyrical Visual previewed in London's, *Dryden Street Gallery*, Covent Garden, in April 1999. The exhibition was an attempt to combine my work as a songwriter, musician and artist, through the use of abstract illustration music and lyricism. Starting with a collection of ten original songs, it aimed to portray the various aspects of human emotion, from jealousy and hate to love and happiness.

The interrelationship of art and music is something many musician/artists have explored. Brian Eno, during his long career as musician, composer and producer, acknowledges the use of metaphor and image in the construction of new musical work. From Eric Tamm's book, *Brian Eno* (1989). Eno explains how he sees certain synthesised sounds as "synthetic formica", and other acoustic instrumental sounds as, "more like a forest, beautiful and complex at any degree of magnification." Other musicians such as David Bowie and Paul McCartney have successfully combined their creative work in music and art. In a book about his paintings, *Paul McCartney* (2000) McCartney writes about the connection of his visual and musical work, " There is a big crossover between the two things, one the visual and one the sound. Space is very good. Negative space, (in music and art) is very important.

Lyrical Visual, is an exhibition of abstract graphic art. It is somewhat personal; my own visual interpretation of my own songs. Ever since the age of fourteen, (at the time my scribbled poems developed tunes) I have seen colours and shapes emerging in each new song composition. *Lyrical Visual* is an attempt to connect the aural with the visual. I have tried to paint the shapes that form from music and rhyme. Anchored to my teaching of advertising, it relates to the potential of word and image to influence and communicate human mood and emotion. Each artwork aims to evoke feeling. The audience are asked to relate their own experiences to what they see and read in each image.

The abstract images in *Lyrical Visual* start as drawings or hand-crafted collages using natural objects such as bark and grass. The final compositions are scanned into a computer, manipulated and re-coloured. The results are a combination of the organic and the man-made.

The following images and lyrics are taken from the exhibition. I hope they provide food for thought and raise questions for critical debate.

Amen

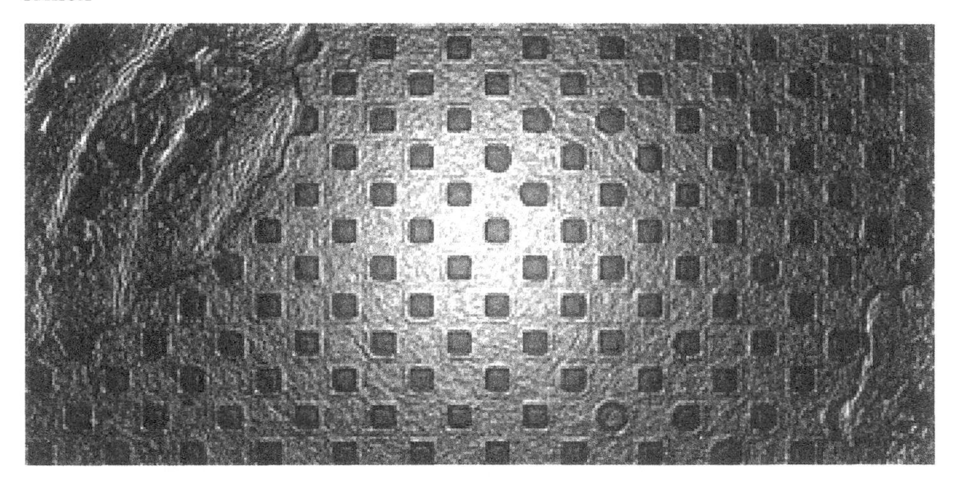

What's it like to be old,
what's it like to be old.
Losing your control,
doing what you're told.
What's it like to be young,
what's it like to be young.
Having so much fun,
Daddy hit your mum.
What's it like to be you,
what's it like to be you,

doing what you do,
seeing things as you.
What's it like to be there,
what's it like to be there,
in the world you share,
rocking in your chair.
What's it like on the edge,
what's it like on the edge,
suffering the pain,
needle in your vein.
What's it like to be poor,
what's it like to be poor.
Wanting so much more,
evil at your door.
What's it like on the edge,
what's it like on the edge,
kneeling on your bed,
rope around your head.
What's it like to be you,
what's it like to be you,
to be waiting for the day,
when at last to pray. AMEN.

Silver Wave

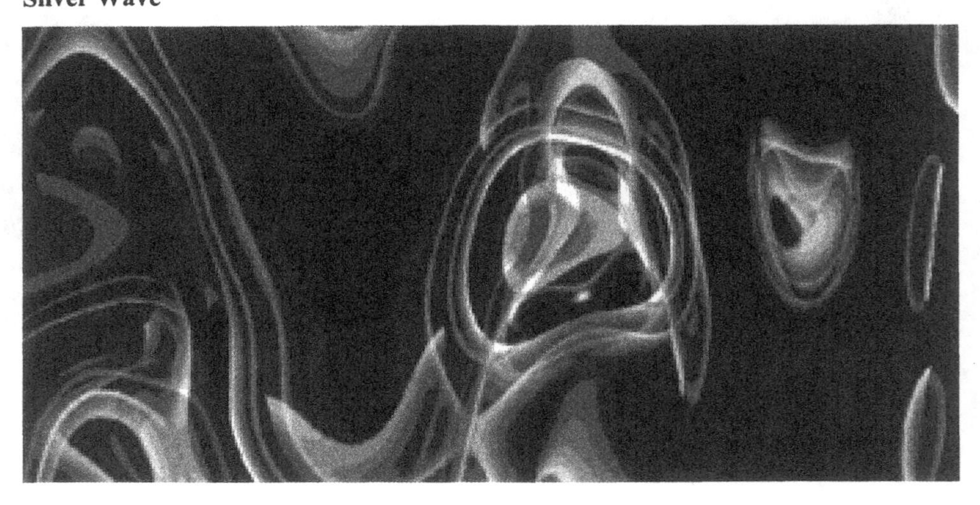

Hey Little angel,
kick off your shoes,
and dance me a rainbow,
around my blues.

Climb on your halo,
spread a little light,
and shine on my sin,
cover me baby,
let me in.

I wanna learn,
I wanna turn,
ride on that silver wave.

Hey little baby,
you blow me away,
with the truth of your innocence,
and the words you say.
Come ride on my shoulder,
come sailing on the dark side of within,
cover me baby,
show me now, let me in.

I wanna learn,
I wanna turn,
ride on that silver wave.

So Down Low

Did you go there, where it's not good,
did you stay there, hoping you could...
Oh yeah, oh yeah.

On the way down, did you look back,
on the way down, did you think that...

Oh yeah, oh yeah.
Well, so how does it feel now?
Blamed, shamed,
hurt digging at the pain,
I bet you're so down low.

Do you sometimes, think about me,
and do you sometimes, wish it could be.
Oh yeah, oh yeah.

Well, so how does it feel now?
Blamed, shamed,
hurt digging at the pain,
bound, found,
trippin' up and down,
lame, tame,
fear you go insane
I bet you're so down low.

REFERENCES

Heylin, C., 2000, *Bob Dylan: Behind the shades*, (London: Penguin).
McCartney, P., 2000, *Paul McCartney paintings*, (New York: Little, Brown and Company).
Tamm, E., 1989, *Brian Eno. His music and the vertical color of sound*, (London: Faber and Faber).

Dynamic interactive aesthetics: facilitating affective input from the audience in the creative design process

A Bennett
Rensselaer Polytechnic Institute, USA

R Eglash
Rensselaer Polytechnic Institute, USA

INTRODUCTION

Traditionally, the role of the graphic designer is to translate information for a target audience using cultural signs and symbols (Friedman, 135; Aneceshi, 8-9). S/he creatively designs various types of conventional and unconventional communication artifacts to transport messages to their respective audience. The communication artifact transmits information to the target audience through its visual language and aesthetics comprised of emotive signs and symbols, graphics, visual treatments, and other types of image-based information (that the graphic designer chooses traditionally). In this traditional context, information exchange occurs at the intersection between the communication artifact and the audience's perception of its aesthetic makeup (i.e. its graphic design). First, the graphic design attracts the audience's attention. Second, the target audience interacts with the communication artifact in order to retrieve a message. If the audience lingers to read the information in its entirety and/or returns for more information, it is then that effective communication occurs. Therefore, a communication artifact composed by the graphic designer succeeds in transporting the message to the target audience when the visual information emotionally induces the target audience to read it.

Today, the role of the graphic designer is broadening to that of a facilitator of participatory design processes that engage target audiences from the conception to the production and distribution of the communication artifact. Thus, as communication artifacts become increasingly multicrafted (that is, multiauthored and multidesigned) works, technically facilitating audience participation becomes problematic when the participating audience is remote. A graphic design that employs Interactive Aesthetics (Steinhauer, 2000), however, can engage a remote audience in the design decision making process of static communication artifacts. Interactive Aesthetics (Steinhauer, 2000) describes a facilitative infrastructure (based on the use of interactive technologies) that enables members of the target audience as well as the client and the graphic designer to engage collaboratively in the graphic design process. The application of IA also allows graphic designers to work with the target audience (as well as with the client and other collaborators) whether local or at a distance. IA permeates geographical, linguistic, and cultural boundaries through the use of interactive technologies that convey the graphic designer's intentions (in regard to the graphic designer's choice of visual treatment like their choice of cultural signs) and also facilitates audience feedback and input.

IA is intended for static sketches though—like documents and other two-dimensional sketches and prototypes. Static forms that appear sequentially in the same space, such as walk/don't walk signs, can also be treated by the IA process. For dynamic sketches and prototypes though (such as online animated faces, humanoid robots, flexible architecture) a difficult and different set of challenges arises for the graphic designers who seeks affective input from the audience. The participatory design process for dynamic sketches is problematic because dynamic aesthetics morph in time, transforming shapes across a continuum over a period of time. As a result, the collaborators (i.e. participants) need to be able to experiment with these dynamic parameters, altering the ways in which the dynamic sketch moves from one state to the next in a fluid manner (e.g. changing the expression of Kismet's face). Herein we introduce the term Dynamic Interactive Aesthetics (dIA) to describe the application of interactive aesthetics to the making of dynamic communication artefacts.

INTRODUCING DYNAMIC INTERACTIVE AESTHETICS

Systematically, the graphic designer may start the graphic design process with a thumbnail sketch or storyboard that s/he distributes electronically to his/her collaborators. Therein begins the participatory design process as the client and members of the target audience interact with the sketch in order to retrieve information and submit information about how the dynamic communication artifact should look and function. Since these forms are often a crucial interface for projects in "affective computing," electronically mediated communications, online avatars, and other human-machine interactions, facilitating audience participation in their graphic design is an important step in their utility.

The graphic designer, who may be formally trained in graphic design history, theory and practice, encodes the message using visual language from her/his cultural reservoir. What results is a graphic design of signs. These signs transport complex social and cultural messages that can be emotionally evocative. Employing these signs and symbols that represent cultural information is a useful approach to configuring a graphic design that simultaneously engages and reflects a target audience that may be globally situated and multicultural. One advantage of analog representation is that the signifier literally resembles the signified (Meggs 8). The change from a smile to a frown is understood globally. In cases where there is an arbitrary relationship between the signifier and the thing signified, communication is entirely dependant on previous knowledge and experience (8). Signs can convey their message to only those individuals who have learned the sign or the sign system. For example, Chinese calligraphy, pictographs, roman numerals and the English alphabet, are pre-arranged, coded systems that are understandable only to people who have learned the language (Meggs 2). Thus, history, culture and tradition all play a role in the interpretation of the arbitrary or digital sign, and a three way interdependent relationship results between the signified, the signifier and the interpreter (6, 8). Hence cross-cultural emotional communication exchange is constrained, and the graphic design process must take this into account.

In the special case of a dynamic communication artifact that engages and reflects diverse users, the signified is the thing that is represented (i.e. the emotional state). The signifier is an emotive sign that represents the signified; and, it is comprised of a multicultural aesthetic that includes words, images, lines, shapes, textures and/or colors. The interpreter is the person/audience/user who affectively perceives, constructs and derives meaning from the sign[ifier] (6). However, this three-way relationship has to be contextually based in order for the author of the message to yield the proper response from the interpreter(s). That is, a visual sign or symbol that the target audience decodes in one cultural setting may be barely recognised and acknowledged by a different target audience in a different cultural context.

The given signifiers may include the author's message(s) visually translated in a prototype of the communication artifact with preliminary aesthetics chosen by the designer. The variable signifiers—that the audience would have the option to edit—could include components of the chosen aesthetics including graphic elements that signify culture like colors, typestyle/size and spatial orientation within the boundaries of the dynamic composition. A communication artifact, even a dynamic one, has culturally specific interpretations that depends on the context in which it is interpreted. Furthermore, interpreting it the way that the graphic designer intends depends upon the target audience's visual literacy. Therefore, the incorporation of interactive aesthetics that teach representatives of the audience is needed in order to ensure that the audience will extract the intended message from the multifunctional and multicultural signs that graphic designers use to aestheticise dynamic sketches and prototypes. Such an infrastructure will not only teach the audience the visual language and how it is used/can be used to convey meaning but it will also evoke an active emotional response from the audience. Thus, dynamic interactive technologies like analog representation of emotional state that we present in this paper may present a better alternative to the physically arbitrary signs and symbols commonly used as conduits for information exchange.

We posit that the stimulus for this dynamic participatory design interaction is implicitly contained in the aesthetic makeup of the communication artifact's graphic design. The visual language of the communication artifact can deliver information to the target audience; and, if it is interactive, the visual language is also capable of soliciting creative input from the target audience. Comprised of variable and given signifiers, dynamic interactive aesthetics become digital interactive technologies that when placed in the hands of the author or maker, users and interpreters at the same time facilitate the participatory design of dynamic communication artifacts. Although we use digital technology, in the sense of computer chips running binary code, at the level of human interaction, dynamic forms are better described as analog representation. In some ways that's true for all physical existence—we exist as a collection of discrete atoms and electrons at the smallest scale, but at the scale of our perceptions we experience a smile, a hand gesture, or a dance as a continuous movement through time—what is often referred to as analog. An online avatar's smile or robot's eyebrow wiggle exists as a digital series of ones and zeros at the level of silicon, but for the purpose of communication design it is better treated as an analog waveform.

Dynamic Interactive Aesthetics stipulates that the collaborative design of dynamic communication artifacts requires the application of interactive aesthetics through analog representation. When applied to the dynamic communication artifact, dIA enables the client, the audience, and the creative designer to give input across a range of contrasting states of being. For instance, the participant may slide a handle (by clicking and dragging the mouse from left to right or right to left) that makes a humanoid robot's (e.g. Kismet's) expression, for instance, morph from sadness to happiness (See Figure 1 for a sketch of what this handle might look like among a series of other hypothetical analog controllers that would allow members of the target audience to collaboratively design a dynamic prototype and give affective input).

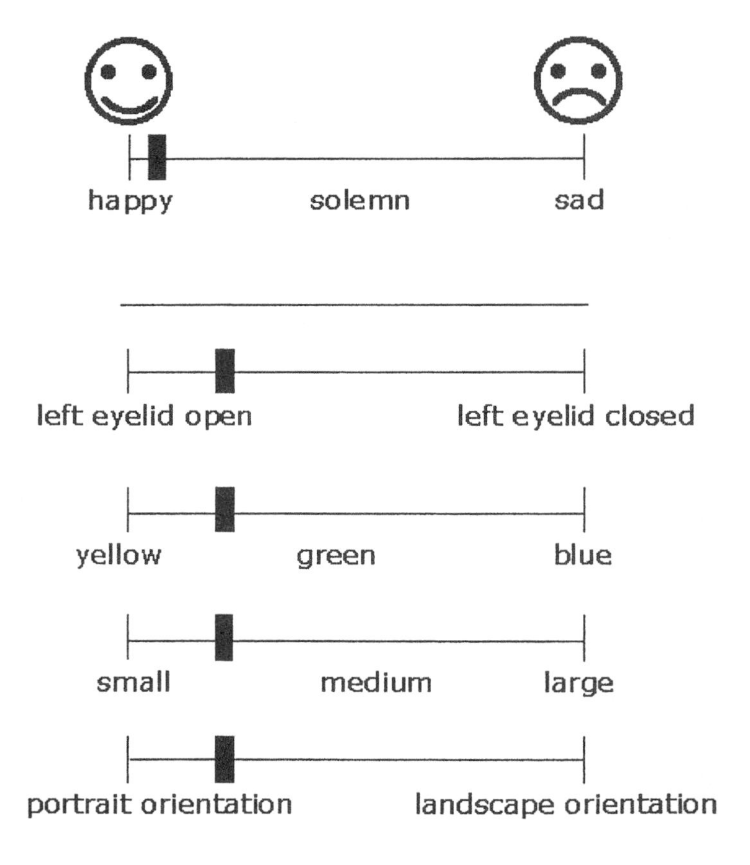

Figure 1 Examples of analogue controllers that facilitate audience input for dynamic prototypes

This allows the user to see how the curve of the smile will change in response to the emotional change. It is this relation between the physical parameter (degree of curvature of smile) and the informational parameter (degree of happiness) that defines an analog representation. Once the user has an intuitive feel for this relationship, they can then limit or expand the dynamic range. For example, children often respond better to a broader dynamic range; they like to see a cartoon character's grin literally expanded from ear to ear. Adults usually prefer a more subtle range of expression. The dIA process allows individuals to custom fit the dynamic range to their personal preferences (which might change depending on the context, the user's mood, etc.). This is one way (among others) to technically engage the target audience in the process of designing when a sketch or prototype is dynamic. This paper begins to answer questions about participatory design processes that involve dynamic communication artefacts. Our focus is on promoting and facilitating audience participation in the conceptualisation, design, and utilisation of aesthetics in sketches and prototypes that are dynamic.

Aneceshi, G., 1996, Visibility in Progress. *Design Issues*, 12:3, pp. 3-13.

Friedman, M., 1989, *Graphic Design in America: A Visual Language History*, (Minneapolis: Walker Art Center).

Meggs, P. B., 1992, *Type & Image: The Language of Graphic Design*, (New York: John Wiley & Sons).

Steinhauer, A., 2000, Transcending Space and Culture: Interactive Aesthetics that Facilitate Remote Participation in Graphic Design Processes. In Adjunct Proceedings of CoDesigning 2000, edited by Scrivener, S.A.R., Ball, L.J., and Woodcock, A., (Coventry: Coventry University), pp. 155-160.

This article has originally appeared in the Journal of Design Research 2002, Vol 2. Issue 2, http://jdr.tudelft.nl

Colour and emotion in design

Sheila Baker
De Montfort and Loughborough Universities, UK

INTRODUCTION AND BACKGROUND: AUTHOR'S PHILOSOPHY

This paper describes and discusses an approach to teaching and learning developed by the author on the basis of teaching at De Montfort (DMU) and Loughborough (LU) Universities. The paper firstly describes the author's personal philosophy of colour and its relation to emotion, based on Vedic philosophy. This is then illustrated by four simple case studies of work at the two universities with first year students. These are followed by a discussion and conclusions are drawn.

The author believes that colour is a fast track to emotion: a hot line to instinct. It is powerful, evocative and mood changing. It has energy. Colour is sensual. When the senses are involved the effect is heightened. The vocabulary of product semantics includes texture, scale, form and line, but in the author's experience colour often has the first impact as well as the most memorable and lasting effect.

Cultural background, personality, life experiences and fashion trends affect responses to colour. In ancient times colour was symbolic and had great meaning. However today this symbolism is less potent. The first decorative rattle or mobile that a baby sees starts the desensitising process. We are surrounded by a cacophony of colour chaos, which may be responsible for the clichéd responses often found in students' colour work. Colour is the interface between a product and the designer; a product and the buyer. It is not skin deep and superficial. Both designer and buyer can be influenced by colour, but the product can be most affected - it can change its personality! Colour can be awesome and scary, it can make or break a design. No wonder that grey is so popular!

The product has character; the designer and consumer each has a personality and emotions; all part of the equation and all are influenced by colour. Designs can have mood and character, an emotional energy, and the designer has alchemical powers to affect others. The mood of the consumer also plays a part. However, the main theme is that the designer also has emotions and that this is often ignored in design education.

There are many useful models about perception and emotion to be found in Vedic philosophy. One idea has inspired and influenced this particular theme of colour and emotion affecting the designer, the product and the consumer. This concept proposes that there are three interwoven mental states which are: energy, inertia and clarity and that we all fluctuate between degrees of these states. These three qualities are given colours. This a refreshing angle and not the clichéd red is hot, blue is cold approach. Energy is symbolised by red, inertia by black or dark blue and clarity is light and colourless (Baker, 2000). Red mental energy can move through agitation, anger, passion, excitement towards pleasure, enthusiasm, humour. Black mental inertia can move through deep dark depression, sadness, lethargy, boredom, fatigue to sleepiness and relaxation. Clear mental clarity has no degrees, it just *is*, still, light, balanced.

Emotionally, as designers, we need the interplay of these states as we strive for equilibrium and balance. Products also have similar qualities. Designs may have a certain red energy, a passion , sexiness or aggression. Some designs may manifest a dark energy by being solid, heavy, soothing, safe or downright boring. Other creations may be light, clear, harmonious and uplifting. This idea can be generalised by saying that all strong, bright colours, can be energising, all dark and dense colours may be heavy and grounding and that clear, light, colour is soothing.

METHOD

The author teaches Interior Design at DMU and Industrial Design at LU and has used her philosophy of colour in developing and structuring projects. These projects have been a means helping students look at and express such emotions from a personal perspective whilst attempting to give their designs mood and character. This approach offers students an opportunity for self-reflection, a creative opportunity to explore personal feelings in relation to the design process. Students are encouraged to involve all the senses, but particular emphasis is paid to the power of colour.

All projects are interlinked by these concepts. In this paper four projects are presented which show how these ancient concepts may inspire current design thinking. Each project was taught using a similar structure: an introductory lecture with slide presentation, group discussion and interaction on the project including brainstorming, mood boards, exploration and investigation via photography, group tutorials, presentations and peer feedback on effectiveness of communicating emotions. At DMU Interior Design students carry out 4 projects which focus specifically on colour over a 12 week module. At LU Industrial Design students carry out a range of projects over semester 1, but none are specifically focused on colour. In this case inputs on colour are integrated into other projects.

The Entrance Project DMU

Students design an entrance which will evoke a certain feeling or expectation in the viewer. This entrance will represent the student themselves, how they feel they present themselves to the outside world. Students are asked to show a glimpse of the interior within, which captures a hidden self. Lectures on the symbolism of archetypal shapes and the emotional influence of colour are given (Baker, 2000). Students prepare a mood board which captures the essence of themselves, their energy. This is a lateral rather than a literal representation. The use of colour, scale, shape, media and method of entrance all add to the character. One to one tutorials are invaluable as students often find this self-reflection challenging. The project brings an awareness of the subliminal messages of a facade, the manipulation of colour and media to communicate character and how the external may not always represent what is internal. The project provokes a sense of enquiry and new reference points...*the self.*

Evaluation shows that students are stimulated by the challenge of capturing personality in a non-figurative manner. Subsequent projects expand on the energetic content of spaces and interiors and how they affect emotions.

Installation project - DeMontfort University

This project involves more directly the Vedic model of colours being categorised into 3 qualities of emotional energy, red (active), black (inert) and light (clear, balanced). Students work in teams, each given a quality to explore in the context of a space, place or local environments. Feedback involves each student listing a group of words which capture the emotional essence of the chosen environment. The red teams often choose markets, shopping arcades, swimming pools. The black teams explore underground car parks or underpasses. The white teams try churches, parks, temples. Groups identify how the colours, shapes, lighting and their interpretation by the individual's senses and perceptions have affected their emotional reactions. A non figurative and non literal installation is then designed and constructed using the colours, textures, smells, sounds, lighting, scale etc to recreate the emotional energy originally experienced.

The installations are presented and the year group explores each environment writes down 6 words which describe the emotional effect each space has upon them. These words are then compared to the original qualities recorded. Evaluation of this project shows that students become more sensitive to their own responses and see formulae for recreating emotional experiences and that individuals may not always respond in the same way.

Colour mood boards - Loughborough University

This project introduces students to the use of colour as a way of evoking emotions as, at this point, with their background, it is generally perceived as simply decorative. It focuses on the subliminal message of a product. For many students this is the first time they have been exposed to the effects and meanings of colour.

Students work in teams to produce a mood board, inspired by a given emotive word: *depression, passion, tranquillity, neutrality, aggression.* They are encouraged to avoid using stereotypical colour ranges and to explore colour combinations to capture the essence of the mood. This is supported by lectures on colour symbolism and product semantics. Individually students use their colour range and mood to restyle an existing product. Feedback sessions are used in which staff manipulate digital images of the original mood boards to change the colour ranges of the moodboards. Projects such as this are about confidence building in the use of colour and mood.

The red, black and white project - Loughborough University

This was a voluntary, two day rapid response project . Students selected a simple geometric shape to symbolise a quality of energy and to represent a product. This was reproduced and mounted on numerous coloured and textured backgrounds. Students explored how the mood of the original colour and form can be enhanced, reduced or destroyed in relation to its environment or context. Continuing excercises involved adding a permutation of colours and shapes to represent buttons and switches and to further influence mood and message.

DISCUSSION AND CONCLUSION

The central theme of the paper is that there is an interaction between the user, the product and the designer's emotions, and that colour is one of many keys to unlocking these. At DMU the context is an Art and Design department with art trained staff. The projects are in a 12-week module dedicated to the exploration of colour. DMU takes 45 undergraduates each year, 95 percent had attended an Art Foundation Course and most took an 'A' level in Art. DMU student reaction is generally open. Most show enthusiasm and even excitement to what may be perceived as an unusual angle. A minority may find the ideas challenging. Attendance on this module is always high, with grades above average. Students readily expand on the ideas and involve other senses - often adding sound, taste, temperature or touch to their work. Feedback shows it is a module rarely forgotten.

At LU the context is Industrial Design. Here the team of staff mostly have a background in technology. Projects are interwoven within a Design Practice course which stresses many technological considerations. The university takes 120 students per year, mainly direct entry, with 'A' levels in Design and Technology and two others. Five percent of students at LU previously attended an Art Foundation course and twenty five percent took Art at 'A' level. LU student reaction is mixed. For some it is new and exciting, for others it is "arty farty", whilst some find it scary, unexpected and difficult to adjust to. For many this is a first exposure to using colour, not to mention emotions and oneself. The work produced ranges from naïve to very sophisticated and exciting. Those with open minds, who "feel the fear and do it anyway" create stimulating and original solutions.

This approach is being developed in order to balance intellect with feeling.

Feedback from students and graduates indicates that any initial resistance to these ideas tends to develop into acceptance as subsequent projects unfold. Such influences are difficult to quantify as the seeds sown will germinate at different times. Feedback was sought from both students and graduates on their experience from such projects (Baker et al., 2000). This summarised findings as "it seems that the approaches were either loved or loathed by the students, but never forgotten and often appreciated long after the event."

In conclusion it seems that undergraduates on more art-based courses are generally more immediately receptive and imaginative, whereas those with a background in technology are often initially more inhibited and cautious. However, the learning curve for these students is sharp and rewarding. The author feels that the opportunity for students of all backgrounds to explore their own reactions and moods as part of their design education can be a colourful and rich experience.

REFERENCES

Baker, S.L., 2000, *Power Diagrams*. Website accessed 01/2000 (http://www.transientspaces.net)

Baker, S.L., and Wormald, P.W., McDonagh-Philp, D.C., & Denton, H.G., 2000, Sthira, Sukha: Introducing industrial design undergraduates to softer design issues, *Proceedings of the International Conference on Design Education*, Swann, C. & Young, E. (Eds), Curtin University of Technology, Perth, Australia, pp 94-102

Colour, design and emotion

Jacquie Wilson
UMIST, UK

Simon Challis
UMIST, UK

INTRODUCTION

Many elements make up a textile design; the construction method used, the yarns and fibres used and the way the fabric is coloured or patterned. Repetition is an important element in most textile designs, apart from some "one off" pieces, and it is by repeating one section of the work that the pattern is formed. There are variations in the way sections that make up one repeat within a design can be repeated, from simple straight repeats and half drops through to repeats that follow complex mathematical formulae.

RULES AND THEORIES FOR DESIGNING

There are many rules and theories as to how designs should best be constructed. Within nature many mathematical rules can be found and these rules are often echoed in designs made by humans, either deliberately or intuitively.

The Golden Section

One such rule that has been used by humans for many centuries is the Golden Section, a ratio that occurs again and again in nature. The ancient Egyptians were the first to use mathematics in art and it seems almost certain that they ascribed magical properties to the Golden Section and used it in the design of their great pyramids.

Fractals

Another construction method for design uses fractals, which are defined by the Oxford English Dictionary as 'curves or geometrical figures, each part of which has the same statistical character as the whole'. This type of patterning is mirrored in nature; for example the construction of a leaf is as a miniaturised version of a twig, a twig of a branch and a branch of a tree.

"I wonder whether fractal images are not touching the very structure of our brains. Is there a clue in the infinitely regressing character of such images that illuminates our perception of art? Could it be that a fractal image is of such extraordinary richness, that it is bound to resonate with our neuronal circuits and stimulate the pleasure I infer we all feel? P. W. Atkins (Bourke, 2001)

Other rules
Owen Jones in his famous treatise "The Grammar of Ornament" (1856) gives a whole host of rules to help produce "good" decoration or ornament. He feels that much should be based on architecture.

"... there are no excrescences; nothing could be moved and leave the design equally good or better."

"As in every perfect work of architecture a true proportion will be found to reign between all the members that compose it, so throughout the decorative arts every assemblage of forms should be arranged on certain definite proportions; the whole and each particular member shall be a multiple of some simple unit."

Jones, 1856

REPETITION AND VARIATION

In his book "The Pleasures of Pattern" (1968), William Justema talks at length about pattern. On the simplest of patterns - polka dots – he says "they make no demands on us, which is doubtless the reason we find them charming." We are comfortable with the familiar – repetition is important to us, but there seems to be a need for visual imperfection too. Is this a reflection on the human condition? Justema continues, "Although repetition is what makes a pattern, Variation is what makes it rewarding to look at ... Variation is that principle of design that relieves the ... mechanical regularity of a pattern".

Repetition occurs when elements that have something in common are repeated. When a design consists of shapes that are exactly alike, repeated in a uniform and regular manner, then that design tends to become very formal. By varying the shapes and the spaces between them a more informal interest is created that seems to have a greater appeal. Repeated shapes make patterns. Many textile designs, because of the manufacturing processes, will automatically repeat when made as a length of fabric.

William Morris and a hatred of machine made pattern

William Morris hated the ability of the machine to dehumanise labour and to debase design. The first company he was involved in setting up in 1862 was to be a co-operative of artists producing their own designs for limited hand production.

He hated mass-production. Morris combined historic structure with natural forms to produce his designs, extensively using the fabric collections of the Victoria and Albert museum for inspiration. He organised trees, flowers, birds and animals within pattern sequences he had seen to emphasise the craft of decoration.

Re-humanisation of print design

Subsequently, other designers have revolted against the dehumanisation of design. In the 70s and 80s a common design technique was to place a black line around the shapes making up a print design (by far the majority of Viyella prints at that time had these black outlines). Susan Collier and Mary Campbell, a prominent team of sisters producing printed textile designs, rebelled against this technique much as Morris had done against the designs of his time. The print designs Collier Campbell produced were new and refreshing in comparison with so much of the print design of the time.

Handcrafting – the importance of variation

Hand-crafting, or at least the appearance of hand-crafting, seems to be important to the human psyche – we like our patterns to have repetition so we feel comfortable, but not so exact as to feel overly formal, clinical or cold. Is this why there is a timelessness to many ethnic designs that would have been originally made by hand? The enduring nature of many ethnically derived patterns is due in part to the subject matter, often florals or geometric patterns that have no negative significances. The execution of these ethnic patterns however also would appear to be important; computer-generated versions of simple ethnic patterns do not appear to have the same appeal as the originals.

PERCEPTION

What we see is determined not just by the eye but by also by the brain. Gregory (1966) talks of a painting as having a double reality. "The painting itself is a physical object, and our eyes will see it as such, flat on the wall, but it can also evoke quite other objects – people, ships, buildings –lying in space".

How we perceive is determined by our life experiences. An artist can make us see an apple, and even a naked woman, by the skilful use of only a few lines; but for us to see these we have to have experience of the actual objects. This experience in our world also impacts on the value we give an object or design. It is said that any object can be given the value and status of a work of art. Take any everyday item, place it on a white pedestal in an art gallery and it assumes an importance that it didn't have before.

Still with art; a photocopy of a piece of artwork or a design on glossy paper can give a more professional, "better" feel to the work than the original has itself. This is due, at least in part, to the fact that much of our appreciation of art is now through books, beautiful extravagant coffee table works rather than seeing paintings and art 'live' alongside other works. The photocopy truly 'two-dimensionalises' original artworks.

The emotional response

Our perception of objects, artworks and designs often involves an emotional response. Our hypothesis here is that when a drawing, pattern or design is 'good' most people are able to acknowledge this. It may be difficult to articulate exactly why, but it is not difficult to recognise.

Colour and Emotion

"Emotions can be stirred by colour"(Tucker, 1987). It is well known that some colours make us feel excited, while others give a feeling of calm and restfulness; much research has been carried out in this area. But what happens when colours are combined?

In textile design, what causes us to like some colour combinations better than others? In any range of colourways for a patterned fabric, there will usually be one that outperforms the others in terms of popularity and sales. (A colourway is a version of a design using a set of colours. An alternative colourway is a further version of the same design in different colours.)

Using a colour solid model to map the co-ordinates of hue, saturation and chroma for each of the colours in a successful colourway, can further colourways be produced with the same mathematical relationships between the constituent colours? And will these have the same level of appeal as the original "best" colourway? Is there a pattern in the colour relationships that exist in best-selling colourways? Research at UMIST is being carried out to find answers to these questions.

Textile designers accrue a large knowledge of successful colour combinations over a period of time, but lack the grammar of colour or the tools to be able to communicate their know how. As a result, imprecise, often abstract similes are used to describe a mood or 'story'. So why is colour so difficult to talk about, quantify and analyse? To an experienced design eye, good colourways or mistakes in a colourway are easy to spot. Why these are good or bad is however much less easy to explain.

Under normal light conditions the human eye can determine well over 10 million different colours. Through history the study of colour and the development of colour theories were frequently undertaken by artists and designers and many artists such as Seurat have spent their lives attempting to understand colour (Hughes, 1980).

Some colours are associated with cold (blues and greys) and some with warmth (reds and oranges). Colour can convey the time of day, weather conditions and temperature and even the time of year. Colours can be designed to blend in with the environment or to stand out. Frank Lloyd Wright perceived architecture as growing directly from the earth on which it stood and therefore used the colours of the surrounding areas in his buildings while Richard Rogers and Renzo Piano, the designers of Paris's Pompidou Centre, used colour, not to blend in, but to code the different pipe systems for heating and other services. This emotional use of colour extends to clothing as camouflage. Studies in fashion adoption recognise the need to blend into our surroundings (often for reasons of safety) (Batterberry and Batterberry, 1982 and Laver 1982). It is not surprising therefore that in cities we wear urban building colours and in the country we wear the colours of the earth and flora. The colour of an item influences the perception of it and this is used extensively in marketing. It is usually the first thing that attracts a consumer to a product, and this is particularly true of clothing.

Many people believe a red car will drive faster than a white one. A series of trials showed that coffee served in a red mug was preferred to the same coffee served in a yellow mug (considered too weak) and in a brown mug (too strong). Foodstuffs claiming to be pure and unadulterated often use blue and white packaging to communicate purity. Products communicating strength adopt vibrant and contrasting colours with the greater the contrast giving stronger associated power (Porter, 1980).

The colour of the universe

In January this year it was revealed by two American researchers Dr Karl Glazebrook and Dr Ivan Baldry that the average colour of the universe is a greenish hue halfway between aquamarine and turquoise. This is the colour that would be seen if all the visible light of the universe was mixed together. They say that earlier, when young stars dominated the universe, the average colour was bluer. Perhaps it is no co-incidence that blue is a calming colour, that some research has found it to be the most popular colour and that it is the colour for the Virgin Mary.

REFERENCES

Batterberry, A., and Batterberry, M., *Fashion, 1982, a Mirror of History*, London, Columbus Books

Bourke, P., *Fractals*, http://astronomy.swin.edu.au/~pbourke/fractals/

Glazebrook, K and Baldry, I., *The Cosmic Spectrum and the Color of the Universe, http://www.pha.jhu.edu/~kgb/cosspec/*

Gregory, R. L., 1990, *Eye and Brain: the Psychology of Seeing*, - 4th ed. - London: Weidenfeld and Nicholson.

Hughes, R., 1980, *The Shock of the New*, London, Thames and Hudson

Justema, W., 1982, *The Pleasures of Pattern*, New York; London: Van Nostrand Reinhold.

Laver, J., 1982, *Costume and Fashion, a Concise History*, London. Thames and Hudson

Porter, T., 1980, *The Language of Colour*, Slough, ICI.

Tucker, J.V., 1987, Printing, psychology of colour, *Target Marketing*, 10 (7), 40-49

Vasagar J., Jan 11 2002, And the colour of the universe is ... , *The Guardian*

Emotional responses to solid shape

John Willats
Loughborough University, UK

INTRODUCTION

Shape and colour are two key factors in design, and in order to think and talk about the shape and colour of a product during the course of the design process effective ways of describing both are needed. A comprehensive vocabulary for describing colour is available and there is a well-established link between colour and the emotions. But what words can be used to describe shape, and are there particular emotional responses to different kinds of shape?

DISCUSSION

According to Marr (1982) any scheme for representing solid shape must have at least two design features: its *primitives* and a *coordinate system* for relating these primitives to each other in space. The primitives of a representational system are the smallest units of information available in the system, and may be zero-, one-, two- or three-dimensional. For example, a shape description of an object such as a table can be built up of zero-dimensional primitives or *points*, and the positions of these features and the spatial relations between them can be defined using Cartesian coordinates. (Using this system the positions of points on the surface of an object are referred to three axes at right angles whose directions are defined relative to the principal axes of the object itself.) This method of defining the shapes of objects is of great importance, because it currently forms the basis for defining and manipulating shapes in computer programs. However, representing solid shape in terms of Cartesian coordinates is not, for most people, a natural way of thinking about or responding to the shapes of objects.

One alternative is to define the shapes of objects in terms of primitives that are one-dimensional instead of zero-dimensional, so that instead of using points, the shape of a table might be described in terms of the directions of its *edges*. The problem with this approach is that it can only be used for objects that *have* edges, and it is therefore not very useful for describing objects with smooth or irregular surfaces – and this includes many objects made by humans as well most objects in the natural world. It is also possible to build up descriptions of objects in terms of two-dimensional primitives, so that the shape of a table could also be described in terms of its *surfaces* and the spatial relations between them. However, this approach is still limited to objects having surfaces with regular shapes.

Finally, it is possible to build shape descriptions in terms of three-dimensional or *volumetric* primitives. According to Marr (1982), the primary function of the human visual system is to take the images available at the retina and use them to compute what he called the 3D-model: shape descriptions based on volumetric primitives. These volumetric shape descriptions are stored in memory and used to recognise objects when we see them again from a different direction or under new lighting conditions. But while the idea of building up shape descriptions in terms of points, edges and surfaces is relatively familiar, using primitives consisting of whole volumes is perhaps more difficult to grasp. Such descriptions can, however, be developed using what is called the *extendedness principle.*

Using the extendedness principle, all three-dimensional objects can be classified in terms of their relative extensions in three-dimensional space (Willats, 1992). In this way all objects can be described as either *lumps, sticks* or *slabs*, or combinations of these shapes. *Lumps* are volumetric primitives that are about equally extended in all three dimensions, *sticks* are primitives that are extended in one direction only, while slabs or discs are primitives that are extended in two directions but not the third. Thus a table can be described simply as a slab standing on four sticks.

One of the most important features of this approach is that shape descriptions of objects based on the extendedness principle can be given independently of the shapes of their edges or surfaces. Cubes, balls and irregularly shaped objects such as potatoes can all be described as 'lumps', and pencils, fence posts and the trunks of trees can all be described as 'sticks'. As a result, this method of shape description is exhaustive: all three-dimensional objects, whether natural or made by humans, can be classified in terms of their extendedness. Another advantage of this approach is that it is more economical than other methods: it would take thousands of point primitives to describe the shape of a table, whereas a volumetric description given in terms of the extendedness of its component parts requires only five primitives: four sticks and a slab. This method of classification can also be applied to two-dimensional images, so that the two-dimensional primitives in a picture or an image on the retina can be classified as either *round regions* or *long regions*. In this form this method of shape description is very restricted, but its scope can be enlarged by using what are called *shape modifiers*. For example, a banana can be described as a stick with the addition of the shape modifier 'being bent', while an ice-cream cone can also be described a stick, but with the addition of the modifier 'being pointed'.

The evidence that human beings actually use this way of classifying shapes comes from four sources: adult languages, children's early speech and drawings and adult pictures. Rosch (1973) gives examples from experiments with colour perception to show that categories such as 'pure blue' are not arbitrary but are 'given' by the human perceptual system. As a result, she argues, the meaning of such categories is 'universal across languages' (p. 112). Much the same seems to be true of the extendedness principle in relation to shape. In adult language extendedness is described using what linguists call 'noun classifiers': obligatory additions to nouns that classify properties such as shape or gender. Perhaps one reason for the relative unfamiliarity of the concept of extendedness is that English is relatively impoverished in noun classifiers for describing shape properties, although phrases such as 'a lump of wax', 'a stick of wax' and 'a slab of wax' act rather like classifiers. English is, however, exceptional in this respect: in nearly all other languages, from Bantu to Burmese and Japanese to Eskimo 'extendedness is a widespread and perhaps universal principle of noun classification' (Denny 1979).

In Japanese, for example, the classifier for shapes that are about equally extended in all three dimensions is *ko*, but this does not simply mean 'round' because it is used for square-cornered objects such as blocks and boxes, as well as apples and oranges. Moreover, the scope of this method of shape classification is usually extended by the use of shape modifiers called 'sub-variables'. In Cree, for example, an American-Indian language, the root *wawiye* is used for genuinely round shapes like a basket ball, while *nonim* is used for shapes that are both round and long such as the back of a Volkswagen Beetle.

The same principle can be seen at work in children's early speech and drawings and artists' pictures. Children regularly overextend the meaning of their first words so that they act like classifiers, using *doggie*, for example, to mean all animals. However, the commonest overextensions children use are those used to describe *shape*: a child will use a word such as *baw* (ball) to mean all round shapes such as apples, and *tee* (stick) to mean all long shapes such as a ruler or an umbrella (Clark, 1976). Similarly, in their early drawings children will use round regions to represent all kinds of round volumes, including square shapes such as cubes and houses as well as round shapes such as people's heads, while lines or long regions are used to represent long volumes such as arms and legs (fig. 1). At a later stage of development children use shape modifiers such as 'being bent' as a way of differentiating between legs and feet, or add short lines or little 'hats' to round regions as a way of representing the edges and corners of cubes. It is not until the age of about 12 years that children use lines rather than regions as shape primitives (Willats, 1997).

Figure 1 Drawing of a man by a five-year-old boy.

Adult artists have also used the extendedness principle as a way of representing three-dimensional shapes in pictures. In Picasso's *Rites of Spring*, for example, the heads of the figures in silhouette are represented by round regions, the arms and legs are represented by long regions, and the horns of the goat are represented using the shape modifiers 'being bent' and 'being pointed'. A similar approach was used by the early Greek vase painters (Willats, 1997).

If the extendedness principle provides a natural way of classifying shapes, how did it come to be built in to the visual system? It seems likely that this may have come about as the result of evolutionary pressure. There are simple mathematical relations between lumps, sticks and slabs in real scenes and the round and long regions that these shapes project to the retina. Lumps, whatever the detailed shapes of their surfaces, must always project round regions as images on a retina, and sticks will project images in the form of long regions except when they are seen end-on. Slabs and discs, on the other hand, will project round regions if they are seen from above, but long regions when they are seen from the side. Given certain assumptions, it can be shown that the chances of a round region being a view of a lump are better than evens – in fact, about 3 in 5 – and the probability of a long region being a view of a stick is about the same. In contrast, the probability of either a round region or a long region being a view of a disc is rather low – about 1 in 3 – and as a result it is difficult to see discs (such as the tambourines held by the dancing figure in the *Rites of Spring*) in static silhouettes (Willats, 1992). However, it can also be shown that there is a simple way of detecting slabs in changing images. Detecting the differences among lumps, sticks and slabs in the real world can thus be achieved using a coarse retina with very few receptors, so primitive organism whose survival depends on being able to distinguish among mates, prey and predators in the environment in the form of lumps, sticks or slabs will be driven by evolutionary pressure to develop a visual system that can distinguish between long regions and round regions, and will learn to associate this distinction with the appropriate responses.

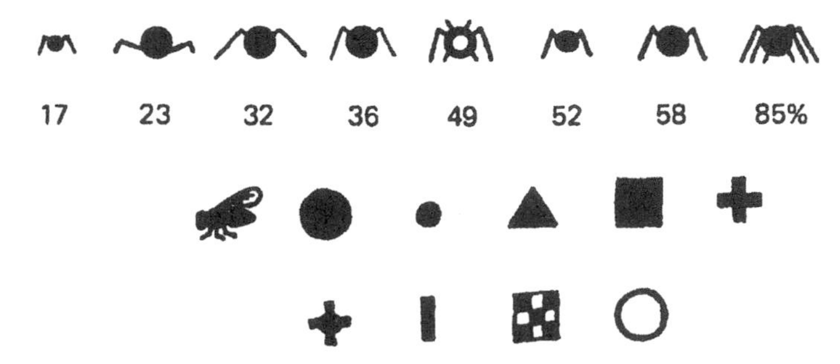

Figure 2 Drawings used to evoke courtship behaviour in jumping spiders (top) together with the variety of shapes that jumping spiders will attack (bottom). The numbers show the percentages of courtship responses (Land, 1990). Figure originally published in Z. Tierpsychol. 9:169-207, (1952) Reproduced with permission from Blackwell Verlag GmbH.

This kind of association can be illustrated by the behaviour and visual system of the jumping spider. Jumping spiders do not make webs, but stalk their prey like a cat stalking a mouse. To do this they are equipped with four pairs of eyes. One pair is vestigial, two pairs detect movement, and the remaining two eyes have coarse, rotating retinas that are designed to detect long regions in relation to a central round region. Jumping spiders will attack most things that are moving and of about the right size, but will only attempt to mate with objects that appear to have the right numbers of legs in the right places. Figure 2 shows the percentages of courtship responses to different combinations of long and round regions, together with the variety of shapes that jumping spiders will attack.

Although the jumping spider is not one of our evolutionary ancestors this example shows how the shape property of extendedness can become associated with the fundamental emotional properties of fear, attraction and desire. It also demonstrates that responses to shape can be tested. It is possible that humans may also, in the course of evolutionary history, have developed similar links between the emotions and certain solid shapes, and that these links can also be tested.

CONCLUSION

Marr (1982) argues that the primary function of the human visual system is to derive volumetric shape primitives from the ever-changing images available at the retina. These volumetric primitives can best be described in terms of the extendedness principle, and this provides a way of representing solid shape that is widely used in both natural languages and pictures. The extendedness principle thus provides the basis for a real rather than artificial system for describing shape comparable with the natural categories we have for describing colour. The emotional responses to colour have been the subject of widespread research, and the results are widely used in design. It seems likely that there may be similar links between the emotions and the extendedness of shapes, and if this proves to be the case it could have important implications for the design process.

REFERENCES

Clark, E.V., 1976. Universal categories: on the semantics of classifiers and children's early word meanings. In *Linguistic Studies Offered to Joseph Greenberg on the Occasion of His Sixtieth Birthday*, Vol. 1, edited by Juilland, A. (Saratoga, Ca.: Anma Libri), pp. 449–462.

Denny, P.J., 1979. The 'extendedness' variable in noun classifier semantics: universal features and cultural variation. In *Ethnolinguistics: Boas, Sapir and Whorf Revisited*, edited by Mathiot, M. (The Hague: Mouton), pp. 97–119.

Land, M., 1990. Vision in other animals. In *Images and Understanding*, edited by Barlow, H., Blakemore, C. and Weston-Smith, M. (Cambridge: Cambridge University Press), pp. 95–212.

Marr, D., 1982. *Vision: A Computational Investigation into the Human Representation and Processing of Visual Information* (San Francisco: Freeman).

Rosch, E., 1973. On the internal structure of perceptual and semantic categories. In *Cognitive Development and the Acquisition of Language,* edited by Moore, T.E. (New York: Academic Press), pp. 111–144.

Willats, J., 1992. Seeing lumps, sticks and slabs in silhouettes. *Perception*, 21, 481–496.

Willats, J., 1997. *Art and Representation: New Principles in the Analysis of Pictures* (Princeton: Princeton University Press).

Colour preference and colour emotion

Li-Chen Ou
University of Derby, UK

Ming Ronnier Luo
University of Derby, UK

INTRODUCTION

Colour plays an important role for customers in making decisions on what they like or dislike. It evokes different emotions such as exciting, energetic, and calm. These emotional feelings are called colour emotions. There have been a number of attempts to define colour emotion factors such as Kobayashi (1981) and Sato *et al.* (2000). However, these studies focused merely on single-colour emotions. In our daily life a colour is rarely seen in isolation. Designers are keen to know how to make a colour combination that generates a particular emotion. It is therefore necessary to develop an integrated model for both single colours and combinations.

Colour preference research has long been focusing on single colours only. A few studies on colour combinations are found in the field of psychology, such as Guilford (1931), Lo (1939), and Hogg (1969). These were aimed to predict the preference value of a colour combination by its constituents. Unfortunately, none of the attempts were successful. In the present study, the relationship between colour preference and colour emotion is investigated. A colour preference model for combinations based on colour emotions is also developed.

METHODS

Experiments were conducted to investigate the ten colour emotions expressed by word pairs: warm-cool, heavy-light, modern-classical, dirty-clean, active-passive, hard-soft, tense-relaxed, fresh-stale, feminine-masculine, and like-dislike. The word pair "like-dislike" represents the colour preference. All of the word pairs were selected according to three primary factors given by Osgood *et al.* (1957): "evaluative", "potency", and "activity".

Thirty-one observers, including 14 British and 17 Chinese, took part in the experiment. Each assessed 20 colour patches and 190 colour pairs against a mid-grey background in a viewing cabinet illuminated by a daylight simulator, the illuminant D65. The 20 colours were selected from an entire range of hue, lightness, and chroma, so as to give a good coverage in the CIELAB uniform colour space. This colour space was recommended by the CIE (Commission Internationale de l'Eclairage) in 1976 (Hunt, 1998). In Figure 1 are shown the 20 colours plotted on the hue plane of the CIELAB space. Each colour patch had a size of 3 by 3 in. The 190 colour pairs were all combinations of the 20 colours.

Each observer answered a series of questions, "which word is more closely associated with the colour/colour pair presented, warm or cool?" For each colour and colour pair, the ten word pairs were asked in a random order.

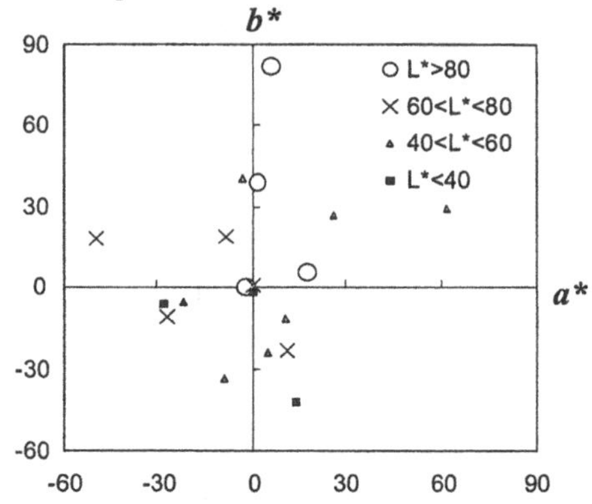

Figure 1 The 20 colour samples plotted in CIELAB a* b* diagram (hue plane).

RESULTS AND DISCUSSION

Experimental results were analysed according to Thurstone's Case V of Law of Comparative Judgement (Thurstone, 1927). The frequency of the first word chosen from each word pair was converted to the z-score. The z-score indicates the extent to which the observers agreed with the first word chosen from a word pair to the colour or colour pair presented.

COLOUR-EMOTION DIMENSIONS

The factor analysis is able to reduce a large data set from a group of interrelated variables into a small set of independent factors. In this study, the z-scores of the ten emotional word pairs were used for the factor extraction, so as to reveal the colour-emotion dimensions for single colours and colour pairs.

The results from British and Chinese observers were found to have no significant difference, and therefore the z-scores obtained for all observers were used in the factor analysis for both single colours and colour combinations.

In the case of single colours, three word pairs, "tense-relaxed", "warm-cool", and "like-dislike" were excluded from the extraction procedure, because poor agreement was found for "tense-relaxed" and "like-dislike" between the two cultures, and that the word pair "warm-cool" is completely independent from others. Two dimensions were extracted from remaining word pairs. The third dimension was determined from the scale "warm-cool". This scale has been widely studied as one of the most important colour emotions (Wright and Rainwater, 1962; Sato *et al.*, 2000). The three dimensions for single-colour emotions are *colour activity* ("active-passive", "fresh-stale", "clean-dirty", and "modern-classical"), *colour weight* ("hard-soft", "masculine-feminine", and "heavy-light"), and *colour heat* ("warm-cool").

For colour-combination emotions, on the other hand, three dimensions were identified using the same methods. These were found similar to those developed for single-colour emotions, and were also labelled "colour activity", "colour weight", and "colour heat". The three dimensions agree well with those obtained by Kobayashi, i.e. "clear-greyish", "soft-hard", and "warm-cool" (Kobayashi, 1981), and Sato *et al.*, i.e. "activity", "potency", and "warm-cool" (Sato *et al.*, 2000), respectively. It should be noted that the two studies were based on Japanese observers, while the present work used British and Chinese data. This suggests that the underlying dimensions of colour emotions are consistent across cultures.

RELATIONSHIP OF COLOUR EMOTIONS BETWEEN SINGLE COLOURS AND COMBINATIONS

An additivity relationship of colour emotions between single colours and combinations was revealed in this study. Based on this relationship, the emotional effect (z-score) of a colour pair can be predicted by the average of two component colours.

Table 1 shows the performance of this model for the ten colour emotion scales together with the three colour-emotion dimensions. The results indicate that the model has good predictive value for all but the scale "like-dislike", with Pearson correlation of 0.37. This implies that the colour preference ("like-dislike") cannot be predicted in such a simple way. The results agree with those found by Hogg (1969). Besides, the three colour-emotion dimensions, i.e. "colour activity", "colour weight", and "colour heat", were found in close correlation between single colours and combinations. This confirms that the colour-emotion dimensions for single colours and combinations have the same framework.

Table 1 The performance of the average model (Pearson correlation)

Emotional Effect	warm cool	heavy light	modern classical	dirty clean	active passive	hard soft	tense relaxed
British	0.80	0.81	0.63	0.83	0.71	0.57	0.60
Chinese	0.86	0.85	0.75	0.78	0.80	0.76	0.63
All	0.89	0.89	0.78	0.87	0.82	0.77	0.69

Emotional Effect	fresh stale	feminine masculine	Like Dislike	**colour activity**	**colour weight**	**colour heat**
British	0.65	0.69	0.33	-	-	-
Chinese	0.79	0.68	0.44	-	-	-
All	0.84	0.75	0.37	0.89	0.84	0.89

COLOUR PREFERENCE FOR COMBINATIONS

By looking at the 190 colour pairs presented on a desk and ranked in order of preference, one may have general feelings from colour pairs on the "liked" side, such as "clean" and "light". Some general feelings may also be found on the "disliked" side, such as "dirty" and "tense". This kind of general impression can easily be obtained by using eyes only. It is also supported by the experimental results of this study.

Table 2 The Pearson correlation between "like-dislike" and the 9 colour emotions for combinations

	cool (warm)	light (heavy)	Modern (classical)	clean (dirty)	Active (passive)
British	0.18	0.18	-0.18	0.15	-0.31
Chinese	0.13	0.40	0.40	0.51	0.11
Both	0.32	0.35	0.14	0.39	-0.12

	soft (hard)	relaxed (tense)	Fresh (stale)	feminine (masculine)
British	0.08	0.61	0.02	0.01
Chinese	0.19	0.35	0.48	0.21
Both	0.09	0.47	0.27	0.14

Table 2 shows the Pearson correlation between colour preference and other nine colour emotions used in this study. Chinese observers were found to prefer colour pairs with emotional feelings "light", "clean", and "fresh", while British observers preferred those with the feeling "relaxed". A colour preference model for colour pairs was derived by the method of linear regression, as follows:

Colour Preference (for British observers)
= -0.095 + 0.748 Relaxed + 0.236 Clean + 0.303 Hard (2)

Colour Preference (for Chinese observers)
= -0.517 + 0.297 Relaxed + 0.350 Clean + 0.212 Hard (3)

The two equations require the input of the z-scores from three colour-emotion terms, "relaxed", "clean", and "hard", so as to predict colour preference. The model indicates that the stronger emotional feelings of "relaxed", "clean" and "hard" lead to the stronger possibility that the colour pair is preferred. "Relaxed" is the most important term for British observers as its coefficient is significantly greater than the other two. For Chinese observers, on the other hand, all of the three terms have equal contribution.

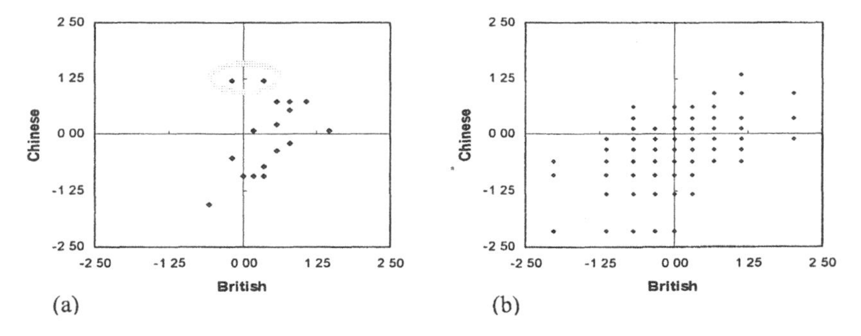

Figure 2 Comparisons between two cultures on "like-dislike" for (a) single colours and (b) colour pairs

The colour preferences were compared between British and Chinese groups. In the case of single colours, the two most liked colours for Chinese observers were found disliked by British observers, as marked in Figure 2 (a). The two colours are vivid green and bright cyan (G-2060 and B30G-1040 in the NCS notation). In general, single-colour preferences agreed quite well between the two groups. For the 190 colour pairs, on the other hand, the colour preferences were also found in close correlation between the two groups, as shown in Figure 2 (b).

CONCLUSION

In this study, three dimensions, "colour activity", "colour weight", and "colour heat", are identified for describing both single-colour and colour-pair emotions. An additivity relationship of colour emotions between single colours and combinations is revealed. The relationship enables the prediction of colour emotion for a colour pair by the average of two component colours. This model gives good prediction for all but the scale "like-dislike".

A colour-emotion based colour preference model for two-colour combination is developed. Three colour emotion terms "relaxed", "clean" and "hard" are included. "Relaxed" is the most important term for British observers, while for Chinese all the three terms have equal contribution.

AKNOWLEDGEMENTS

The authors would like to thank Ms. A. Wright and Ms. A. Woodcock for providing guidance and encouragement for this work.

REFERENCES

Guilford, J. P., 1931, The prediction of affective values, *American Journal of Psychology*, **43**, pp. 469–478.

Hogg, J., 1969, The prediction of semantic differential ratings of color combinations, *Journal of General Psychology*, **80**, pp. 141–152.

Hunt, R. W. G., 1998, *Measuring Colour*, 3rd ed., (Hertfordshire: Fountain Press), pp. 63-67.

Kobayashi, S., 1981, The aim and method of the color image scale, *Color Research and Application*, **6**, pp. 93–107.

Lo, C, 1936, The affective values of color combinations, *American Journal of Psychology*, **48**, pp. 617–624.

Osgood, C. E., Suci, G. J. and Tannenbaum, P. H., 1957, *The Measurement of Meaning*, (University of Illinois Press).

Sato, T., Kajiwara, K., Hoshino, H. and Nakamura, T., 2000, Quantitative evaluation and categorising of human emotion induced by colour, *Advances in Colour Science and Technology*, **3**, No. 3.

Thurstone, L. L., 1927, A law of comparative judgement, *Psychological Review*, **34**, pp. 273–286.

Wright, B. and Rainwater, L., 1962, The meaning of color, *Journal of General Psychology*, **67**, pp. 89–99.

Type, motion and emotion: a visual amplification of meaning

R Brian Stone
Ohio State University, USA

Daniel P Alenquer
Ohio State University, USA

Jeffrey Borisch
Ohio State University, USA

INTRODUCTION

Typographic messages can be analyzed through three different dimensions: semantic denotative representation, color and texture, and shape. These dimensions, when presented to subjects as stimuli, activate a variety of thoughts, images and meanings that are in both semantic and episodic memory systems. Personal and collective representations trigger a complex sequence of reactions known as emotions. We believe that the introduction of motion as a new dimension in the stimuli will change the asymptotic level of learning that subjects can achieve when viewing typographic messages. The perception of the stimuli may also be changed, having impact on the intensity of the associative bonds, amplifying the emotions as responses.

A typographic vocabulary, through the use of time-based composition, sound and animation can broaden the emotional stimulus in users beyond static delivery systems. The application of kinetic media will enable the typographic designer to add motion, scale change, sequence, metamorphosis, and context (mood or emotion) to typographic communication. Through our results of teaching, and by way of a series of qualitative and quantitative analysis (using Plutchik's mood rating scale), we have concluded that kinetic typography, within an appropriate context, has the possibility to evoke emotion, while enhancing visual form, meaning and communication.

The study of "type in motion" and its stimulus effect on emotions, establishes a foundation for more meaningful applications in professional practice. Ultimately, these concepts can be realized in products such as film and television titles, movie previews, commercials, information kiosks, multimedia programs, web sites, and presentations. By bringing "type in motion" studies into our design curriculum we investigate varying type size, weight, spatial relationships, form – counter form, and movement within a word (or words), producing a variety of rhythmic and expressive uses of kinetic typography, that preserve sound typographic principles. We used these projects as the visual stimulus in our preliminary research with human subjects. By objectively proving the effectiveness of motion typography to the communication of a visual message, we hope to make "type in motion" a standard part of the design curriculum. Our test is small in scope and serves to give us clearer questions and methods for future study.

METHODOLOGY

The set of stimuli used in this experiment consisted of four typographic animations, and four still images that corresponded to the typographic animations. They were extracted from student projects produced during the course "Type in Motion" instructed by R. Brian Stone at The Ohio State University.

There were 4 animations that were divided into two categories, a pair of "verbal and visual equations," and a pair of "animated antonyms." The four still images, were modified screenshots of a corresponding animation.

The procedure consisted of showing the stimulus to the subject and asking him or her to complete a quantitative mood rating scale. Subsequent to viewing the stimulus, they were asked to make qualitative comments on what they viewed. Prior to seeing the first animation or still, subjects were asked to complete a mood rating scale, indicating how they were feeling at that moment in time in order to establish an emotional baseline. Test participants viewed the stimulus on a computer screen. Figure 1 shows a typical still image of a typographic stimulus, contrasted by the animated message sequence shown in figures 2–6.

Figure 1 Still image of corresponding animated verbal and visual equation.

Figure 2 The letterform first appears as an abstraction of a paddle.

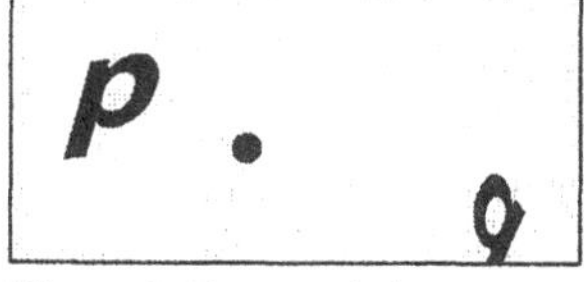

Figure 3 Abstracted elements begin a playful juxtaposition, mimicking a game.

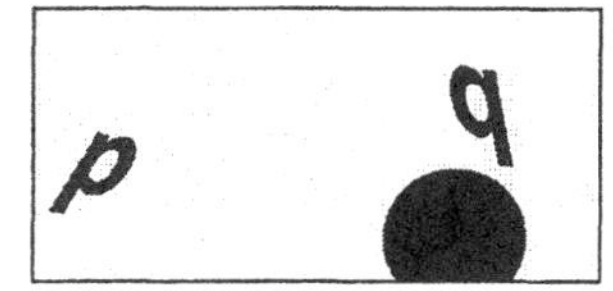

Figure 4 Abstract elements become recognizable as letterforms.

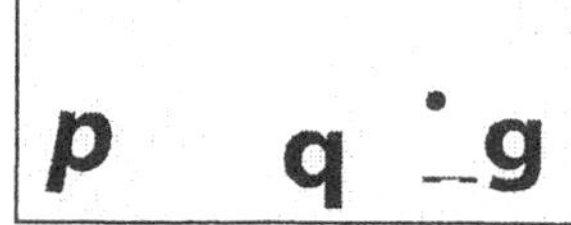

Figure 5 Letterforms take order, forming a word picture. Dot is associated with a ball.

Figure 6 Vertically bouncing movement semantically represent the game.

This quantitative measure is synthesized with qualitative observations of participants. The facial expressions and body language of participants were recorded on videotape. Our experience has shown that research participants may not always fully articulate their feelings through a quantitative scale. What they "say" is not always what they actually "feel." We assumed that some participants might not be fully aware of the emotional condition they are in at a particular time. Additionally, the participant's condition may be transitory enough to not affect the outcomes of the mood rating scales. The questionnaires would give us a structured reflection of the emotional state they believe they were at that time. Their facial expressions and body language would complement the questionnaires, giving us other elements and more information to analyze.

Quantitative Analysis

Analyses of the mood rating scales were used to determine if there were any measurable emotional changes based on the stimulus viewed. A mood rating scale (Plutchik, 1980) was used as the primary quantitative metric for evaluating mood changes in participants. This scale shows eight words that represent the eight clusters of mood terms corresponding to the primary emotion dimensions. These words – Happy, Fearful, Agreeable, Angry, Interested, Disgusted, Sad and Surprised – were to be classified into five different levels of activation: Not at all, Slightly, Moderately, Strongly and Very Strongly.

By completing the mood rating scale, subjects would be able to give us a structured reflection of the state they believed they were in at the time they were exposed to a given stimulus. 10 participants viewed a module that displayed still images first, and animated messages second. 10 other participants viewed the same modules but animations first and still images second.

The nature of the emotional responses varied from the four different stimuli presented, but the differences between animated versus still emotional responses were constant within the same stimulus. Summarizing all the stimuli into one graphic confirmed to be ineffective, as different stimuli trigger different emotions, and one tended to suppress the other. For investigation purposes we analyzed each stimulus individually, and selected to report the stimulus "Ping-Pong" as an example. It shows a representative display of the differences observed on the emotional responses of animation versus still image.

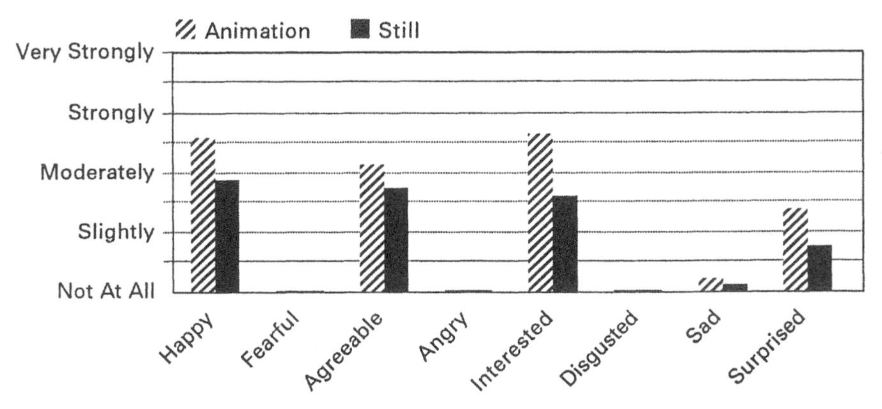

Figure 7 Shows average emotional response from participants for stimulus "Ping-Pong" "Ping-Pong" animation by Bok Young Kim, The Ohio State University

Although participants responded differently when we switched the order of the animated stimuli to the still stimuli, there does not appear to be any measurable difference in how participants reported their emotions. Negative emotions tended to flat line, indicating very little change. Furthermore, the results from the analysis show that the emotion activation is more powerful when the typographic message is in motion than when it is shown in a static state.

Qualitative Analysis

To complement the quantitative data, while viewing the participants via videotape, we applied Schlosberg's (1954) work on the activation theory of emotion. In it he developed the idea that at least three independent dimensions are needed to describe what we know about facial expressions relative to emotion. One of these dimensions is pleasantness-unpleasantness. The second is attention-rejection, and the third is intensity or level of activation. Every facial expression of emotion could be described in terms of these three axes.

"Paul T. Young (1961) began by pointing out that we judge the presence of emotion in another person on the basis of various kinds of evidence. The kinds we use are (1) knowledge of the situation, (2) knowledge of how a person typically reacts to various situations, (3) physical signs of disturbance, and (4) types of behavior used by a person to adjust or adapt to the situation. Based on these kinds of observations we make interpretations or inferences about the presence of emotion in another person" (Plutchik, 1980).

These qualitative evaluations emerged as a valuable portion of this research. A limitation of the structure of the quantitative portion is that the emotional state is sampled at a point between the stimuli. Through our qualitative observations, we saw that it may be more important to measure in some fashion the subject's emotional state while viewing the stimuli. This may involve the use of measuring physiological responses. While viewing the stimuli, bearing these hypotheses in mind, we observed a variety of changes in facial expressions when participants were exposed to the stimulus. Again, these typically favored the positive emotions. Figure 8–10 shows participant #5 viewing the animated sequence shown in figures 2–6. Figure 11–13 shows the participant viewing the corresponding still image (figure 1). The changes in facial expressions while viewing the animated stimulus are clearly visible, while the still image evoked very little change in expression.

Figure 8 Participant appears interested and/or inquisitive

Figure 9 As animation progresses, eyes are focused

Figure 10 Meaning is revealed; participant smiles or agrees

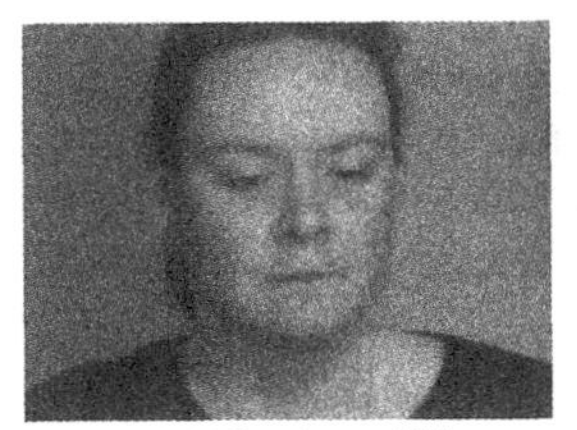

Figure 11 Viewing still image, participant appears passive

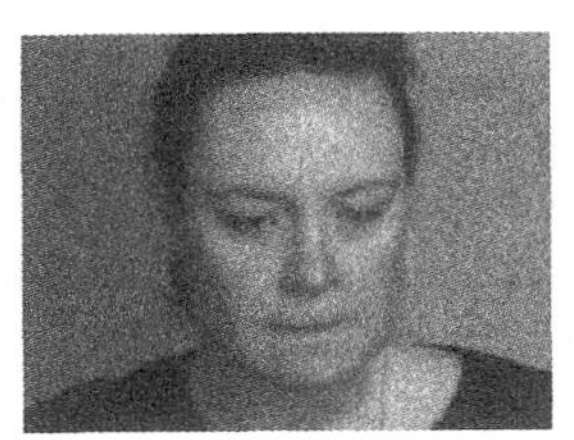

Figure 12 Participant shows small changes in expression

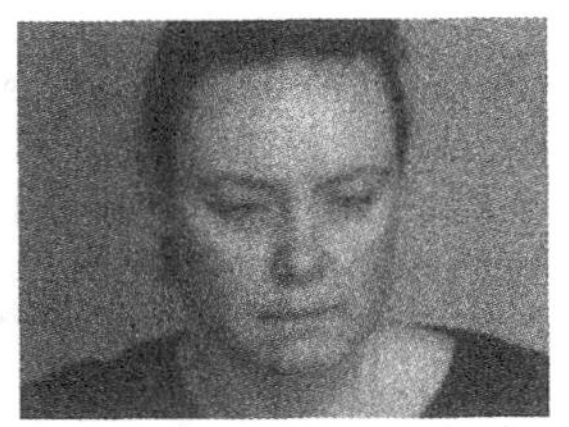

Figure 13 As time passes participant appears pensive

CONCLUSION

We believe that "type in motion" or kinetic typography that is specifically designed with the intent of enhancing meaning, can evoke emotional responses. Figure 7 shows that the emotion activation is more powerful when the typographic message is in motion than when it is shown in a stationary manner. This quantitative result, synthesized with our empirical data was typical in the majority of participants tested. Further studies are needed if the medium is to move beyond arbitrary applications of "flying type." Our preliminary results appear to support the idea that motion typography has the ability to influence reactions within the viewer. These reactions, whether they are excitement, delight, humor, agitation, or tension, make typographic communication a richer, more memorable experience, even if these reactions are momentary.

Further Study

Our preparatory research has indicated that "type in motion" studies and its effect on emotions, may yield more meaningful kinetic applications. Although emotions have different elements (namely, feelings, behaviors, and physiological changes), they are still unitary phenomena. We intend to carry this research further to develop theoretical insights relative to the inherent features of "type in motion." To accomplish this task, the research will include the messages' emotional effect, by incorporating the following:

- Improve the control of stimuli variables
- Develop new methods of testing and measuring emotions
- Investigate aspects of feelings and behavior through different methodologies
- Investigate the physiological changes in participants
- Test with a larger sample and a randomization of the stimulus

REFERENCES

Plutchik, R., 1980, *EMOTION, A Psychoevolutionary Synthesis*, (New York, NY: Harper & Row).

Schlosberg, H., 1964, Three dimensions of emotion. *Psychological Review*, pp. 61, 81–88.

Young, P.T., 1961, *Motivation and Emotion*. (New York, NY: Wiley)

Animated emotion: human figure movement and viewer emotion

Valentijn Visch
Free University, Amsterdam and Industrial Design Delft
The Netherlands

INTRODUCTION

Although a lot of feature films have been produced in the last century, little empirical research has been done on the viewer impression of the bodily movements of the characters. Questions concerning the emotional, aesthetic, narrative or generic impact of actor movements in film have only appeared at the sideline of studies from very diverse research areas such as perception research, ethnology, communication studies, psychology, theatre and film theory. This research will try to bring these theories together as it aims to identify which parameters of bodily movements in a filmic scene are responsible for what cognitive and aesthetic viewer impressions. It is supposed that once these parameters have been identified, a specific configuration of them will result in a predictable viewer impression.

The knowledge of these genre/emotive parametric configurations could be applied both in productive, and in recognition computer programs. For instance in animation programs: when a basic character action-scene is animated, it should be possible to change it with one press on a button into a dramatic, a comical or a realistic non-fiction scene. The research's outcomes could also be applied in recognition programs: a decoder could be able to identify which genre or emotion belongs to which cinematic scene.

OUTLINES OF THE PROJECT

The main question addressed by the project is: what movement properties of actors are determined by film genres. There are typical action scenes that occur in many different genres. For instance *arrivals, departures, chases, happy reunions, love at first sight* can be found in comedies, drama film, action film and non-fiction film like documentaries or news-items. These scenes portray goal-oriented actions by at least two characters. They are crucial in a narrative, in the sense that they end in only one or a few outcomes that have relevance for the characters. The outcomes are interdependent for the characters involved; X succeeds in his/ her goal or doesn't succeed and X's success is related to Y's. For our ends, the scenes need to meet further requirements. While we differentiate between three fiction genres, action, comic and drama, the scenes that will be used should occur in at least these three genres. Moreover, the scene should occur in reality as well, because one of our basis hypotheses is that genre variations in movements are variations of realistic movements.

To summarize, we are selecting scenes in which the action-content or plot-structure remains constant over all genres, including non-fiction, whereas the realisation in terms of bodily actions varies in each genre. Through comparative analysis we attempt to find the parameters that are responsible for the variations in action realisation.

METHOD OF RESEARCH

Newtson *et al.* (1977) have empirically proven that the perception of ongoing actor behaviour is organized into perceptual units of movements, marked by breakpoints. The scenes to be researched will serve as gross action units that may be divided into smaller units. As breakpoints convey more information for a subject, in the sense of actions, than non-breakpoints (Newtson and Engquist, 1976), subjects who make fine units will perceive more differentiated impressions and have a greater tendency to perceive the actions as causal than subjects that make gross-units (Newtson, 1973). The amount of perceived breakpoints in the realisation of a cinematic action scene will be used as a first genre predictor. A running chase in an action movie will for instance convey more breakpoints than a running chase in a drama film; breakpoints will tell the preference of genre for a scene.

At present we have worked on the genre-crossing scenes *running chase* and *happy reunion*. We collected about 200 examples of these scenes, ordered in genres, in an electronic database. Each of these scenes have been analysed from its most common description to its detailed genre dependent actor movements. The analysis of the happy reunion runs as follows:

1. abstract description of the gross action:
$X \Leftrightarrow Y$; with the restriction that character X and character Y are both willing to reunite.

2. description of successive finer action units:
I recognition
II approach
III physical contact
IV informative explanation

3. schematic plot representation of these fine action units, in terms of causal links and affect states of the characters (based on Lehnert, 1981).

4. overview of possible intentional variables of the characters involved and the setting of the scene, e.g.:
X2: X knows in advance he/ she will meet Y.
Y3: Y doesn't recognize X.
S4: The happy reunion takes place at a dangerous location.

5. detailed description of the bodily movements of the actors, e.g.:

X and Y walk with increasing speed and big steps towards each other. When they are two steps apart they decelerate and take two or three smaller unbalanced steps. They grasp each other's heads. Kiss three times. Push their heads and trunks together, and lift each other slightly from the ground, while rotating about 30 degrees to left and right. Than they quickly walk away with arms around shoulders.
(description of a news-item where two brothers meet each other at the airport accompanied by press. Two days before the meeting one of them was said to be dead.)

From these analyses we discerned the following set of *kinesiological variables*:

- position/ spatial movement of limbs
- velocity of movements and locomotion
(action actors move faster than drama actors)
- proportion of smooth/ jagged movements
(dramatic actors move more smoothly than comic or action)
- ground-stability of the character
(action actors have higher ground stability)
- proportion of occupied horizontal/ vertical body space
(comical actors occupy more horizontal space - spread arms and legs - than action actors)
- degree of perceptive details
(non-fiction actors show a higher degree of fine detailed movements)
- realistic efficiency of movements
(non-fiction actors, like athletic sprinters, run more efficient than action actors)

In a 3-D computer-animation program we will build animations of the genre-transient scenes. Subsequently we will compute the kinesiological variables into sliding parameters. By varying the configuration of parameters it should be possible, on basis of subject experiments, to build for each scene a prototypical genre-specific realisation of the scene. Prototypes are in this sense defined as being "the clearest cases of category membership defined operationally by people's judgements of goodness of membership" (Rosch, 1999).

As we are interested in the effects of the cinematic body movements, we will try to eliminate any other cinematic effect as much as possible. The characters will appear faceless and their bodies will appear as abstract line figures against a blank background. There are however some cinematic properties that can't be negated or eluded, partly because they can't be separated from the cinematic medium (such as camera-usage), and partly because they will have a considerable effect on the subject appraisal of the scene. Therefore the following cinematic scene properties will also be treated as a second set of variables in the experiments:

cinematic variables:
- duration of the scene
- duration of scene-units
- amount of breakpoints
- completeness of the scene-units
- order of scene-units
- genre
- narrative context
- camera-usage

In a series of experiments the effect of these kinesiological and cinematic variables and their relation will be tested on two kinds of subject response-levels:
- cognitive level: questions regarding action intention, character intention, amount of breakpoints and genre.
- emotive level: questions regarding character emotion, subject emotion and aesthetic appraisal.

HYPOTHESES

1. A set of actor movement parameters exists that can regulate viewer response to genre. This set of parameters can be applied to any basic sequence of actor movements.

2. Movements in any fictional genre are systematic transformations of realistic movements. It is supposed that non-fiction actions contain more and more complex movements than fiction actions. A fiction genre selects a set of these realistic movements and transforms them into its own genre-specific movements.

3. Fiction prototypes will be preferred over non-fiction prototypical actions. A first explanation is that non-fiction scenes contain more and more complex movements than fiction scenes. It will thus be easier for viewers to judge fiction scenes as being prototypical. A second explanation concerns the test-material itself; as the tests are performed with animated characters, it will be more difficult for subjects to consider them as being realistic than as being unrealistic or fictitious.

PRELIMINARY RESULTS OF A PILOT-STUDY

Recently we performed a pilot-study to test the function and significance of the hypothesized genre-determination variable *velocity*. 48 subjects were presented with 30 film-fragments, which all contained *happy reunion* scenes taken from feature films and news items. The appearance of all fragments was digitally manipulated in the same way: only the outlines or borders of the objects, i.e. actors, were visible by a blue line against a black background, and all sounds were eliminated. In the pilot-tests the fragments were varied on the factors *genre* (i.e. the film-genres the scenes were taken out, as there are: comic, drama, action and non-fiction), *genre-prototypicality* (prototypical or a-typical) and *velocity* (the speed of the scene was manipulated by accelerating it with 1/3, decelerated it with a 1/3, or its speed was kept original). Subjects were asked to judge on a 5-point scale how well each of the 30 fragments fitted in each of the following four genres: comic, drama, action and non-fiction.

Resulting from MANOVA a main effect of *velocity* was obtained in the judgement of fitting of the fragments in the comic genre (F = 17.3, $p < .001$) and in the drama genre (F = 8.3; $p < .001$). Scheffe's post hoc test showed that accelerated ($p < .001$) and original ($p < .05$) fragments were judged as fitting significant better in a comical genre than decelerated fragments do. Considering the fittingness in a dramatic genre, post hoc test showed that decelerated fragments fitted better in a dramatic genre than original ($p < .05$) and accelerated ($p < .01$) fragments do.

These results confirm our hypotheses that the genre-determination variable *velocity* is most effective in the genres comic, as being rather fast, and in drama, as being rather slow. The influence of velocity in the action- and non-fiction genre is possibly suppressed by other genre-determination variables, whose determination, influence and interdependence will be a main topic of research in this project.

REFERENCES

Lehnert, W.G., 1981, Plot units and narrative summarization. *Cognitive Science*, **4**, pp. 293-331.

Newtson, D., 1973, Attribution and the unit of perception of ongoing behavior. *Journal of Personality and Social Psychology*, **28**, pp. 28-38.

Newtson, D. and Engquist, G., 1976, The perceptual organization of ongoing behavior. *Journal of Experimental Psychology*, **12**, pp. 436-450.

Newtson, D., Rindner, R., Miller, R., and Lacross, K., 1977, Effects of availability of feature changes on behavior segmentation. *Journal of Experimental Psychology*, **14**, pp. 379-388.

Rosch, E., 1999, Principles of categorization. In *Concepts: Core Readings*, edited by Margolis, E. and Laurence, S., (Cambridge: The MIT Press), pp.189-206.

Gone to the wild: using virtual reality for sensory stimulation

Ulrika Westergren
Umeå University, Sweden

Katrin Jonsson
Umeå University, Sweden

INTRODUCTION

The combination of the senses helps us define and relate to the reality we are living in (Ayres, 1988). However, not all people can experience all of their senses. Severely mentally disabled people might have trouble using all of the senses in the traditional way (Ayres, 1988). For these people sensory stimulation is often neglected. However, with the help of Virtual Reality (VR), they can experience new environments through immersion into a virtual one. Technology is also easily controllable and adjustable to the individuals' needs, so that a minimum/maximum of stimuli can be provided without the risk of overexposure.

This paper explores the use of theories of sensory stimulation in designing virtual environments and presents a design proposal for a sensory stimulating environment. While operating on the border of the real/surreal, Virtual Reality is the guide that can be used as the link between the mind and the body, creating a sense of being where senses can be stimulated with the purpose of evoking emotions.

SENSORY STIMULATION

Rudolf Steiner (König, 2000) developed the theory of the twelve senses, all of which he deemed necessary in order for a person to develop fully and to take advantage of all that the surrounding world has to offer. The senses can be divided into three groups, each with a slightly different focus. In the first group one finds the lower senses; the sense of life, touch, proper motion, and balance. These senses give an immediate feedback of the experiences of the human body. The next group consists of the middle senses; the sense of smell, taste, sight and warmth. These senses provide a person with information about the surrounding environment. The final group of four senses is the upper senses; the sense of the meaning of a word, the sense of thoughts, the sense of self-awareness and the sense of hearing. According to Steiner these four senses connect a person to the spiritual world, and form the individual (König, 2000). Followers of Steiner tend to focus on integrating the senses into an entity through the concept of learning by doing, enhancing creativity and stimulating the senses through experiencing (See for example www.camphillvillage.org).

The theory of Sensory Integration stems from the research of occupational therapist Jean Ayres who has studied how the different human senses interact in order for a person to relate to his/her surroundings and develop a composite picture of who, when, where and what. Sensory Integration is a way of organising and coordinating the sensory impulses sent to and through the brain, in order to improve perceptual and motorical skills (Ayres, 1988).

Ayres (1988) stresses the presence of a well-trained therapist who can guide the person and provide challenges that incorporate the brains' ability to respond to various stimuli and the human desire to experience.

Participating in an experience together with a carer is also one of the characteristics of Snoezelen, a theory on sensory stimulation that was developed in the Netherlands in the 1970's. The theory focuses on stimulating the senses by providing an environment conducive to both leisure and activity. Snoezelen has been described as offering a selection of primary stimuli in an attractive environment (Bosaeus, 1997). Rooms are decorated in certain ways with different themes as to stimulate different senses, which encourages the person to explore and take an active part in the environment (Sjösvärd, 1993). For more information see www.rompa.com.

Steiner's theory of the senses, Ayres' Sensory Integration and Snoezelen, all challenge the traditional concept of the five human senses. There is a strong focus on the ability to choose and participate in various activities since all theories believe that an active mind enhances creativity, learning and development. The shift from passive to active is also something that appears to be important in all theories. In Snoezelen there are certain rooms designed for activity, others for relaxation. Sensory Integration stresses carefully monitored activity, so that the person being exposed is neither under - nor overly stimulated.

VIRTUAL REALITY AND SENSORY STIMULATION

Virtual Reality (VR) provides a relatively safe and monitored environment, where the level of induced stress or difficulty can be changed according to the individuals' needs and wishes. The combination of realistic settings and the ease with which procedures can be monitored makes VR an invaluable tool in assessing cognitive ability and also a powerful asset in the rehabilitation of people with disabilities (Andrews *et al.*, 1995). These are key concepts when exploring the use of VR for sensory stimulation.

Galen Brandt (www.virtualgalen.com) believes that combining technology and art is a way to heal in the real world. Her creations focus on immersion into the virtual, which she believes will change the way people will see, and experience themselves. By combining visual and auditory stimuli in a fully immersive environment, Char Davies (www.immersence.com) creates a virtual space called Osmose where the mind and body form a whole. Osmose is described as "a space for exploring the perceptual interplay between self and world, i.e. a place for facilitating awareness of one's own self as consciousness embodied in enveloping space". As shown by the work of Brandt and Davis, a fully immersive experience blurs the notion of conscious self and allows for a new way of sensory stimulation.

The Focus, Locus and Sensus model aims to explain cognitive function during a virtual experience. Developed by John and Eva Waterworth, the model spotlights the shift between presence and absence (Focus), the real and the virtual (Locus), and the conscious and unconscious (Sensus) (Waterworth *et al.* 2001). Breaks in presence refer to a transfer of attention from the virtual environment to inner reflection. The Locus of attention can be seen as the degree of immersion into a virtual environment that a person experiences. A fully immersive environment means that what is virtual is perceived as being real, and that people can identify themselves within that world. The third dimension of the model, the Sensus of attention refers to a transfer of the state of mind from unconscious to conscious being (Waterworth, 2001). According to Waterworth and Waterworth, virtual environments can and should be designed for both absence and presence, and consciousness and unconsciousness, which indicates that VR should be able to stimulate both active and passive participation.

GONE TO THE WILD- A DESIGN PROJECT

Based on the theories of sensory stimulation and the properties of VR, the following design proposal is intended to incorporate these into a meaningful whole. The basis for the design is a virtual environment called Gone to the Wild that stimulates the senses while providing experiences that encourage active participation. The environment is implemented as a three-dimensional (3D) world which can be viewed with an ordinary web browser. Other studies of virtual environments have shown that a higher sense of presence emerges when we are in a secluded area surrounded by the virtual reality (IJsselsteijn *et al.* 2000, Waterworth and Waterworth, 2001). However, in the spirit of Snoezelen and Sensory Integration, it should be possible for a carer or a staff member to participate in the experience together with the person until they feel comfortable enough to experience on their own. We have therefore decided to project the environment onto a semi-circular screen surrounding the person so that they can focus on the experience while being accompanied by someone familiar. The first prototype is going to be evaluated at an activity center for severely mentally disabled people in Mörbylånga, Sweden.

To find out how the persons' body reacts to being immersed in the virtual environment we are using objective measurements such as heart rate and blood pressure. However, we are also conducting interviews with accompanying carers who have first hand knowledge about the person and who are able to interpret their reactions while immersed. With their help we can obtain a subjective evaluation as a complement to the objective methods. By using these methods of evaluation we strive to learn more about the actual virtual experience.

The overall concept for the design is a canoe ride along one of three rivers through the wilderness, each with a different theme: the Meadow, the Forest, and the Fjord. The canoe ride simulates an actual experience with the benefits of the virtual reality control mechanisms. Paddling a canoe down a river implies taking an active part in steering it, thereby making conscious choices. The virtual travel starts at an island from where one can choose to follow one of the three rivers. The three themes range from passive to active participation, with the Meadow being the most passive environment and the Fjord the most active. The Meadow includes grazing cattle, green grass and a richness of flowers. Images of the forest have been shown to have a relaxing effect on stressed persons (Clay 2002) and in Gone to the Wild the Forest is designed to be a moderately active path where one can calmly sit back and enjoy the lush vegetation but also encounter deer and foxes and hear the sound of birds. The Fjord is a path with high mountains and fast flowing water. Rocks may fall from the mountaintop and the canoe can tumble down a waterfall.

The more rapidly flowing river of the Fjord builds up tension and increases the level of presence, as the person must stay alert in order to control the chain of events. The quietly flowing water of the Meadow encourages relaxation and absence, shifting the focus from active participation to inner reflection. In Steiner's theory of the senses, this would mean shifting from a focus on the lower and middle senses, to the upper senses, where one can be in touch with the spiritual self. As suggested by the concept of the Sensus of attention, it is important to provide for both conscious and unconscious activity as this provides input that stimulates both the middle and the upper senses. Not only are the people consciously processing information about the surroundings, they are also an integrated part of the environment and can as such unconsciously experience being a part of a whole, opening the door for reflection.

All previously described theories on sensory stimulation stress the importance of active participation. A shift from passive immersion to self-motivated involvement in an activity stimulates brain activity and inspires a person to explore. A richness of different activities and sensory input will further increase interest. It is therefore possible not only to chose one of the three rivers to travel along, but also to shift from one river to another. The ability to control which is stressed in both Sensory Integration and Snoezelen is implemented both in allowing for a choice of activity and a choice of motion, that is steering the canoe.

Both Steiner and Ayres mention the sense of balance and proper motion as a key to sensual awareness. This implies that it should not be possible to fly around anywhere in the virtual environment as this can upset the sense of balance. In Gone to the Wild the person always follows a guided path, grounded. The canoe is in constant motion as it follows a river, but the speed can be altered depending on the current and on the needs and wishes of the person. By providing tactile stimuli, one can simulate the use of a paddle to steer the canoe and stimulate the sense of touch. Since a virtual environment takes place on a screen one must provide physical props to reach this sense. In this first version the participants navigate through the use of a joystick.

CONCLUSIONS

Gone to the Wild combines the theories on sensory stimulation and the benefits of a controllable and adjustable virtual environment into a virtual world of sensory stimulation aimed at severely mentally handicapped people. Based on earlier research done on both Virtual Reality and on sensory stimulation we conclude that a sensory stimulating environment should include the ability to control, means for both active and passive stimulation, and a choice of various activities. We also want to stress that when designing for severely mentally handicapped people, it is especially important to include the ability to control, so that the virtual environment can be adjusted to the individuals' need and wishes. By doing so the amount of stimuli can be monitored and the risk for overexposure is decreased.

REFERENCES

Andrews, T K, Rose, F D, Leadbetter, A G, Attree, E A and Painter, J (1995). The use of virtual reality in the assessment of cognitive ability In: I. Placencia Porrero & R. Puig de la Bellacasa (Eds), *The European Context for Assistive Technology: Proceedings of the 2nd TIDE Congress, Paris, France* (Amsterdam: IOS Press), pp 276-279

Ayres, J (1988) Sinnenas samspel hos barn, 2nd ed, (Stockholm: Elanders Gotab)

Bosaues, M (1997) Med öppna sinnen i vita lekrummet, (Lycksele: Nya Tryckeriet)

Clay, R.A. (2001) Green is good for you in Monitor on Psychology, Vol. 32, No.4

IJsselsteijn, W A, de Ridder, H, Freeman, J, and Avons, S E, (2000) Presence: Concept, determinants and measurement. *Proceedings of the SPIE*, Human Vision and Electronic Imaging V, 3959-76. San Jose, CA

König, K (2000) Sinnesutveckling och kroppsupplevelse, (Järna: Telleby Bokförlag)

Sjösvärd, A-M, (1993) Sinnenas Gym, (Göteborg: Graphic Systems AB)

Waterworth E L (2001) Perceptually-Seductive Technology - designing computer support for everyday creativity, (Umeå: Solfjädern Offset AB)

Waterworth, E L, Waterworth, J A, and Lauria, R (2001) The Illusion of Being Present. *Proceedings of Presence 2001*, 4th International Workshop on Presence, Philadelphia, May 21-23.

Waterworth J A and Waterworth E L (2001) In Tent, In Touch: beings in seclusion and in transit. Extended Abstracts of CHI 2001 Conference on Human Factors in Computing Systems. Seattle, Washington, USA, March/April 31-5.

Internet Resources On Sensory Stimulation:

Camphill Ass. of N. America: www.camphillvillage.org Visited: 2001-12-09

Char Davies official homepage: www.immersence.com Visited: 2001-11-29

Galen Brandt official homepage: www.virtualgalen.com Visited: 2001-11-29

Snoezelen Official Homepage: www.rompa.com Visited: 2001-11-27

The gourmet foodstuff: packaging our impulses

Liz C Throop,
Georgia State University, USA

INTRODUCTION

The number of packaged foods available to U.S. consumers is enormous – estimated at 40,000 kinds of items in the average supermarket (Trout, 2000). Between the general supermarket and the gourmet store an urban American will encounter dozens of choices in almost every food category. While items like medical insurance and safe housing are out of the reach of many Americans, foods are available to suit every budget, and multiple flavours and styles of food are presented within each price range. This surfeit of choices forces consumers to make purchasing decisions based on more than a package's contents alone.

Because American consumers make complex and fine-grained distinctions about their food purchases on a frequent basis, gourmet packages constitute an excellent subject of study when observing the intersection of affluence, taste, and design.

For this discussion I will focus on how the formal design qualities of food packages help us distinguish foodstuffs on the basis of perceived "gourmet" qualities, or "good taste."

FOODS IN GOOD TASTE

Preoccupation with good taste arises in periods of increased abundance, when plenty of food or other commodity is available to everyone. Elite classes must form new ways to characterize membership within their group in order to maintain exclusivity. The conventions that characterize such membership are known as good taste, and are constituted through various forms of restraint, by simplicity rather than frivolity, by rational discrimination rather than hedonistic indulgence. Bourdieu (1984, p. 32) calls this the "...refusal that is the starting point of the high aesthetic, i.e., the clear-cut separation of ordinary dispositions from the specifically aesthetic disposition." Those considered to have good taste are those who appreciate the very best by discerning and refusing that which is less than best. While terms such as "good taste" and "elite" were once considered at odds with America's supposedly classless society, the term "gourmet" has become popular for luxury food products that appeal to the select tastes and budgets of an elite group (Holt, 1998).

The successful package for luxury foods balances contradictory appeals to the consumer. It offers a rarefied aesthetic experience and at the same time some base pleasure gained from consuming the food. When purchasing a gourmet product shoppers must be made to feel that exercising their good taste sets them apart from the ordinary, yet in some way must also be reassured that, once purchased, the food inside will provide sensuous, bodily pleasure. The contents should taste at least as good as the cheaper, more vulgar offerings available at the supermarket.

Gourmet products accomplish this complex task through a limited array of materials, typefaces, colours, and illustration styles. One may imagine that designers of upscale merchandise have free reign, but in fact they are bound by a visual and symbolic vocabulary that distinguishes their packages from cheaper mass-market goods.

Because each individual food purchase is relatively inexpensive and transitory, it is seen as of little consequence. Thus each purchase is made relatively without self-consciousness. It is because each individual food purchase seems trivial that the sum of these trivialities, what Bourdieu (1984) would call "habitus", constitutes a system of distinction known as good taste. "The schemes of the habitus, the primary forms of classification, owe their specific efficacy to the fact that they function below the level of consciousness and language, beyond the reach of introspective scrutiny or control by the will." (Bourdieu, p. 466)

As trivial as each individual food selection may seem, such selections are based, at least in part, on the design elements of the food package. These elements include colour, symbology, typography, and materials. The gourmet package and the mass-market package must each, through these formal elements, appeal to their respective audiences, who vary not only by level of affluence but also by reading skills (in the expanded sense) and attitude toward food and pleasure.

COLOUR

Colour is recognised as a powerful element that affects how people perceive foodstuffs. The colours on food packages for sweets must be appetizing for that kind of food, and will differ markedly from appetizing colours on frozen vegetable packages. Yet beyond the use of this or that colour for a particular kind of food, we can observe another, broader variation in colour usage between "gourmet" and mass-market food packages.

In the United States certain pure, intense colours such as black, white, red, yellow, green, and blue are learned in early childhood and can be identified by most everyone. Debates rage as to whether any of these constitute a truly universal set of "focal colours" (Berlin and Kay, 1969), but in the U.S. even people with limited education are comfortable with conventional, easily-named colours. These colours are memorable (in the sense that one easily recalls that Coca-Cola packages are red and Pepsi packages are red and blue) and attention getting. It is not surprising that they are used lavishly in mid-range and inexpensive food packages.

In addition to these conventional colours, one or two colours will become very popular for a season or two in fashion-related industries. Mass-market food packages use fashion colours, but use them sparingly and in combination with conventional colours.

Gourmet food packages also use conventional colours and fashion colours, but use them in more subtle combinations. Intermediate hues such as blue-violet or grey-green appear more frequently on gourmet packages, as do very restricted palettes, such as black and white with small red highlights, or white against off-white. Complex colour schemes that involve many colours, or that vary from simple complementary relationships, are markers for the gourmet status of the products.

Designers with extensive training in the use of colour may delight in the chance to create gourmet packages, using the chance to explore advanced concepts of hue, saturation and value. Yet the designer of gourmet packages faces certain restrictions. The food must not look common or cheap. He must avoid the clashing, garish colours so widespread in the main aisles of the supermarket for his product to successfully reach "gourmet" status.

ILLUSTRATION TECHNIQUE

Illustrations are common on both gourmet and mass-market labels. Here we can discern two broad categories of images – one descriptive, the other stylish.
In the sociologist Pierre Bourdieu's (1984) classic study of taste and social class, *Distinction: A Social Critique of the Judgement of Taste*, he observes that members of the working class are not interested in pictures as ends in themselves, but instead in pictures that serve some other purpose.

> "...working-class people expect every image to explicitly perform a function, if only that of a sign, and their judgements make reference, often explicitly, to the norms of morality or agreeableness. Whether rejecting or praising, their appreciation always has an ethical basis." (Bourdieu, p. 5)

This observation is confirmed by the different styles of art on food labels in supermarkets and gourmet stores. The images on the fronts of supermarket foods often cannot be easily classified as photographs or illustrations, *per se*. Instead one imagines they began as photographs, then were cleaned up, exaggerated, and heavily retouched to the point of becoming the sort of illustration that owes everything to a photographic ideal of verisimilitude.

These images serve an easily understandable function of symbolizing the contents of the package. A jam-jar label has a clearly identifiable image of strawberries, suggesting the contents feature strawberries, rather than grapes or raspberries. The mass-market label shows the contents in an idealized manner that is easily identifiable and makes the food look as appetizing as possible. In all cases the use-value of the image is straightforward and easily understood.

The labels on gourmet foodstuffs differ both formally and ethically. The image on the label is as much about an artistic style or technique as about the food itself. Woodcuts, engravings, pastels drawn on coarse-textured paper, all render the contents through abstract, less literal interpretations. These kinds of images suggest a psychological distance from the immediate concerns about the food itself. Instead they suggest the leisured world of artistic appreciation, the world of a person of good taste.

SYMBOLOGY

Not only the style, but also the subject matter of an image demands more interpretation in gourmet food labels. Those with little formal education might not want, or not be able to, decipher why a nun with her eyes cast down in front of a church would signify an especially mild variety of chili sauce, as in "Desert Pepper Trading Company's 'Salsa Divino' " (Lowrey, 1998). Further, members of Bourdieu's working class might object to such indirect signification on ethical grounds, thinking a picture of sweet peppers would be a more honest or practical approach to signifying the jar's contents.

When gourmet food packages do show the food itself, it often appears within a larger landscape that features planting, growing or harvesting. The gourmet shopper is given a chance to appreciate the abstract and remote pleasure of contemplating the rural way of life s/he sustains through purchase. Other themes may be conveyed by the artwork, such as the exoticism of ethnic foods, or the food's historical associations. When food is shown on the label, the emphasis is often on its decorative qualities. For instance a gourmet olive oil label displays a rhythmic pattern of olives and leaves, whereas a moderately priced olive oil shows oil being poured onto salad. In all these cases the base satisfactions of quenching hunger and of tasting food are deferred.

TYPOGRAPHY

Mass-market foodstuff labels, although garish, are surprisingly legible. Upper and lower-case letters are carefully combined from typefaces with generous x-heights to create distinctive word-shapes that ensure shoppers can read them from across the supermarket aisle. (Carter *et al.*, p. 85) Designers create custom ligatures in order to compress the food's name so they can make it as big as possible on the package. The lettering is often so visually distinctive that it serves as a sort of logo within a larger food company's brand. Product names stand out from backgrounds through strong colour contrasts, drop shadows, and outlines. By Bourdieu's (1984) analysis, the mass-market consumer finds such a type treatment appealing in an ethical sense: s/he appreciates that the manufacturer worked hard to make the label easy to read.

In contrast, lettering on gourmet packages, like the illustrations they accompany, do more than just identify the contents of the package. Type appears to be more adventurous than formulaic, although to be perceived as "gourmet" its adventurous treatment is a necessary signifier.

Shoppers for gourmet goods are generally better educated and more literate than their mass-market counterparts. All-capital letters, words letterspaced or rearranged into abstract shapes, or type that is very small or broken into multiple lines are all hard to read by those with low literacy, but successful in signifying the gourmet nature of the product. In fact, up-market consumers may find this hard-to-read type amusing or enjoyably challenging. In all cases these treatments offer the gourmet consumer a reprieve from the banality of most food labels.

On the gourmet package, the type treatment is not so restricted by legibility issues, and consequently whatever treatment is chosen carries greater symbolic weight. Letters may suggest handicraft or folksiness through calligraphic treatment, evoking the rural way of life mentioned earlier. Or they may suggest sophistication and urbanity through highly geometric forms. In all cases type must be in keeping with the stylistic tone of the illustration to maintain an overall artistic effect.

MATERIALS

Large food producers in the U.S. can take advantage of economies of scale to package things however they want. Iridescent foils and embossed box fronts are just the latest in once-exotic containers to make their appearance in the mainstream supermarket. This trend toward more elaborate mass-produced packaging provides a challenge to packagers of gourmet foods, who want their products to appear in some way better than mass-market goods, even though they can never compete in terms of per-unit cost. It is probably for just this reason that bits of raffia around the neck of a bottle or jar remain a popular cliché of the gourmet package: the tying of the cord demands handwork.

The better mass-market producers become at creating shiny packages that reach out and grab the consumer's eye, the more that once-humble uncoated papers come to signify gourmet cachet. Traditional uncoated stock can look understated and can convey freshness and rusticity – however uncoated papers and boards don't appear in supermarkets because they have high spoilage rates from staining and scuffing. Gourmet package designers face a significant challenge in creating durable containers that hold up to the retail environment, yet look like they have been handled with care.

PRISON HOUSE OF TASTE

There are distinct visual clues that mark the difference between packages of upscale and moderately priced foods. Gourmet food packages signal their exclusivity through subtle and unusual colour combinations, artistic type and illustrations, and understated materials. Mass-market food packages, even when they imply they are "as good as" gourmet foods, always reassure their audience that they provide a straightforward experience that is easy to appreciate.

Many Americans today pay dearly for distinctiveness by buying "upscale" products such as Starbucks coffee, whereas their parents may have been more comfortable buying moderately priced Maxwell House coffee in order to seem like everyone else. Heller and Fink's (1996), *Food Wrap: packages that sell*, notes that, "In the 1950s virtually all supermarket brands, regardless of the product, looked basically the same." (Heller and Fink, 1996, pp. 12–13). They remark optimistically that today, a wider range of designers, many of whom are non-specialists, design packages and consequently "the old school conventions have been challenged and taboos busted." This is probably true to some extent. But the aesthetic range we see today may have as much to do with different class aspirations and a wider distribution of income than with the triumph of design. This discussion points out that the system of differences known as "taste" is much larger than our own design production. Like Jameson's "prison-house of language" (Jameson, 1972) the prison-house of taste is inescapable in a commodity-centred, stratified Western culture.

REFERENCES

Berlin, B. and Kay, P., 1969, *Basic color terms: Their universality and evolution*, (Berkeley and Los Angeles: U Cal Press).

Bourdieu, P., 1984, *Distinction: A social critique of the judgement of taste*, (Cambridge: Harvard University Press).

Carter, R., Day, B., and Meggs, P., 1985, *Typographic design: Form and communication*, (New York: Van Nostrand Reinhold).

Heller, S. and Fink, A., 1996, *Food wrap: Packages that sell*, (Glenn Cove, NY: PBC International, Inc.).

Holt, D. B., 1998, Does Cultural Capital Structure American Consumption? *Journal of Consumer Research*, **25**, pp. 1 – 25.

Jameson, F., 1972, *The prison-house of language: A critical account of structuralism and Russian formalism*, (Princeton: Princeton Univ. Press).

Lowrey, T. M., 1998, The effects of syntactic complexity on advertising persuasiveness. *Journal of Consumer Psychology*, **7**, pp. 187 – 206.

Trout, J., 2000, *Differentiate or die: Survival in our era of killer competition*, (New York: Wiley).

Emotions in action: a case in mobile visual communication

Esko Kurvinen
University of Art and Design, Finland

INTRODUCTION: EMOTIONS AND INTERACTIVE PRODUCTS

Emotions play an important role when consumers adopt new technology into their everyday life. Understanding emotional expectations and emotional responses that emerge during use will help us design usable, enjoyable, and therefore successful products. While the design community widely accepts the importance of emotions in consumer preferences, their interactional properties are poorly examined.

Usability and user centred design (UCD) typically places a lot of emphasis on the first confrontation between the user and the machine, including e.g. the learnability of the structure or the success of the terminology of the interface. As such, this orientation is unable to capture what happens in the course of time when products are gradually domesticated into our everyday life.

The aim of this paper is to study emotional responses as a socio-interactional phenomenon. I will present data from an experiment that prototyped one wide-visioned feature of the so-called third generation mobile phones (3G). That is, capability to take, send and receive photographs using a mobile communication device.

3G AND MOBILE VISUAL COMMUNICATION

According to product development scenarios, mobile imaging will be the key feature of the mobile Internet. In the 3G mobile phone visions, the small portable devices of the near future are versatile in production, editing and sharing of images.

Though all the key players in this area pursue technological development, there are only few studies from the perspective of the end user. While the commercial breakthrough of SMS (Short Messaging Service) has only recently awakened us to understand the potential of social use, it remains to be seen if the key paradigm of 3G is also going to change. According to Pantzar (2002), the consumers will make the final selection between *tools* and *toys*.

METHODS

This paper studies the production of mobile messages that are somehow emotional. Instead of extensive discussion on messaging in terms of emotions, I focus on the turn-by-turn making of meanings in messages. First, I will show examples of messages where users use photographs to represent their feelings or emotional state. I will also discuss the role of text in these messages. Secondly, I will show how in the context of communication, the meanings of a message do not arise from its content or intentions of the sender only, but are *achievements* in a co-operative activity.

My analysis draws from ethnomethodology (Garfinkel 1967) and conversation analysis (e.g. Hutchby and Wooffitt. 1998), both methodological orientations that focus on the sequential organization of social events.

DATA

To get a glimpse on the possible uses of future mobile phones, we used today's off-the-shelf technology. We gave a Nokia 9110 Communicator and a Casio infrared-capable digital camera to four user groups and followed the way in which they used this package. Our study concentrated on consumer grade technology and everyday uses, not professional photography.

The equipment enabled our subjects to take photos and send them instantly as email attachments. With the Communicator, subjects were also able to receive and view photos and comments while on the move. In practice, they most often ended up reading the messages on a PC because of the slow GSM connection and small black and white screen of the Communicator.

In 1999-2001, four groups participated in the study: a pilot group, a group of young adult males (born 1973-75), a group of young adult females (1973-76) and a control group. Each group had five members. The examples in this paper are from groups 2 and 3. For a full description of the data, see Koskinen *et al.* (2002)

Emotional Mobile Images

Some of the messages sent during our study were clearly charged with emotions. In the example below, Kirsi sends a picture of herself to her boyfriend, who is studying abroad.

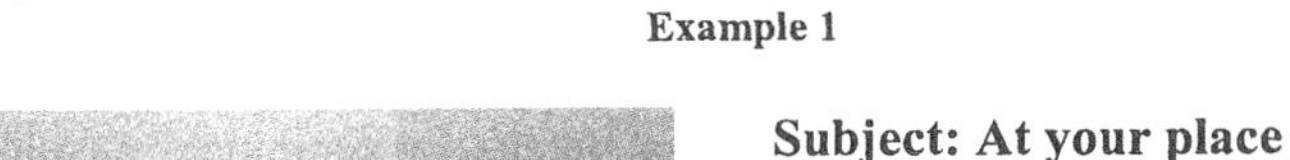

Example 1

Subject: At your place

I came here to check out the storage room, which is really small by the way - and to give Toni the keys. So strange being here without you.

Kiki

In the message, Kirsi's longing for her boyfriend is communicated with a subtle combination of visual and textual means. She is intensively looking at the camera, or rather, mediated by the image, the receiver of the message. In the text, she is using her pet name and telling that her being at this particular place is not the same without him.

Interaction with Image and Text

Practically all of the messages sent during our study contained both images and text. Even in cases where text was missing, the author had otherwise quite explicitly linked the image to the previous message(s) or conversations outside email. In a prime example below, from the male group, text explicates both the content of the image and context in which it was taken.

Example 2

Subject: dishes

HELLO!
Dudes check out this pile of dishes; oh sheet.

Some statistics: almost forty glasses, twenty mugs, a bloody load of plates. What a fantastic way to start Saturday morning.

Gentlemen, things are not fine right now!

- Ike

Typically, the sender provides the recipients at least some instructions on how to interpret the message. Most often these instructions were given in the text. As a result, the text-image pair of a message builds a frame of reference against which subsequent messages can orient to. Of course they need not do so, but if they are to continue on the same topic and especially if they are to discuss the first message or image in more detail, they often take it into account. Next example shows the reply that followed 20 minutes later.

Example 3

Subject: Re: dishes

Me too, the dishes hit the fan and just look at all those damn crappy plates... only 4 h to go, I am already parched...

The replying message was constructed using both visual and textual references to the first one. The replying photograph did not only present a visually similar view, it was also not taken until the first photograph was received and provided guidance on how to take a similar one. Style-wise, the text of the second message follows the first one, its strong first-person perspective and swearing projecting the unpleasantness of the situation.

Subsequent messages can also draw from the previous ones without having to entirely agree with them. In example 4, Lisa continues a monologue she has started on her childhood neighbourhood. In example 5, a replying message from Minna oriented to what was presented earlier but turned it into something new.

Example 4

Subject: at the corner store

Continuing the Friday afternoon series... pictures from my childhood home store Valintatalo.
Part1, Sami and a Baguette

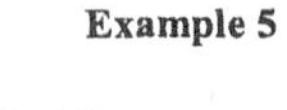

Example 5

Subject:

I follow the series started by Lisa (man and a baquette) and challenge others to join. Here is a man and a paddle. Minna

In Finnish, *baguette* (patonki) and paddle (mela) are both slang words referring to the male sexual organ. Though this type of interpretation was by no means necessary, Minna's invitation for others to join the series was built on this sexual tone.

The replying message altered the meaning of the original in two ways. First, in her message Minna utilized the subject of the photo, Sami. Since other recipients have no photos nor the desire to discuss him, a more general frame, men, still applicable to Sami, yet more interesting, had to be initiated. Secondly, Lisa's monologue may be uninteresting as such, but its established serialness is a legitimate justification for subsequent messages, even though herself Lisa never called upon this kind of theme.

Both of these transformations were done in order to better facilitate an invitation. Yet, this not simply a case of abuse or intentional misinterpretation. Looking at the dates of the photos, one notices that Minna's photo was taken before she received Lisa's message from the store. Despite of the photo of a man and a paddle being aesthetic, funny and useful as such, it was not sent until Lisa's message made it relevant in this context. It can be concluded therefore, that the two photographs, instead of one being inherit from the other, mutually defined each other in the turn-by-turn process of group communication

DISCUSSION

The successful devices, services and interfaces of tomorrow's imaging phones may not be *themselves* emotionally charged. As I have shown above, each visual message is a potential resource for not only taking new photos and building new messages, but also for making sense of the preceding ones. Therefore, the meanings, fun and pleasure of communication that leads to our engagement in future mobile imaging services is by no means a take-it-or-leave-it. Rather, it is based on a continuous chain of retrospective-prospective comparisons between messages.

Mobile imaging services should support users' activity at hand. Paradoxically, this is best done by giving users rich access to the local history of their personal and group-specific messaging, as well as storage space for images waiting for the suitable moment to be sent. Sending images to our friends is not a need awaiting fulfilment; it is a process that needs to be nourished.

Garfinkel, H., 1967, Studies in Ethnomethodology, (Englewood Cliffs: Prentice-Hall)

Hutchby, I. and Wooffitt, R., 1998, Conversation analysis: principles, practices and applications, (Cambridge: Polity Press)

Koskinen, I., Kurvinen, E. and Lehtonen, T., 2002, Mobile Image, (Helsinki: Edita/ IT press)

Pantzar, M., 2002, 'You Press the Button, We Do the Rest' – Writing Prospective History, in *Mobile Image*, Koskinen, I., Kurvinen, E. and Lehtonen, T., pp. 105-118

Shop image and interaction: the use of senses in consumption space

Dion Kooijman
Delft University of Technology, The Netherlands

INTRODUCTION

The history of the commercial consumption space has seen two important moments of change: a) the introduction of established retail trade, already some two hundred years ago; and b) recent changes in the previous twenty years that are aimed more at emotion and 'leisure'. In addition, the development of the shop presentation has two ingredients, these are: a) the shop image, and b) the interaction between shop and consumers. The first moment of contact is particularly connected with the importance of the visual impression made by the shop. The recent changes have particularly to do with a deeper interaction.

At the start of the established retail trade the image of the shop became important because of two reasons. The scale increase of retail, and as a consequence the necessary enlargement of the service area, did make it necessary to communicate with consumers in a different way. Advertisements in newspapers could bridge the increased distance between shop and consumer. The function of the image had also to do with the quality of mass production and mass consumption. Of the luxurious fixtures and fittings of shops, it has been said that they must compensate for the 'poverty' of the articles and products on the shelves. Sennett (1993/1977) has called this phenomenon 'adjacent attraction', following Walter Benjamin.

The image is, however, passive and in the end it falls short of the ability to retain the relationship with the viewer or the customer for any length of time. Therefore other forms of involvement are necessary in order to maintain the relationship between shop and client. The retail trade is looking for a more encompassing relationship, in which the customer participates in the sales function of the retail trade. The current competition between suppliers - this is referred to as a displacement market - makes the necessary contribution here. There is a continued search for new possibilities of gaining a (often temporary) head start on the competition.

On the basis of the recent changes I would like to speculate about the return in a modified form of an oral type of society.[1] Management books and retail books are full of an emotional retail environment. The intended environment is, however,

---

[1] The hypothesis is derived from my own work, *Machine en Theater: Ontwerpconcepten van winkelgebouwen* [Machine and theater. Design Concepts for Shop Buildings], 1999. In that study, I analyse the architectural history of shop-buildings, together with a few aspects of modern business administration and supply and demand factors. Here my focus on the image remains, but the subject has become more elaborated in an analysis of the aspects of social psychology.

limited to the external appearance, full of temptation and theatre. This appearance is at the same time the product of a rational decision-making process, of efficient logistics and machine, of 'shopping as a science'. At present, consumers only experience the shop presentation. The logistics and management are much less emphatically present. This combination of 'deep' business rationality and superficial emotionality can be indicated as 'secondary orality' (Ong, 1982).

Furthermore, there is also the reverse tendency to which Benjamin referred earlier: the commercial shop space will become characterised by stressing the distinctive features of the public space. During the nineteenth century public space became uniform and anonymous. Now, two hundred years later, this public space become more individual, varied, and differentiated. E.g. kitchens and sitting rooms appear in the retail environment with a striking parallel of social meaning and psychological functioning. Therefore this secondary orality in retail is to my opinion a special one.

Interaction between shop and consumers

Today the interactivity and the relationship between shop and consumers shop environment are more important than ever. The interactivity of the new medium, Internet, seems to offer an alternative for the merely visual, and as a consequence, passive interaction. This interaction is active because the participation of the user is a part of the 'product'. Similar forms of interactivity can be found today in many types of shops, dealing with self-service and all kinds of events.

Participation is not stimulated merely through visual impressions, but also by exciting all the senses. For a long time there has been a visual and participatory relationship. After all, self-service has always been more than just looking (just looking happens in shops such as the Dutch Kijkshop and the British Argos, in mail order companies and Internet shops). Currently: touch, hearing, smell and flavour are all called into action. This kind of participation implies a new connection between consumers and retail trade. The relationship is less detached and anonymous. Consumers participate in the logistic process (through self-service), in the quality management of the shop stock (through an institutionalised form of client reactions[2]) and even of the production process (through of the cooking school). In the last example, the consumer 'completes' the 'product' or the 'service' by making his or her own personal contribution.

Of all the forms of participation, the cooking school in the supermarket is at this moment the best illustrative example. In the cooking school, consumers/students learn to prepare the recipes published in the supermarket magazine (Albert Heijn[3] has, for example, AllerHande) to produce meals using

---

[2] Superquin, a supermarket in Ireland, pays consumers for their services if they help to improve the functioning of the supermarket. Tips concerning faults in plates, products past their sell-by date, a cooler that does not maintain the correct temperature, are rewarded with an apple tart, etc. 'This goes beyond self-service – the shopper is selling quality control services to the store' (Davis and Meyer, 1998: 57).

[3] Albert Heijn is a Dutch supermarket chain that currently owns approximately 650 shops and has a market share of almost 30%. The supermarket business is part of the internationally oriented Ahold concern, with interests in Europe, the United States, Brazil and South-East Asia, among other places.

products from the same supermarket. By offering services as well as the necessary products, this enables the supermarket to provide a combination of participation and variety. The fashion show or the furniture exhibition often still have the character of a demonstration that points to a future use. Participation in a cooking school is an immediate participation; a participation in the present time.

The cooking school represents integration in a new social arrangement: it is representing the kitchen outside the private home. The departments Chill-Out and Yes-or-No Shop of the Dutch department stores the Bijenkorf and Vroom & Dreesmann, The Lab Antimall in California (USA), and even IKEA's Kinderland all have an individual target group that is coupled with a specific part of the shop. This time the homely atmosphere of the shop is referring to another part of the private home: the sitting room. Through the formulation of 'target groups', two mechanisms become combined: the commitment to the product or the service and the isolation in a particular group. This isolation also brings a separate space with it. The spaces of department stores become separated, so that they all become small, specialized shops within one large store, increasing more individuality, variation, and differentiation.

RATIONALITY AND EMOTIONALITY

McLuhan and others have speculated about lettered and oral societies (McLuhan, 1995/1964). Orality is 'the whole of the unwritten form of communication' (Mostert, 1998). Spoken word, attitudes, gestures, smells, flavours and non-verbal messages belong to the domain of orality. The historic development is, grosso modo, one from unlettered to lettered, from oral to literacy. In large parts of social life this unlettered characteristic still exist, not just as a residue of a former social order but also as reinvented social places.

Mostert does not see oral and written forms of communication as opposites. Forms of the spoken word continue to be sounded in the written word. 'Printing remains directly and indirectly connected with the oral forms' (Mostert, 1998). Also, he asserts that: 'Watching television itself requires no schooling, but nevertheless the message that the medium carries is clear: the modern cultural ideal is lettered and educated' (Mostert, 1998). In this connection the concept of secondary orality is important; this includes all those forms of non-written communication that in the end depend on writing (Ong, 1982; Mostert, 1998).

Perception and reflexive consciousness do not appear to be constants but seem to be dependent on historic dynamism. It is noticeable that in the commercial consumption space a multitude of consumer senses is currently being stimulated. In addition to visual aspects, touch, hearing, smell and the sense of taste are in the areas of interest. In as far as visual aspects are concerned, the figurative and associative images are particularly dominant. 'Complicated' texts are taboo in the shop presentation. For the retailer/entertainer, the efficient deployment of what Packard (1957) has called the 'hidden persuaders' count to an increasing extent.

It is not the rationality, but the emotionality of the consumer bond that is central to current sales methods. The Enlightenment must have nothing of sensuality (smell). The western cultural ideal of education is based on this. That was not always so. Romanticism on the other hand saw the importance of, for example, the sense of smell, also in the science of business (Vroon et al., 1994). Today we can experience a new Romanticism. The extensive stimulation of all the senses in the current shops is part of this. Reaching a condition of 'flow' should be the ultimate consumption experience. The question however is whether and to which extent this stimulation can be effective.

SMELL AND THE DETERMINATION OF BEHAVIOUR

Psychologists have explained the 'schizophrenic' situation of distant rationality and emotional involvement through the unconnected structure of the brain (Vroon, 1978). The limbic system is primarily responsible for the emotions. In the neo-cortex we find the seat of reason. The neo-cortex is, seen from an evolutionary viewpoint, the product of a recent period. This part of the brain received its form in a relative short time. The neo-cortex has developed more or less independently of the older limbic system. This uneven development may explain the unconnected workings of the brain and, as a consequence, our behaviour.

One of the favourite senses in retail and advertisement is smell. Both areas speculate about the great effectiveness of the scent. Today retail often has even her 'corporate smell'. The advertisements of Axe and Fa show irresistible allure for men or touch the woman under the spell of the (smell of the) man. The products supposed to have the capacity to regulate behaviour. A specific scent should lead to a specific action. There are certainly possibilities for conditioning. But the effect of that which has been conditioned can be other than that which was originally intended.

Advertising and shop image are not the result of the emotionality of a 'creative' or an absent-minded architect. The appearance of present-day shops has a clear basis in the 'machinery' of logistics and marketing. 'The game with pictures, images and identities is played in a more differentiated and complicated way', as Mommaas (1993: 115) has also remarked. The concept 'secondary orality' of Ong (1982) and Mostert (1998), expresses this new combination of rational management and oral emotionality well.

Whether the image will disappear in the short term, is still an open for discussion. There are shops on the Internet e.g. that have been exhaustive in the creation of a virtual 3-D world. Other virtual shops are satisfied with a global representation that presents a strong likeness to a comic strip. The first category is certainly well drawn but nostalgic in content. The second category is vague and blurred, but may well represent the future. The clear picture is just nostalgic of the last few decades. One of the latest examples in the retail trade, in which a clear picture is also aimed at the future, is that of self-service in the supermarket. The mechanical machine and the conveyor belt were the source of inspiration for this.

The question whether the extended consumer participation and the related shop design is good or bad is not easy to answer. From historical perspective there is in de last decade or so a trend from rational to emotional shop environment. The metaphorical character of the shop image (the theatre for the department store, the market for the supermarket, the town centre for the shopping mall etc.) is extremely helpful in the emotional participation of the consumer. But this metaphorical characteristic is helpful too for the innovation process in retail. Today there is again much discussion about client relationships, one-to-one marketing, and service aspects. At the same time it also represent a possibility for quality improvement beneficial to the consumer. Retailers are often stimulating the consumer participation. Consumers are sometimes after convenience shopping, sometimes after leisure, sometimes both. Shoppers looking for a effective and efficient bargain will experience too much leisure as an (over)stimulating of the senses. For shoppers spending their free time leisure will be a desired seduction. What matters here is the match between the supply side and the demand side.

REFERENCES

Davis and Meyer (1998) *BLUR, the speed of change in the connected economy.* Reading, Mass.: Addison-Esley.

Kooijman (1999) *Machine en theater. Ontwerpconcepten van winkelgebouwen.* Rotterdam: 010 Publishers.

McLuhan (1995) *Understanding media. The extensions of man.* Londen: Routledge (first edition: 1964)

Mommaas (1993) *Moderniteit, vrije tijd en de stad: sporen van maatschappelijke transformatie en continuïteit.* Utrecht: Van Arkel.

Mostert (1998) *Oraliteit.* Amsterdam: Amsterdam University Press.

Ong (1982) *Orality and Literacy: technologizing the word.* New York: Methuen.

Packard (1957) *The hidden persuaders.* London: Longmans, Green & Co.

Sennett (1993) *The fall of the public man.* London/Boston: Faber & Faber (first edition: 1977).

Vroon (1978) *Stemmen van vroeger: ontstaan en ontwikkeling van het zelfbewustzijn.* Baarn: Ambo.

Vroon, Van Amerongen and De Vries (1994) *Verborgen verleider, psychologie van de reuk.* Amsterdam: Ambo.

EMOTIVE EFFECTS OF THE OTHER SENSES

Sensory interaction with materials

H Zuo, T Hope, M Jones, P Castle
Southampton Institute, UK

INTRODUCTION

The visual aesthetics of a particular material and the surface texture information are signalled through visual and tactile feedback. This will contribute to how the user perceives the 'material representation'. There is little evidence to suggest that a wealth of information exists about the sensory and aesthetic characteristics of materials.

This research is concerned with people's perception of the inherent and perceived properties of materials within a holistic environment and system, which is referred to as 'material representation'. It is defined as "the perceived images, properties, meanings, and values of a material in the human-product interface under a specific set of environmental conditions" (Zuo *et al.*, 2001). It emphasises the sensory interaction between materials and their users. Because of various variables, a visual narrative matrix is being developed to systematically explore material representation. Both the theoretical study and the experimental findings are needed for construction of the matrix.

The results of our previous research have made it possible to identify the way in which people subjectively describe a material texture through touch. Subjective texture description of materials has been classified into four dimensions: geometrical dimension, physical-chemical dimension, emotional dimension, and associative dimension. Each dimension consists of a primary and a secondary group of lexicons (Zuo *et al.*, 2001).

This paper details some experimental results about people's scaled sensory responses to materials' texture by touch and the correlations between these responses have been investigated.

EXPERIMENTAL RESEARCH METHOD

The controlled experiments were conducted to assess a person's response to different materials with particular surface textures. Information was obtained through both visual touch and blindfold touch.

In this experiment, 29 different material samples from four categories (steel, plastic, thermoplastic elastomer, and aluminium) were adopted as samples. A population of subjects consisting of design students, psychological students, engineering students and staff served as participants.

Subjects were asked to make scaled evaluations of 13 items of description to a material texture. The environmental condition was set up as: temperature 20±1.5°C; humidity 50%; and under TLD70W lighting condition.

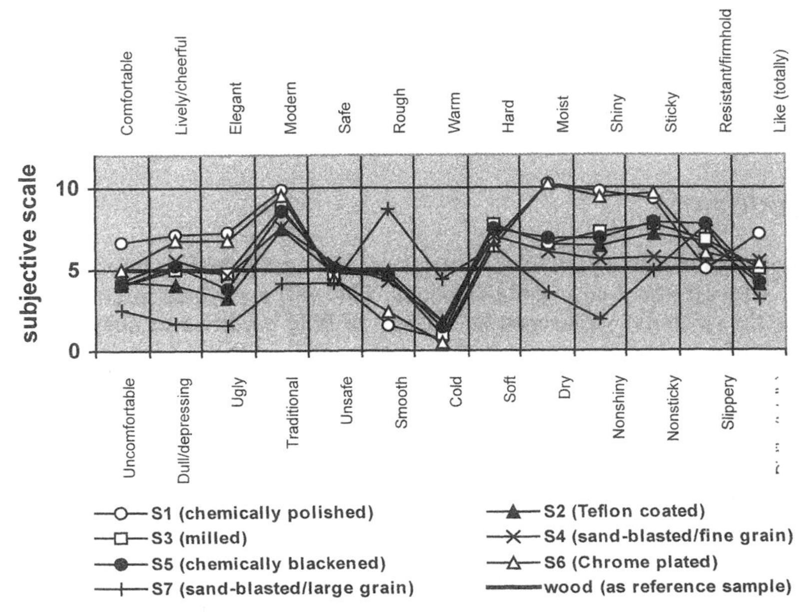

Figure 1 Subjective scaled responses to steel samples with different textures by blindfold touch

EXPERIMENT RESULTS AND DISCUSSION

The subjective scaled evaluation of texture through blindfold touch on 7 steel samples with different surface finish is shown in Figure 1. It is shown that there is only a slight difference in subjective response between the 7 samples when evaluating the surfaces in relation to 'soft-hard' and 'unsafe-safe'. The other sensory responses are obviously related to the 7 different surface textures, and the largest spread of responses occurs on 1 geometrical item 'smooth-rough', 3 emotional items 'dull/depressing-lively/cheerful', 'ugly-elegant', 'traditional-modern', and 3 physical-chemical items 'shiny-nonshiny', 'dry-moist', 'sticky-nonsticky'.

According to the subjective evaluation, it seems that the 7 steel samples can be clustered into three groups: group1 consisting of S1 (chemically polished), S6 (Chrome plated); group 2 consisting of S2 (Teflon coated), S3 (milled), S4 (sand-blasted with fine grains), and S5 (chemically blackened); group 3 consisting of S7 (sand-blasted with rough grains). On the other hand, there seems to be some correlation between the response items underlying the data shown in Figure 1. Correlation analysis and factor analysis confirmed these assumptions.

Table 1 confirmed that except for soft-hard and slippery-resistant responses, all other evaluation items have a certain correlation with the evaluation of roughness. Especially, the correlation between roughness and emotional responses are comparatively stronger. Factor 1 includes these emotional responses plus the smooth-rough item (see Table 2). Factor 2 includes four of the physical-chemical items, i.e. dry-moist, nonsticky-sticky, nonshiny-shiny, plus cold-warm (although cold-warm could also integrated into factor 1). Because of no significant correlation with other items, the slippery-resistant response and soft-hard response separately represent factor 3 and factor 4 respectively.

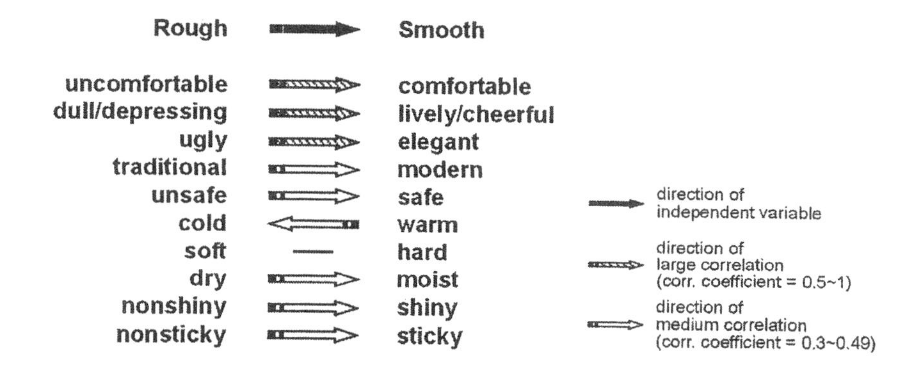

Figure 2: Correlation between other subjective evaluation items and subjective roughness through blindfold touch on steel. When the subjective response steel surface changes from rough to smooth, the correlation changes in the direction of arrows.

Results suggest, when blindfolded, people tend to give more positive emotional responses such as lively/cheerful, elegant, modern, comfortable feelings to a smooth surface than to a rough surface. Similarly, the correlation between the items within factor 2 suggests the perceived surface will be either 'cold-moist-shiny-sticky' or 'warm-dry-nonshiny-nonsticky'.

It's also worth looking at a relationship between factor 1 and factor 2. Within factor 1, only roughness (smooth-rough) is primary information directly from the material surface, while other items are all emotional or a kind of secondary, evoked information. Interestingly, the absolute value of the correlation coefficients between smooth-rough and all the items within factor 2 all exceed 0.4, whether positive or negative. This middle correlation supposes that a 'cold-moist-shiny-sticky' response corresponds to a 'smooth' surface, while a 'warm-dry-nonshiny-nonsticky' corresponds to a 'rough' surface.

Therefore, it seems reasonable to suppose that the roughness could be the key variable in the interface between material surface and human sensory response when people are blindfolded. The relationship between roughness and all the other response items via blindfold touch can be summarized in Figure 2.

As to the subjective evaluation of the surface grip characteristics slippery-resistant, the situation seems a bit more complicated. This can be reflected by checking the standard deviation of data. The mean standard deviation for slippery-resistant evaluation data is 2.56, 66% higher than the lowest deviation for the evaluation of warmth and shininess (1.5 and 1.52), especially in the case of polished and chrome plated surfaces. It could be explained that the individual skin characteristics, e.g. the sweatiness, more strongly influence the interface between skin surface and material surface. This influence is greater for a smooth surface than for a rough surface. Dryer skin interprets smooth surfaces as quite slippery, and for sweating skin, the surface feels resistant.

Table 1 Correlation Matrix

	uncomfortable-comfortable	smooth-rough
uncomfortable-comfortable	1.000	-.576
dull depressing-lively cheerful	.532	-.652
ugly-elegant	.456	-.637
traditional-modern	.349	-.423
unsafe-safe	.201	-.443
smooth-rough	-.576	1.000
cold-warm	-.466	.485
soft-hard	-.159	.104
dry-moist	.300	-.432
nonshiny-shiny	.262	-.457
nonsticky-sticky	.190	-.417
slippery-resistant	-.138	.165
dislike-like	.517	-.605

(a part of the correlation matrix)

Table 2 Rotated Factor Matrix*

	Factor			
	1	2	3	4
dislike-like	.805	.335		
dull depressing-lively cheerful	.792	.340		
smooth-rough	-.724	-.312		
uncomfortable-comfortable	.604		-.317	
ugly-elegant	.586	.526		
unsafe-safe	.579			
traditional-modern	.496	.360		
cold-warm	-.420	-.396		
dry-moist		.870	-.359	
nonsticky-sticky		.748		
nonshiny-shiny	.339	.706		
slippery-resistant			.505	
soft-hard				-.663

Extraction Method: Principal Axis Factoring.
Rotation Method: Varimax with Kaiser Normalization.
* Rotation converged in 9 iterations.

It should be pointed out that the results reported here are only derived from blindfold touch on steel samples. In fact, if the material category changes, sensory condition changes (e.g., touch plus vision), and if also the environmental context for perception changes (e.g., temperature and media), the results would possibly change as well. That's why in the introduction section, the matrix of material representation was supposed to be developed as a holistic system.

Similar results using steel were obtained also for the visual touch condition. Also results for plastics and thermoplastic elastomer will be reported later.

CONCLUSIONS (BLINDFOLD TOUCH USING STEEL SAMPLES)

1. Metallic materials with different surface textures produce different tactual responses, with the exception of subjective hardness.
2. Subjective tactual texture evaluation can be further identified using four factors. Factor 1 – emotional responses plus subjective roughness; factor 2 – subjective shininess, subjective moisture, subjective stickiness, and subjective warmth; factor 3 – subjective hardness; factor 4 – subjective grip.
3. Within the tactual response items, subjective roughness is supposed to be a key component, which correlates other responses to texture.
4. Smooth metallic surfaces evoke positive emotional responses such as lively/cheerful, modern, elegant, and comfortable; while rough metallic

surfaces evoke negative emotional responses such as dull/depressing, traditional, ugly, and uncomfortable. But whether this response is linearly proportional to surface roughness needs further exploration.

An example of how a designer might use this information is in the selection of materials for the handle on a consumer product. The material, texture and characteristics selected would be selected in the knowledge that the interface with the user will provide a positive, cheerful response.

REFERENCES

Zuo H., Hope T., Castle P., and Jones M., 2001, An Investigation into the Sensory Properties of Materials. In *Proceedings of the international conference on affective human factors design*, edited by Martin G. Helander, Halimahtun M. Khalid, and Tham Ming Po. (London: Asean Academic Press), pp. 500-507.

Dreamy hands: exploring tactile aesthetics in design

Marieke Sonneveld
Delft University of Technology, The Netherlands

INTRODUCTION

Tactile aesthetics are important in user-product interaction, and therefore, should be a part of the Product Design Education Curriculum. Because little is known about the tactile aspects of the users' experience of products, and even less about how designers should deal with these aspects within design projects, it is a challenge to integrate research and education about tactile aesthetics. To explore the potential of research embedded in a course in tactile aesthetics, a 40 h pilot course 'Tactility' was organised, for third year students of the School of Industrial Design from the Delft University of Technology.

This paper presents the structure and the lesson learned from these pilot studies, and concludes with suggestions to elaborate the course and to assess its potential for research.

EDUCATING THE SENSES

The basis for this course was the assumption that augmenting awareness for the tactile aesthetic aspects of products is mainly a matter of development of the sensitivity of the students themselves, through their own tactile experiences. To illustrate this thought, the analogy with a chef cook is presented. To be able to cook delicious food, s/he has to develop the 'taster' in him/her, has to learn by cooking, and above all, by tasting while cooking. Similarly, the designer has to develop the 'feeler' in him, and to use this sensitivity while designing.

To be able to develop this aesthetic sensitivity, the main task of the tutor will be to present exercises and experiences that will challenge and incite the student to explore their own aesthetic judgements. Gauguin 'proposes the notion of the *eyes that dream* as a way of viewing the world', to experience it (Caranfa, 2001). This notion was translated to the tactile domain: instead of exploring the world with an intelligent hand, the student explores their surroundings with dreamy hands.

THE COURSE 'TACTILITY IN DESIGN'

Theory on tactile experience

During the course, lectures on theory about tactile aesthetics were given, illustrated by examples from design practice. This theory distinguishes between the tactile *perception* and the tactile *experience* of products. Tactile perception is considered to be the determination through active touch of the physical aspects of a product. These aspects are: shape and size, texture, weight, balance, temperature, material properties (like elasticity, plasticity) and dynamics.

Tactile *experience* is considered to be the affective response of the user to the physical aspects of a product, through touch. As in tactile perception, different aspects can be distinguished in tactile experience (figure 1). These aspects were determined in previous research by the author on tactile aesthetics.

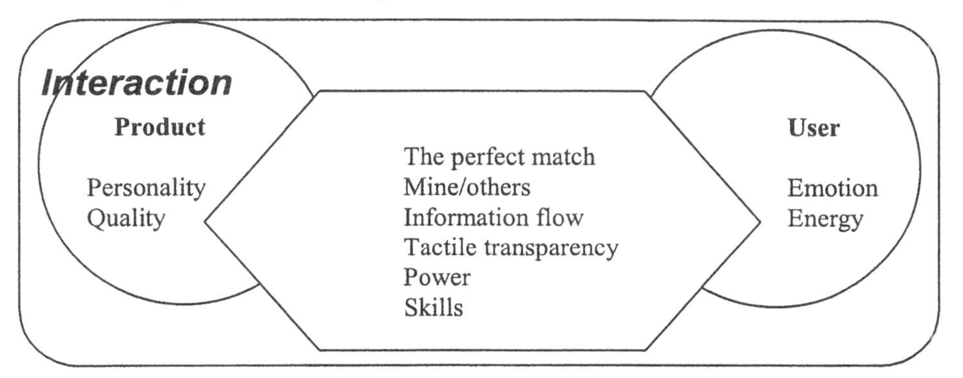

Figure 1 Aspects of tactile experience in user product interaction.

Aspects related to the product

Personality

As with people, we tend to attribute personality characteristics to products when being in contact with them. Within the physical interaction with products, the product can be perceived as elegant, mean, blunt or friendly. Moreover, the personality of people is often characterised by material properties: weak, strong, hard, soft, flexible, rigid, etc. Likewise, product personality can be characterised by the physical properties of the applied materials.

Quality

An important aspect of the tactile experience of products is the assessment of the quality of the product. Our sense of touch tells us whether we can trust the thing or not, whether it is valuable. Aspects like weight, finishing and the feeling of the moving parts are some of the important cues to this quality of products.

Aspects related to the actual interaction of the user with the product

The perfect match

When touching an object, we assess the way it fits us. We enjoy the feeling that something can be perfectly right. Shoes, tools, chairs or clothes can feel as if made for us. This feeling of a perfect match can be obtained through shape, size, adaptability, absorption properties, etc.

Mine/others

Within time, an object may be transformed by the interaction with the user, and therefore become unique. This allows the user to recognise the object among others as their own by touch, or, when this is not the case, as strange to them.

Information flow, tactile feedback

Within the interaction with physical objects, we obtain a lot of information about the state of the physical world. Designers can use this richness, especially when it comes to mediate information that can hardly be conveyed in another way without the use of complicated coding and decoding.

Tactile transparency

People have the ability to feel through objects. Thus, in user product interaction the attention of the user is not necessarily focussed on the contact surface between user and product, but rather somewhere else in the outside world. In a way, objects that facilitate this 'feeling through' become tactually transparent.

Power, control

During the physical interaction with a product, we know how the balance of power is. Who is in control and who is the strongest? This aspect of user product interaction is thrilling. It can be overwhelming to sense the power we have, to feel that we are completely in control (car), but the opposite can be exciting as well (roller coaster).

Skills

Related to the issue of power is the ability of the user to develop skills, and of the product to challenge the user to do so. Once acquired, the skill can be a source of pleasure, of flow. The way skills are acquired in time is an important aspect of tactile experience.

Aspects related to the user

Emotion: Within the user experience, emotion plays an important role. Especially for tactile perception, as the sense of touch is considered to be the channel for affective communication. The emotions reported within tactile experience are 'basic' emotions, the gut feelings. They have to do with feeling secure or feeling threatened, with being attracted or being repulsed, with lust or disgust.

Energy: Closely related to the experienced emotions, but clearly distinguishable is the effect of the experience on the user's energy level. Some experiences may give us energy, even if it the interaction with the product costs some effort, and other experiences may cause an energy leak.

The exercises

The course emphasises exploring one's own tactile experiences through the following exercises. Sharing the experiences between students strengthens the effect of the exercises.

Introduction exercise

The students are blindfolded and study the different tactile aspects of products that are presented to the group by the tutor. They learn to distinguish between touching to learn the physical aspects of objects (the intelligent hand) and touching to experience the object (the dreamy hand).

Second exercise

The students are asked to bring objects from their homes to the classes: 3 objects that are appreciated for their tactile qualities, and 3 objects that are not appreciated. The objects are presented to each other. In a discussion, the students explore the common aspects and the differences.

Third exercise

The students are asked to study, during one week, their encounters with a specific material: wood, leather, rubber, steel, textiles, etc. When do they encounter it, in what circumstances, how do they experience it? They describe their experiences and try to portray the material in terms of affective aspects.

Fourth exercise

To shift from contemplation to creation, the students transform a wooden stick into an object that feels pleasant on one side, and unpleasant on the other side. Again, the results are discussed. The exercise adds the experience of designing to merely exploring existing objects.

Fifth exercise

The fifth exercise adds the context of use to the design challenge. The students redesign a product in order to make the use of the device a pleasant tactile experience for themselves. The exercise had three phases: first, exploration of their actual experience of existing products; second, expressing the desired experience through words, collages and 3D objects; and third designing the new product through sketches but above all, using 3D models. During the exercise, the student uses the aspects of tactile aesthetics as presented during the lectures.

RESULTS

It was clear that the students started with a vague notion of tactile aesthetics. They had a rather uneasy feeling about it. First, due to this unfamiliarity, and second due to the fact that they had to deal in such an intimate way with their own experiences. Gradually, they became more confident, were able to express themselves and were enthusiastic to share their findings.

The end results of the course made the students aware of the fact that like for the visual experience, the quality of the tactile experience is strongly individual, and has little common ground. Although this fact is evident in daily life when it comes to the appreciation of food (taste), music (hearing) and fashion (vision), this discovery is a surprise when it comes to the tactile experience.

The students' evaluation of the course shows that its content answers a wish of the students for courses on experiential aspects of user product interaction, especially with a focus on the non-visual senses.

CONCLUSION

The results of the pilot show that the basis and the structure of the course are fruitful for the development of the students' sensitivity for tactile aesthetics. Both students and tutor/researcher are aware of the difficulties in the design process one is confronted with when dealing with the tactile aesthetics of products.

The course can be improved by concentrating on the exercises, where more detailed exercises may help to explore the tactile experience. Next, sophisticated product examples will help to illustrate the theory on aesthetics of touch.

For researchers in Product Design, the course showed that conducting a course in Product Tactility is an appropriate means for researchers. First, to explore the design implications of the results of their own research, and second, and even more important, to deepen and elaborate these results. The experiences reported by the students are valuable data for the tutor/researcher to elaborate insights on tactile aesthetics. Thus, research and education become one.

REFERENCES

Caranfa, A., 2001, Art and science: The aesthetic education of the emotion and reason. *Journal of Art & Design Education,* **20**, pp 151-160.

Design of sensorial sporting goods: a user-centred approach

Nicolas Bouché
Decathlon Research Center, France

INTRODUCTION

In the last few years, the world of the media has undergone a change. It has passed from the era of the television, where the programme schedules are the same for all, to the era of the Internet, where individual choices can be made giving total independence. This main evolution illustrates the change to a hedonic society, where the customer no longer respects the rules imposed by society and takes privilege in autonomy. Today, for economical reasons, our environment has also become standardised concerning sensations we feel, so individuals need to re-educate their senses to achieve more enjoyment. Therefore, as customers get more sensitive, the sensorial product is now a necessity to fulfil their expectations. The sports practice is a privileged time to relax and take pleasure. It seems obvious then that sporting goods need to induce more sensorial feelings too.

WHICH SENSATIONS FOR THE SPORTING GOODS

Today, our senses are not very solicited and then they are still dull. Therefore, each new feeling will surprise us. The first effect of the sensorial product is thus the creation of surprises, which attract and allure customers. However, this surprising effect has a relatively short lifespan, because very quickly we accustom ourselves to a new feeling. Then, it is necessary to renew unceasingly these surprising effects. Of course, this task is delicate and requires a great creativity. Otherwise, the play side of surprising effects could easily be perceived as a gimmick, from which a lot of customers might turn away quickly.

Many sales leaflets put forward the characteristics and functionalities of a product without giving customers the possibility to check their presence or the efficiency. However, these product features are essential in the selection stages. Therefore, if the customer could check them by himself, the product would certainly be more attractive. Feelings are once again essential here to give customers the opportunity to verify the presence and efficiency of the products' features we promise them. Let me take the case of trekking shoes, for which customers await a good grip. The cramps drawing, position, number and shape, their depth as well as the adherent touch and the mat colour of the rubber are many features which will allure and reassure customers (Figure 1). Here, we deal with controlling what the customers perceive of the quality of a product out of the feelings it induces (Figure 2).

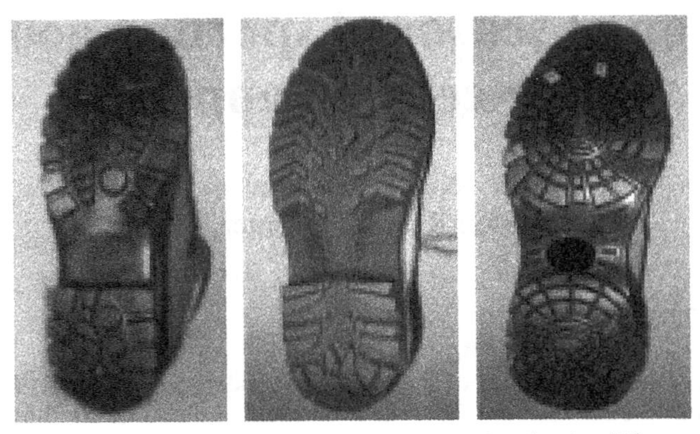

Figure 1 Different design of outdoor shoes outsoles, showing different cramps drawings, positions, depths, mat or bright colours .

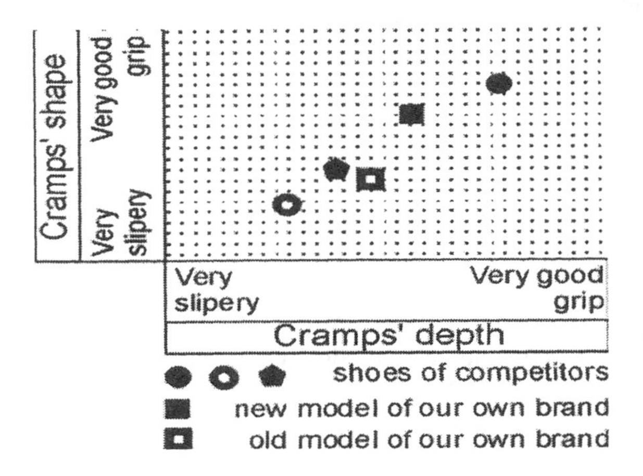

Figure 2 Perception mapping of outsoles considering two parameters: cramps' depth and shape.

Otherwise, a sensation can make it possible to push an invisible functionality forward. Today, many products offer anti-bacterial properties that the customer cannot detect differently without the presence of a leaflet or a distinctive visual sign. Why shouldn't we associate with such a property a little used sense: smell? Indeed, we could create an identical olfactive signature on all products presenting these anti-bacterial properties.

The s ensorial product seduces customers but it has above all to give entire satisfaction to its user, who must not be disappointed by the sensations they experience using it. Let us take again the example of the trekking shoes: they must have a good grip and give the sensation of it during the hours spent in the mountains this in each and every single condition of practice of that sport.

As regards clothes, the thermal comfort they provide must last despite the climatic variations and very changing physical activity levels of a day of trekking. Only at this price will the customers be faithful no more to a brand but to sensations - put to the test in the course of time - that will characterise this brand. Controlling the sensations experienced in the product utilisation stage is thus an extraordinary means to establish customer loyalty.

The main sensation in our modern activities is well being. Products must then be easier to use. Obviously, this implies the reduction of the physical, physiological and mental constraints while using it. A product with an intuitive functioning would significantly increase the customers' well-being. I believe that sensations can allow us to make the products more intuitive. Let us put the example of a bicycle rack that has to be strapped up around the hatchback of the car. It is very hard for the user to know when the straps are tightened enough, which generates a risk to put him or her in a dangerous situation. Nevertheless, if a characteristic sound signal indicated the right tightness of the straps, the product would be intuitive and the danger and customers' worry suppressed. This is thus an example of sensations used to improve the ergonomics of a product.

At last, some fashioned sporting practices and some ways of life lead us to talk about a "glide generation". Glide brings pleasant feelings, sometimes sensational or intense experiences. However, how to define them? How to reproduce them as well in the water as on the snow or in the air? Here is another kind of sensations - a more delicate one to deal with - in relation with sporting practices.

WHAT ARE THE SPECIFICITIES OF THE SPORT MARKET?

First of all, the sporting goods stores offer a great variety of products and there are a great number of brands on the international market as well as on national ones. In front of such diversity, the only problem for the customers is that they have a large range of choice: they turn towards products that offer either an excellent quality-price ratio or a fashionable look or at least a technicality that guaranties the user's security when practising extreme sports.

Second, meteorology and climate, lifestyles and fashion exert a strong influence on this market. That is why it evolves very rapidly with very often renewed product offers. To keep up with those changes, manufacturers need to develop new products always faster. The duration of the cycles of development is thus very short compared with other branches of industry that involve sensorial questions, such as automotive industry, pharmacology, agribusiness or cosmetics.

Finally, the sensation one can experience using a sporting good are linked to the level of physical activity, the state of fatigue and the environment of practice. It seems obvious that a climber caught in a storm will be very demanding and the sensations they feel will lead them to be faithful to a product or a brand. However, it is impossible to reproduce this situation of life with the product in a lab, and even to apprehend correctly remains delicate. Therefore, the best is still to test the product on the field, in its real situation of use, which is unusual in other branches of industry where tests and clinical trials predominate.

SENSORIAL ANALYSIS OR CUSTOMERS' PREFERENCES

Whatever the new product launched on the market, uncertainly remains about its success. How will the customers react? No design or marketing team can say they have a sure answer to this question. Such observation leads me to assert that in each designing process, there are always both a predictable - "rational" - part and an unpredictable - "subjective" - part. In front of this dilemma - which originates an industrial risk we wish to be the lowest possible - every company applies to controlling better and better its own designing process. We have at least two ways of proceeding: either we try to get a better "rational" knowledge or we enlarge the "subjective" offer and proceed by selection.

The sensorial analysis fully comes within the scope of such an approach: it tends to enlarge the "rational" knowledge by intrinsically characterising the sensorial outlines. It brings out the perceptible differences, reveals the sensorial stimuli intensity, and enables to know the customers' acceptance and preferences. However, all the sensorial analysis specialists will agree to say that it is long and expensive to set up such a process. Indeed measurement equipment and sensorial lab cabins as well as the formation of trained sample groups that always need to be kept level; all these represent high investments. Thus, before you launch forth into that sensorial analysis process, I think you had better ask yourself questions about its judiciousness and efficiency in the industrial context of the moment. Some of those questions seem obvious to me: what will the sensorial analysis frequency of use is and will it allow making the expenses profitable? Do we need to characterise intrinsically the sensorial outlines? Is the necessary analysis period compatible with rapid developments of new products? Can we integrate this capitalise the experiences? Must we integrate this know-how or subcontract it?

The specificity of both the sports market and the Decathlon Company do not seem to me to be favourable to the integration of that sensorial analysis process. The great diversity of products we design, our short development time limits, as well as the fact that the way sportsmen & sportswomen perceive the products is linked to their condition, level of practice and outdoor environment. Nevertheless, there is a lot at stake with sensorial sporting goods. We have then to find other methods and designing means to develop these products controlling at the same time the industrial risk. It was soon verified that the reasons why the customers preferred one product - among a set of very diverse propositions - constitutes the most relevant information for the design teams. We attempt thus to characterise the differences and draw the preferences. We centre our effort on the second possible way to control better the designing process - that is to enlarge the "subjective" offer and proceed by selection from the customers' preferences. The main point in this second method lies in the organisation of sensorial preference tests and semantic tests with naive customers from large or targeted populations. That way, we break free from the sample group's formation and level upholding. By addressing naive subjects, we create a new relation between the company and its customers. Such a process allows thus to raise the image of both the company and its brands. The more diverse are the sensorial propositions submitted to the customers, the lower the industrial risk. It is thus necessary to be highly creative to make those propositions very diverse. Let us draw a parallel with the artistic world: a painter will show much more creativity from a rich palette of colours. It is then essential to make up and enrich regularly this palette of primary solutions from which we will be able to create a great diversity of sensorial complexes.

AN OPEN APPROACH OR THE NEED OF STRATEGIC PARTNERSHIPS

The success of this analysis of the preference process lies in the ability to offer a great variety of sensorial stimuli. The richness of the primary solutions palette evoked before constitutes a first factor of success. We have then to update it very regularly being on the lookout for novelties by systematic and organised technological development monitoring. Next, to have preferential relations with some suppliers or transformers allows us - sometimes in an exclusive way - to benefit components and manufacturing processes. Some of those innovations can spring from our different skills put together: on the one hand, the knowledge of the customers' preferences - and so the knowledge of the rejected - and on the other hand, an in-depth knowledge of the materials and processes. It seems thus to be essential to build up those kind of strategic partnerships.

A second factor of success comes from our capability for generating great creativity. Every individual has got numerous ideas but that's a different matter to giving him or her the opportunity to express them. That is why the company has to organise the collection and selection of ideas showing respect, however, for some simple rules in this area: it is important not to judge the individual from the ideas he or she expresses to encourage him or her to go on and keep him or her informed about the evolution of each of them. This organisation will be more efficient if it opens up to the outside to collect more ideas. That way, the academics, suppliers, students, industrials from other branches bring so many different points of view of our activity: they allow us a change of paradigm and to see new opportunities.

Otherwise, though it does not seem judicious to me to invest in the setting up of a sensorial analysis process, it does present many interesting aspects and remains useful. It can then be led in collaboration with academics or specialised providers to determine the intrinsic characterisation of some sensations - such as sliding, grip or thermal comfort, etc...- to be found on different products. Here is a possible bridge between the academic world and the industrial one that would be beneficial. At last, we could think of establishing this same kind of partnerships with various foreign universities to apprehend and then understand some cultural differences linked to the perceived quality of the products.

Finally, even if it remains necessary to develop the sensuality of our sporting goods, it is also essential for the sensations that will constitute the mark of our brands to be constant and guarantied for the customers. Only by maintaining a close relation with our subcontractors will that be possible. Therefore, we would rather involve them in this process than demand of them new controls they would hardly manage to take over then.

Compact disc cover design: transfer of musical style and emotional contents into graphical design

René van Egmond
Delft University of Technology, The Netherlands

INTRODUCTION

For designers in industrial design it becomes increasingly important to consider the effect their product has on the human senses, emotions, and behavior. To state it more briefly designers have to consider—in developing their new concepts—what the complete *product experience* will be. In general, one could state that merely a focus on functional aspects of products will not determine if a product is successful. Of course, a designer still has to compromise between the experience of products and the functional restrictions imposed by the use of a product. One of the major challenges for researchers in industrial design will be to investigate how different senses register the different perceptual aspects of products and how these senses interact, which eventually will lead to the emotional and the aesthetical experiences of products. For this research, we used a domain of products in which designers already have a tradition in employing perceived aspects of one sensory modality (auditory) in their design of another sensory modality (visual).

Imagine one wants to buy a new CD and one walks in a music store or browses the Internet. The only thing you know is that you like to buy a specific style of music or music that evokes a certain emotional feeling in you. The problem is, of course, that the only "information" you can perceive in this specific context is visual. A designer working in the field of CD cover design may want to convey the emotional message of the music and/or the musical style. The resulting question then is, whether designers are successful in the conveyance of these aspects. In this study, this conveyance will be studied using recent CD's. However, a suitable experimental paradigm for this investigation needed to be determined first.

One of the frequently employed paradigms is the matching task, in which a person is instructed to match a sound to a product. This has been used, for example, in studies on portable cassette players and music (Smets & Overbeeke, 1989) and movies and music (Lipscomb & Kendall, 1994). However, this type of paradigm does not produce insight in the underlying perceptual, stylistic, or emotional dimensions. A suitable experimental paradigm to obtain these underlying dimensions is a questionnaire that contains the relevant attributes. We developed a questionnaire that is applicable in the visual and auditory domains. In addition, participants had to group the CD covers and a compilation of the music on the CD on the basis of their similarity. These paradigms allow a comparison among covers and among the music on the basis of their attributes; and the results allow a comparison between music and covers. Because of the space limitations only some results of the cluster paradigm will be presented in this paper.

METHOD

In this experiment the graphical similarity between CD covers and the auditory similarity between compilations of tracks has been determined using a manual cluster paradigm (see Procedure).

Participants

Twenty subjects participated in the experiment (*mean age*= 27.3 years). The participants were selected in such a way that they fitted within the following three groups: graphical expertise (n=5); musical expertise (n=5); amateurs or people with no expertise (n=10). Participants were volunteers and were paid.

Stimuli

An extensive description of the stimulus material will take up too much space. Therefore only a limited amount of information will be presented. Twenty-five recently published rock CD's were selected. The CD covers were copied at their natural size. From each CD 6 tracks were selected. A small representative snippet of 5 seconds was selected from each track. The six 5-seconds snippets were compiled into one track and new CD's were burned with twenty-five 30-second tracks.

Procedure

The experiments for the graphical and auditory stimuli were performed in individual and in separate sessions. The duration of the experiment was one hour for the musical stimuli and half an hour for the graphical stimuli. In general, the procedure was the same for the graphical and auditory stimuli. A person was asked to group the stimuli on the basis of their similarity first on a low level (maximum of 8 clusters; minimum of 5 clusters) and then to group these clusters in a higher order cluster (maximum of 4 clusters). The procedure is probably easily understood for the graphical material (25 CD covers), because the visual stimuli can be seen and moved. However, for the auditory stimuli 25 "blank" cards with track numbers were made to present the participants with a visual means for the grouping procedure. While listening (and skipping back and forth) to the tracks on the CD's, a participant had to group the cards with the track numbers. After each grouping was finished the participants had to indicate on a 10-point rating scale how well the CD's in a cluster fitted together.

RESULTS

In this result section only the analysis of the highest-level clusters (maximum of 4) are presented. In general, the participants indicated that they were (reasonably) satisfied with their clusters. The graphical clusters (*M*=6.33, *SD*=1.92) were rated somewhat lower than the musical clusters (*M*=6.82, *SD*=1.77). The technique used to analyze these clusters is multidimensional scaling (MDS) using the program SPSS, procedure Proxscal. This scaling technique transforms the individual cluster data into a two or more dimensional representation. The analysis was performed separately for the cluster data of the graphical and of the auditory experiments.

Using a specially written program in Director, the coordinates of this solution were used to depict the scans of the CD covers. In Figure 1 the MDS solutions for the graphical clustering (top-figure) and the auditory clustering (bottom-figure) are shown. Note, that the bottom-figure has been based on the analysis of the auditory experiment. In the auditory experiment, the participants did not see the CD covers. The scans of the CD covers are only used here to make the two analyses easier to compare. Four groups are encircled in both figures, Groups *A* (2 CD's), *B* (2 CD's), *C* (2 CD's), and *D* (4 CD's). Before the groups are discussed in more detail, some general trends that can be derived from the analysis will be described.

In the right half of the top-figure—MDS of graphical clustering—those CD covers are presented that all contain face-like pictures or photographs. In the right top-quadrant the CD covers are more veridical (i.e., photographs have been used) compared to the right bottom-quadrant. The designs of the CD covers in the left side of the top-figure become more abstract. It seems that the level of abstraction increases clockwise if one starts in the right top-quadrant of the top-figure.

In the bottom-figure, which depicts the MDS solution for the auditory clusters, a similar type of grouping occurs as in the top-figure. In fact, it almost seems that the top-figure has been rotated 90 degrees clockwise. The CD's in group *A* are mapped very closely together for both the graphical and the auditory cluster data. However, some remarkable differences do occur. In the MDS solution of the graphical cluster data a natural transition appears to occur from group *D*—>*B*—>*C*. In the MDS solution of the auditory cluster data, group *B* has been moved and groups *C* and *D* have been placed next to each other.

DISCUSSION

The most important finding of this study is that certain CD's are clearly grouped together for both the visual and auditory domain. This means, of course, that the designers of the covers somehow appear to successfully "transform" stylistic and experiential aspects of the music into a graphical design. Especially, the indicated groups in Figure 1 appear to evoke a similar experience in people. Furthermore, the groups *C* and *D* are mapped together. Interesting is the behavior of the CD's of group *B* in both domains. The characteristics of the visual design of the CD's map them between groups *C* and *D*, but if these two CD's are judged on their musical characteristics, they are judged more similar to other CD's. These CD's

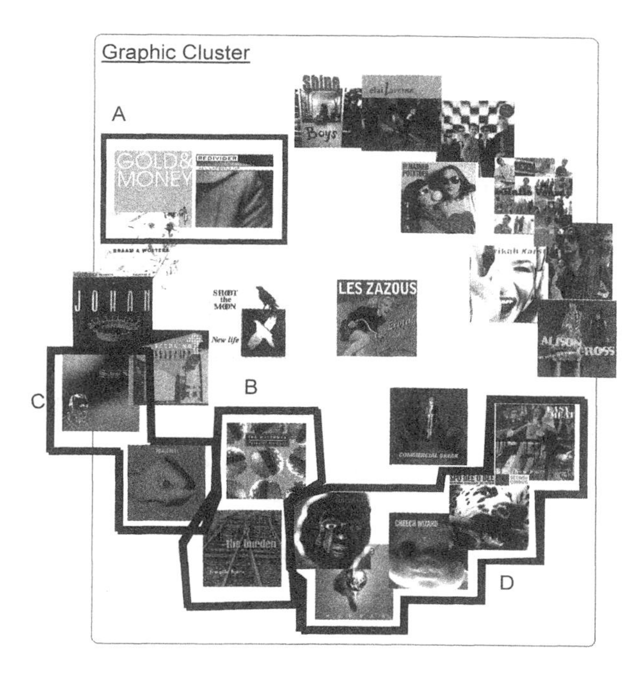

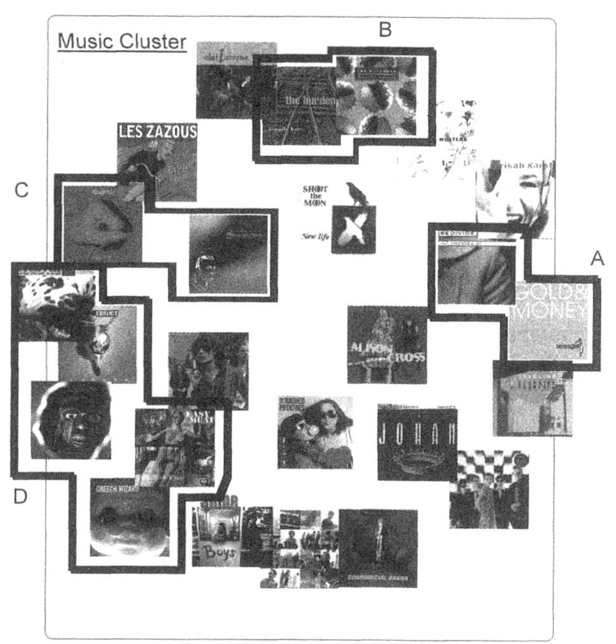

Figure 1 Two-dimensional scaling solution of graphical (top-picture) and musical (bottom-picture) clusters of 25 CD's. Specific groups are indicated by letters.

have been designed using metaphors, which means that a more cognitive translation between the music and the cover design has to be made. This appears not to be the case for the other CD's in which aspects in the music and the design of the covers seem to evoke a similar experience for the auditory and visual perception. In addition, the employed paradigm has a considerable advantage over a matching task: the underlying perceptual and experiential dimensions become insightful. The reason for this is that the *direct* relationships among covers and the music are not investigated. Instead, the music and the CD covers have been tested separately. If, using this paradigm, the same relationships among CD covers and among the musical snippets pop up, it might indicate that the visual design is a good indicator for consumers when they walk around in a shop or browse the Internet (the story with which we started).

Only a very small part of the analysis of the data has been presented here. Several other aspects have to be pursued. One of the first aspects will be to extend the MDS solution to 3 or 4 dimensions, and study if these new dimensions give even more insight in how people judge these CD's. In addition, the MDS solution of the cluster data will be compared with an MDS solution of the questionnaire. Given the fact that this questionnaire is composed of verbal attributes, the obtained dimensions will (probably) be even more insightful in the underlying perceptual and experiential behaviors, which determine the similarity between CD covers and music.

The final step in this investigation is to discuss these results with the designers of the CD covers in order to determine if our findings correspond with their expectations. An intended result of this interaction with the designers will be knowledge, or even a tool, that helps designers to get across their ideas to consumers. For research in industrial design this feedback of the actual designers is essential to come up with relevant research questions in this field. As stated before we have chosen the field of CD cover design because of the already existing tradition in transferring perceptual and experiential aspects of one perceptual domain to the other. It is our aim to use this research approach into other fields of industrial design in which this tradition does not exist or only just has started.

ACKNOWLEDGEMENTS

I gratefully acknowledge Mila Boswijk for her work and discussions, Patrick Groenen for his advice on MDS, Jeroen van Erp for the discussion and CD's, and Pieter-Jan Stappers for the development of a tool to visualize the MDS results.

REFERENCES

Lipscomb, S.D. and Kendall, R.A., 1994, Perceptual judgment of the relationship between musical and visual components in film. *Psychomusicology*, **13**, pp. 60-98.

Smets, G.J.F. and Overbeeke, C.J., 1989, Scent and sound of vision: Expressing scent or sound as visual forms. *Perceptual and Motor Skills*, **69** (1), pp. 227-233.

The emotional townscape: designing amiable public places

Nicolas Beucker
University of Essen, Germany

Ralph Bruder
University of Essen, Germany

INTRODUCTION

Cities are aiming at a personal relationship with their inhabitants. However, cities hardly ever use the potential of design for stimulating the senses. The city as a living space is constituted not only by architecture and urban planning, but also by its design. We all know places we dislike, others we enjoy, some we avoid, and others we appreciate and recommend. What are the local features that make us feel comfortable? How do we identify with the city?

A city is like an organism: its development cannot be predicted. Nevertheless, the ambition of urban design is always to define what can be expected of a location. Architectural elements mark the borders of the urban surrounding and adjacent buildings influence the options for place attractions. Mobile urban elements, devices for access, information and orientation, safety, comfort and adornment are defining the functional options in which a place can be used. Thus, they become the manual for public spaces. The urban design also tells stories about the city's identity. Following the functional needs of past society, space arrangements and street patterns evolved that still determine the image of the city. Obviously, architectural styles and spatial compositions remind us of the influential periods in the history of a city. In addition, design details can reveal the former identity of a location, though the purpose of the place has undergone dramatic changes.

URBAN IDENTITY AND IDENTIFICATION WITH URBANITY

Urban identity is based on the consistency of all aspects that are usually included in a corporate identity (Kutschinski-Schuster, 1993). But, applying the concept of corporate identity to the urban context is not enough. Good city design is not only responsible for the appearance and legibility of the city's outer image, but also for the identification of the population with the city's character (Kil, 1995). Identification with a location depends on the use and acceptance of a setting. Design of urban identity has to evoke individually accepted mental maps of the city. If cities want to compete in their attractiveness, they will have to enhance the townscapes with an emotional appeal. If a city as an institution has the courage to develop an independent profile (Beyrow, 1998), it will have to consider corporate design in a broader sense than implied by mere graphic design.

Design Is Story Telling

Design details tell a story about the historic and present use of the city, about its topography and its inhabitants, and a story that reveals the consciousness for a quality of life provided by the city. In this context, public design becomes much more than just ornamental decoration. It is performing a crucial task for the image of the city. It takes over a responsibility Kevin Lynch already called for [in] 1974, that a well designed public street will "engage and delight" people (Lynch, 1996).

In a survey conducted in 1999 at the University of Essen, information was collected on the ways in which design details convey the image of the city. It was a study focusing on the perception and importance of street furniture. It was undertaken to generate a more precise knowledge of the important aspects of urban elements for a concise image of the city. Slides showing townscape details with furnished streets were shown to town dwellers. They were asked to say whether they know the place shown in the picture, and by which detail they recognise the location. The survey was accompanied by an open interview of approximately 40 minutes in which the participants were asked about the influence that urban details have on their mental map of the city. Emotions were articulated concerning the unsuited design of benches, bollards, and detailed street clocks or fountains. Every interviewee mentioned at least one personal preference for a certain location.

Although the individual perception of urban details does of course differ, some elements shown in the survey were recognised by everybody. Obviously, location, size and quality of design characterise the significance of urban elements for the townscape. News stands, bus shelters, air exhausts, show-cases and other elements, that are in at eye level and often placed in the way of pedestrians, were always remembered. Only those elements that have a distinguished form were definitely linked to a specific location – either they are unique or they belong to a family of elements that furnish a certain place. Pure functional elements with an everyday appearance hardly indicated a certain location. A striking finding was that elements obtaining emotional connotations were remembered because of the whole context that was connected to them subconsciously. For example, some bollards are only used in the old town and were therefore not known in detail. Nevertheless, their form and the pattern of the cobblestone pavement around referred to a historic tradition of street adornment that the test participants easily linked with a specific location in the old city. According to the semantic density of urban design, legibility of the city is increased. Many people were astonished that they did not realise before how much the townscape is dominated by urban details.

An example of synergetic urban design

One way in which design practice can include emotional aspects is demonstrated in a study of an interdisciplinary urban design concept in which town planning, architecture and design perform synergetically. In 2000 a diploma in urban planning and architecture (Julia Kapp, Philipp Dechow; Technical University of Stuttgart) and in industrial design (Nicolas Beucker; University of Essen), presented a concept to redevelop the old Neptun dockyards in Rostock, Germany (2000). This aimed to reconnect the city with the waterfront at the Unterwarnow.

Since city life in Rostock was separated from the water for several decades, due to the inaccessibility of military areas close to the city centre and the belt of ship building industry north of the city, the major goal was to design an attractive location in the vicinity of the water. The focus of the concept was the installation of an experimental opera in the main factory halls of the dockyard. It was an aim to establish a new cultural centre at the water and continue a line that would connect the inner city with the outer suburbs. The old slipway was used as an open-air theatre, a long seaside promenade stretching from downtown to the suburbs in the north with the old dockyards in the middle. Various plazas were planned for the new districts that should transport the waterfront atmosphere into the city. Each detail of the design was based on one theme: Rostock, city at the waterfront.

The concept was proposing situations that should make it possible to perceive the city with all our senses, connecting it with the waterfront. By opening the access to the river the experience of life close to the water was encouraged. Wind, water, and an extension of sight introduce a new quality of perception. In the plans, the view from the promenade and other recreation points near the river open a totally new view on the city and its topographical situation. To bind the existing scenery to the newly developed district, for example all lamp posts designed for the river promenade are painted white, thus perpetuating the image of the white masts of the sailing boats in the harbour.

The image of the existing site was naturally dominated by the impressive industrial style and size of the old dockyards. Cranes, floodlights, and a ssembly plants form the skyline of Rostock. The historic importance of the “Neptun” ship yard was the pride of the city for many years and its most important employer.

Redesigning the area means influencing the image of the city within the collective memory of the inhabitants. Therefore, the project proposed to keep the most impressive buildings and continued to use the industrial style for design details. It was important to give a human scale to the industrial district. To a large part, this was done by the careful application of design details in public space. Street furniture was implemented to define the use of certain areas. The rough and bulky style of the industrial surrounding was maintained in the furniture, but contrasted with details dissolving the massive forms into scaled human proportion.

COMPREHENSIVE IDENTITY

Urban identity is a dynamic system, which depends on the changes of society, nevertheless it can be influenced by design. Coherent concepts enrich requirements of a functional city with emotionally striking patterns. The following aspects can be noted as a background of emotional design to challenge urban design.

Public space as a homage to the pedestrian

Obviously most urban elements can be perceived in detail most easily by pedestrians. The walking citizen is the one who really explores the culture of city life (Boesch, 2001). Indeed, urbanity in its original sense – as a place to foster tolerance – can only exist where people can meet personally. We do not have to be nostalgic and recall again the romantic plazas of renaissance Italy as *the* model of good city design (Rudofsky, 1969), but still we should consider the street as a community room (Kahn, 1991). Emotional public design is mainly for immediate experience, it is design for citizens on foot.

Immediate city life

Emotional urban design encourages immediate encounters with diverse scenarios within the city. The experience of sitting next to a river without any separating fences (Cullen, 1996) constitutes the identification with the city as well as sitting on the stairs of a building, watching the street life, following a game of old French boule players, watching children play or listening to street musicians.

Regulations for the design of the urban environment - as much as they would help for a concise identity – must be imposed carefully and in a way that allows participatory action and spontaneous modification.

Atmospheric townscapes

Architects, designers, psychologists, sociologists, and especially poets mention how emotional design creates distinctive atmospheres (Böhme, 1998). Böhme refers to all sensuous experiences from the olfactive memory of different districts in Paris to the acoustic of street pavement and the holistic enthusiasm of August Endell, who described *the beauty of the big city* (1995) as a synesthetic adventure. Sophisticated urban design should be aware of sensuous compositions.

The hidden physiognomy of the city

As already indicated, design often traces the hidden past of a city. A famous example is the Rambla in Barcelona, which used to be a dried-up bed of a seasonal river. This historic situation survives in the street name *Las Ramblas* that is derived from the Arabic word *ramla*, which means *dried-up bed of a river*. Furthermore, the pavement of the street with its wave-like design is a reminder of the water that was flowing there several hundred years ago. Most towns provide similar examples of detailed semantic connections that refer to particular situations. In a coherent story, the urban context describes the present situation as detailed as the historic.

OUTLOOK

Though the image of the city can be read as the setting of the stage on which life takes place, city layout cannot pretend to be distinct from what was originally there. Emotional design is crucial for the urban identity and can modify the existing grammar of the city slightly, but nevertheless a city has to maintain its authenticity.

At the University of Essen the discussion of urban identity and the importance of design within the grammar of the city will be continued in further practice related research. It will result in a PhD thesis in the design sciences. Finally, one day designers may arrive at a stage, when they will be able to virtuously compose emotional-functional solutions for urban design projects. We might explore cities as fascinating as the invisible cities described by Calvino (1972).

REFERENCES

Beucker, N., Dechow, P., and Kapp, J., *<http://www.baunetz.de/arch/diplom>*

Beyrow, M., 1998, *Mut zum Profil*, (Stuttgart: avedition GmbH).

Böhme, G., 1998, Die Atmosphäre einer Stadt. In *Neue Stadträume- zwischen Musealisierung, Medialisierung und Gestaltlosigkeit*, edited by G. Breuer, (Frankfurt am Main: Stroemfeld Verlag), pp. 149 – 162.

Boesch, H., 2001, Die Kultur des Langsamen. In *Hans Boesch, Die sinnliche Stadt – Essays zur modernen Urbanistik*, edited by E. Pulver, (Zürich, Nagel & Kimche AG), pp. 72-81.

Calvino, I., 1972, *La città invisibili*, 10. rist. (Torino: Einaudi)

Cullen, G. 1996, *Gordon Cullen, Visions of urban design*, edited by D. Gosling, (London: Academic Group Ltd), pp. 49-50.

Endell, A., 1995, *Vom Sehen – Texte 1896 – 1925 über Architektur, Formkunst und „Die Schönheit der großen Stadt"*, edited by H. David, (Basel: Birkhäuser Verlag), pp. 163-208.

Kahn, L., 1991, *Writings lectures interviews*, edited by A. Latour, (New York: Rizzoli International Publications), p. 296.

Kil, W., 1995, Identität entsteht durch Aneignung. In *Risiko Stadt*, edited by Schwarz, U., (Hamburg: Junius Verlag), pp. 144-145.

Kutschinski-Schuster, B., 1993, *Corporate Identity für Städte – Eine Untersuchung zur Anwendbarkeit einer Leitstrategie für Unternehmen auf Städte* (Essen: Verlag die Blaue Eule).

Lynch, K., 1996, Designing and managing the strip. In *City sense and city design: writings and projects of Kevin Lynch*, 3rd ed., edited by T. Banerjee, and M. Southworth, (Cambridge, Massachusetts: MIT Press) p. 597.

Rudofsky, B., 1969, *Streets for people: A primer for Americans*, (Garden City, N.Y.: Doubleday).

FROM DESIGN TO EMOTION

An accessible framework of emotional experiences for new product conception

Carl DiSalvo, Bruce Hanington, Jodi Forlizzi
Carnegie Mellon University, USA

INTRODUCTION

As people become more sensitive to dimensions of products that go beyond traditional aspects of usability, the need to create emotional resonance between people and products increases. However, designers lack a shared understanding and language for emotion within the context of design. This paper describes research towards creating a framework of emotional experiences that is usable by designers for new product conception. We base this research on three prominent approaches to emotion and experience expressed by Dewey's *Art as Experience*, Carlson's *Experienced Cognition*, and Csikszentmihalyi's *The Meaning of Things*. These approaches represent the fields of philosophy, cognitive science, and social science, three fields that converge in design research and practice. By extracting points of intersection across bodies of research, we have begun to synthesize these approaches into a single framework that will help designers to *discover* and *understand* opportunities for new products that stimulate, enhance, or change emotional experiences.

WHAT IS EMOTION?

Both Dewey and Carlson split the structure of what we generically call emotion into two types of responses. Carlson makes a differentiation between emotion and mood (Carlson, 1997). Emotion is defined as short, sharp waves of feeling, arising without conscious effort or reflection, usually accompanied by increased activation of the autonomous nervous system — physiological changes in heart rate and respiration. In contrast, mood is defined as a longer-sustaining, less intense emotional effect.

Dewey differentiates between emotional statements and emotional expressions (Dewey, 1934). An emotional statement is a momentary descriptive response that may seem expressive, but is an automatic action — for example, laughing at a cartoon in the daily newspaper. In contrast, an emotional expression is an ordered response that references the emotions of previous experiences — for example, cherishing a special family heirloom. While Carlson and Dewey use different terms, what is important is that they share the recognition of two types of emotion, one that is *short and reflexive* and another that is *sustained and reflective*.

HOW DOES EMOTION RELATE TO EXPERIENCE?

Dewey does not equate an emotional statement to an experience. When he speaks of an emotional experience he is only referring to an emotional expression, a sustained and reflective response. Dewey regards emotion as a pervasive quality of an experience that serves to unify and shape the experience. This emotional quality cannot be attributed to a single emotion or object, it is a gestalt formed by multiple emotions felt in the moment, from the past and interactions with the environment.

The theory of experienced cognition as presented by Carlson is based on what he terms the co specification hypothesis, or the idea that the experiencing self and the objects in the environment are simultaneously specified in arrays of information processed by the brain and the nervous system. This view of conscious experience is based on an ecological view of perception (e.g., Gibson, 1979). According to Carlson, a purely emotional response (short and reflexive) affects information that is about the self more than the environment. In contrast, mood affects information about the environment and the objects within it more than the self. The state of arousal that comes with emotion and mood shapes the environmental information that we perceive, and its positive or negative influence shapes how we relate ourselves to the real or imagined state of the world.

Although these two approaches come from different traditions, a commonality in understanding the role of emotion and experience exists between them. From both approaches, we can state that a sustained and reflective emotional response (whether we call it "mood" or an "emotional expression") is highly dependent on *the environment*. In contrast, a short and reflexive emotional response is not dependent on the environment but instead to *singular elements or objects*. This response is not an emotional experience.

WHAT DO THE THINGS IN OUR ENVIRONMENT MEAN?

Although Dewey refutes the notion that single specific characteristics of an object can be the cause of an emotional experience, we can extract three ways in which products can function to contribute to emotional experiences: products can function as *stimuli* for *new* emotional experiences; products can function as *extenders* of *existing* emotional experiences; and products can function as *proxies* for *previous* emotional experiences.

To understand how products fulfil these functions we look to the work of Csikszentmihalyi (1981). Original research asked people to identify objects they deemed "special" in the home environment. While the definition was left open to interpretation by participants, the results have implications for emotion and product conception, by providing us with a set of dimensions useful for understanding how human-product interactions create meaning, including emotion.

Dimensions may be extracted from both explicit and implicit (extrapolated) design connections. Most objects in the study that are within the realm of design are categorized as "action" objects, or "those objects whose use involves some physical handling, interaction, or movement," contrasted to contemplative objects (Csikszentmihalyi, 1981, p. 270). Furthermore, "things that produce ordered stimulation, either auditory or visual," may help "focus and objectify emotions."

These features correspond to Dewey's *stimuli*. Additional object meaning categories include experiences (enjoyment, ongoing occasions/activities, release/escape); intrinsic qualities (physical attributes, e.g., size, color, texture, material); style; and utility. Collectively, these categories may parallel Dewey's *extenders*.

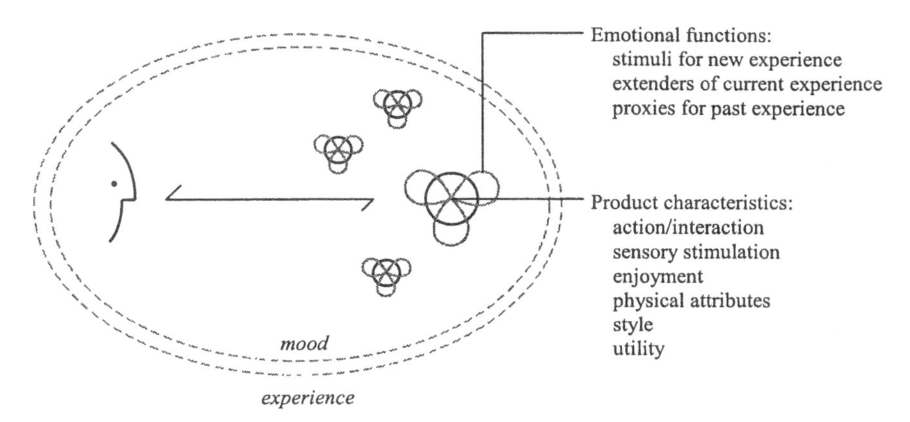

Figure 1 A framework of user-product experiences for emotional product conception.

Beyond formal and interactive qualities, *associative* object qualities may contribute to meaningful and emotional experiences, and correspond to Dewey's *proxies*. These include such things as intrinsic qualities of craft whereby the maker is known, and memories emoted through family heirlooms. While not within the domain of traditional design, these qualities suggest emotional value in evidence of the hand-made and object longevity, with possible design implications for product materials and quality. The work summarised here, supported by our own in-class exercises (Hanington, 2001), have helped us to form an initial list of object qualities contributing to emotional experience.

A FRAMEWORK FOR EMOTIONAL PRODUCT CONCEPTION

From the analysis and synthesis of these perspectives on emotion and experience, we are able to suggest the following framework for discovering and understanding emotional experiences (Figure 1).

1. There are two types of emotional response:
    - short and reflexive (emotion)
    - sustained and reflective (mood)
2. Only sustained and reflective responses (mood) constitute an emotional experience
3. An emotional experience is dependent on the relationship between the individual and the environment in which it takes place

4. The environment is constituted of objects that can function as
   - stimuli for new experiences
   - extenders for current experiences
   - proxies for past experiences
5. The qualities of objects that function as contributors to an emotional experience are:
   - action/interaction
   - sensory stimulation
   - enjoyment
   - physical attributes
   - style
   - utility

NEXT STEPS: A PLAN FOR EVALUATION

Much of the research into design and emotion has focused on short and reflexive emotional responses rather than emotional experiences. Models such as the circumplex of affect (Feldman, 1995) and tools built upon that model (Desmet, 2000) are excellent at measuring momentary descriptive emotional responses, but are not easily applied to emotional experiences. Additionally, these tools are more evaluative and less applicable in the early stages of product conception.

We have stated that designers cannot craft *an experience*, but only the conditions or levers that might create an intended experience (Forlizzi, 2000). Similarly, we cannot design products to generate specific emotional experiences. However, according to Dewey, Carlson, and Csiksentmihalyi, if we understand the environment in which an emotional experience occurs and how objects function as emotional levers within that environment, we may be able to discover opportunities to design new products that have an effect on the emotional experience. This effect could take the form of stimulating or enhancing the emotional experience, or changing the emotional experience to a more desired one. We believe that our initial framework may help designers *discover* and *understand* opportunities for new products that fulfil needs and desires for emotional experiences.

To evaluate this work in progress we propose a study in which we will define a context and collect an initial inventory of emotional experiences and object images using experience sampling and visual diaries. Using these materials, we will then conduct two workshops with a small team of practising designers. In the first workshop, we will present our findings from the experience sampling and visual diaries along with the framework, and ask the design team to define (*understand*) the varieties of emotional experiences. In the second workshop, we will ask the design team to use our findings and the framework to generate (*discover*) product concepts that stimulate, enhance, or change, as appropriate, the emotional experience. The activities will be documented and evaluated with post-session surveys and discussions. If the designers rate the process as effective for generating innovative and feasible design solutions with an emotional component, we will feel confident that our framework will function as a useful tool for practising designers. While future studies would naturally involve responses to design solutions by consumers/users, the current focus is placed on product conception by designers.

Many challenges remain for enhancing the understanding of emotion as it relates to design. This initial work synthesizes findings from three prominent bodies of research into a single generative framework. We hope that this framework will prove to be a useful tool for understanding emotion, with practical implications for design application. As work continues in this area, we can look forward to developments that increase our understanding of the relationship between emotion and design, and in discovering opportunities for new product conception.

REFERENCES

Brandstatter, H., 2001, Time Sampling Diary: An Ecological Approach to the Study of Emotions in Everyday Life Situations. In Brandstatter, M. and Eliasz, A. (eds.) *Persons, Situations, and Emotion. 20-52* (Oxford, Oxford University).

Carlson, R., 1997, *Experienced Cognition.* (Mahwah, NJ: Lawrence Erlbaum).

Csikszentmihalyi, M. and Rochberg-Halton, E., 1981, *The Meaning of Things: Domestic Symbols and the Self.* (Cambridge: Cambridge University Press).

Csikszentmihalyi, M., 1993, Why we need things. In Lubar, S. and Kingery, W.D. (eds.), *History from Things: Essays on Material Culture.* (Washington: Smithsonian Institution Press).

Desmet, P. M. A., Hekkert, P., Jacobs, J .J., 2000, When a car makes yousmile: Development and application of an instrument to measure product emotions. In: S.J. Hoch, and R.J. Meyer (eds.), *Advances in Consumer Research*, **27**, pp. 111-117.

Dewey, J., 1934, *Art as Experience.* (New York: Penguin Putnam Inc.)

Feldman, L. A. 1995, Variations in the circumplex structure of emotion. *Personality and Social Psychology Bulletin*, **21**, pp. 806-817.

Forlizzi, J. and Ford, S., 2000, The Building Blocks of Experience. DIS2000, 17-19 August 2000. In Dan Boyarski and Wendy A. Kellogg (eds.). New York, NY, ACM Publishing.DIS00 Conference Proceedings, 419-423.

Gibson, J.J., 1979, *The Ecological Approach to Visual Perception.* (Boston: Houghton Mifflin).

Hanington, B., 2001, Factoring the human in design education. In Helander, M., Khalid, H. & Tham, M.P. (eds.), *Proceedings of the International Conference on Affective Human Factors Design.* (London: ASEAN Academic Press).

Mobile phone games: understanding the user experience

Hayley Dixon
Gary Davis Associates, UK

Valerie A Mitchell
Loughborough University, UK

Susan D P Harker
Loughborough University, UK

INTRODUCTION

Mobile gaming is viewed by the mobile communication industry as one of the 'killer applications' for future mobile services. Fuelled by the success of games such as Nokia's Snake and the continuing popularity of online and console gaming, the drive is to develop ever more sophisticated and engaging gaming experiences for mobile users. However the current mobile gaming experience in terms of graphics, interaction mode and content more closely resembles that presented by personal computer games of 20 years ago than anything evoked by today's console based offerings. Despite such limitations the appeal of mobile games continues to grow. Market research conducted by Nokia estimates that 85% of people with the game 'Space Impact' on their phones have tried it out and 45% play it everyday (Robens, 2001).

Mobile gaming research has predominantly focused on the "mobility of gaming" (Kuivakari 2001). Such research seeks to exploit the entertainment potential of ubiquitous technologies and augmented reality, making both the proximity of others and the mobile environment itself part of the gaming experience. (See for example Bjork et al (2001), Brunnberg (2002)).

The research reported here aims to provide insight into what motivates people to play existing mobile phone games, despite their limitations, and seeks to identify elements of the current mobile gaming experience that should be preserved within future games. The continuing convergence of computer, consumer and communications technologies within mobile devices is raising many unknowns about how users will perceive these devices and therefore how best to design appropriate form structures and user interfaces (Sacher and Loudon 2002). This research examines the existing convergence of game playing and telephony within the mobile phone and provides early indications of how people may approach future converged devices.

METHOD

Interviews, focus groups and video analysis of game playing were used to elicit user requirements for mobile gaming in relation to the mobile context of use. In total 47 experienced mobile gamers – defined as those who played games on their phones for at least, in total, an hour a week – participated in the study. The age range was between 18 and 28 years of age. The study population consisted of 32 males (21 students, 11 non-students) and 15 females (10 students, 5 non-students). Motives for game playing were extracted from the interview and focus group data using content analysis techniques. Individual game players were videoed playing their favourite games and then asked to talk through the game playing experience as the tape was replayed. The video based interviews focussed particularly on how well the mobile phone's hardware matched the participant's gaming needs.

RESULTS

The data shows that mobile phone games are played in a diverse range of use contexts and for varying motives. In different contexts of use, users demand very different experiences from mobile gaming. The motivations for playing mobile games identified from the content analysis are described below and, where appropriate, their implications for handset and mobile game design are discussed.

Relieving boredom

Not surprisingly, participants were using mobile games during 'bored' times, during travelling, in cars, trains or situations where users were presented with periods of time that lacked any familiar social interaction or an interesting environment. Many users said that they first started playing games when in these situations. Mobile games are particularly suited to travelling and in such circumstances, users may become quite engrossed in the gaming experience. This is likely to be a popular context for playing more sophisticated console style games on a mobile. However this context of use may also require game playing to be abruptly stopped as, for example, the bus arrives. The ability to pause games quickly and resume at the same point was expressed as a key requirement in this context. The demand placed on the battery life in the travelling context is also a critical issue often accentuated by the need to keep the phone switched on through out the entire journey. Participants reported heavy draining of batteries when gaming and therefore require non-obtrusive – e.g. in between levels – alerting before battery levels become critical.

Enhancing social interaction

Today, mobile gaming is almost entirely a solo player experience. However the study revealed that mobile gaming activities were frequently used to facilitate social interaction. Games were being used as common ground for people to start up a conversation. In a social context mobile games were also used to elicit praise and acknowledgement from others. A common practice, reported by participants, was attempting to put a 'highest score' on another's handset with or without their permission. Even the low complexity games currently available were played very competitively within social groups. Recent developments allowing scores to be posted on websites hosted by phone manufacturers (e.g. Club Nokia) will further encourage the competitive side of mobile gaming. Two player and multi player games will also provide new opportunities for socialising with friends and strangers.

Avoiding social interaction and embarrassment

Conversely games were also used to avoid unwanted social contact. Female game users in particular used games to avoid eye contact when on their own, for example when waiting for friends or dates in social environments. One male participant reported playing mobile games when waiting for his girlfriend in a lingerie department. Embarrassment was adverted by keeping his eyes firmly on his mobile phone screen!

Meeting personal entertainment needs

Even when not in a social environment, gamers expressed personal achievement as an important motivation for game playing. Users were game playing on mobile phones for sometimes very extensive periods of time. This was normally in private, in circumstances where they can play without interruption and play well. The expressed goals in such circumstances were often intrinsic – to improve ones performance for personal satisfaction, as well as external – hoping to later impress other players. In these circumstances the limitations of current mobile handsets as a platform for gaming became most apparent. Many participants felt they were pushing the performance of the available games to their limit and were concerned about the effects of heavy gaming upon the phone's keypad. The more experienced gamers became, the more frustrating they found the lack of control enabled by the keypad.

Providing personal comfort

Mobile phones are examples of what are often referred to as 'living objects,' conveying meaning and significance to the user beyond the functional utility of the product (Jordan, 2000). The results of this study suggest that mobile gaming, particularly for female users, is part of this intimate relationship between product and user. Female gamers reported far higher incidences of gaming alone at home than their male counterparts. Comfort seeking use of games was reported in the context of working through emotions after, for example, an argument with a boyfriend and when playing mobile phone games last thing at night. Little attention was paid to the game itself in such instances. It was the interaction with the mobile phone that provided comfort. The low complexity nature of existing mobile games suits this type of interaction. Both genders viewed current mobile phones games as "cheesy." However, their simplicity was perceived by many female participants as fundamental to their emotional appeal. This was confirmed in a later related study which looked specifically at gender differences in mobile gaming (Tyler 2001).

DISCUSSION

Blending mobile telephony and gaming requirements

Amongst participants there was a large diversity of user requirements, some wanting far more gaming capabilities, others expressing satisfaction with current games. Although participants made extensive use of mobile games and many were frustrated with the limitations of the phone keypad for game interaction, they did not want gaming features to detract from the aesthetics of the handset or the efficient performance of voice telephony and text messaging tasks. Similarly, part of the allure of mobile phone games is their ready accessibility within a pocket-sized product. Increasing handset size to improve gaming performance would be incompatible with the spontaneity of use present within many of the gaming contexts and contrary to mobile telephony needs. Improving the playability of mobile games must come therefore through handset enhancements that do not hinder fulfilment of these design priorities. However keypad qualities chosen to withstand the rigours of gaming will also suit the demands of text messaging. Similarly high resolution screens will enhance not only the gaming experience but also improve the readability of text and mobile internet content.

Future games – keeping the essence of mobile gaming

Mobile phone gaming is rapidly evolving. The gaming experience is being enriched through provision of colour screens, and advanced sounds and graphics. JAVA™ based mobile game platforms will enable faster interaction speeds, console game style animation and special effects. Opportunities for two player and location based games are already emerging. However mobile game developers and handset manufacturers must be careful to continue to service the wide range of gaming needs identified. Users wish to maintain the flexibility inherent to successful mobile games where current controls allow both one handed and two-handed use of the same game. Successful computer games present a clear mission so that players are quickly aware of what they are trying to achieve (Clanton, 2000 ***or 1999**). This is of paramount importance for mobile games as users will often be unable or unwilling to spend time figuring out controls or game play motives.

Participants within this study were frustrated by the absence of what they considered quite basic gaming features within mobile games. For example they wanted to easily pause and resume games and place names by high scores. As future handset technologies move the mobile gaming experience closer in character to that experienced on game consoles, users may well become less tolerant of shortcomings in mobile game functionality and control. As user expectations increase, mobile phone designers will need to find new ways of meeting the demands of mobile gaming without compromising handset design to the detriment of either voice or text communications. Increasing the sophistication of mobile games should not increase their complexity. Multi player games in particular raise many novel usability issues. The process of linking players must be intuitive, as must procedures for re-establishing links when the vagaries of the mobile environment cause the link to be broken.

This study suggests that current mobile games play an important role within the personal relationship existing between many users and their mobile phones. They become a source of comfort in times of need and a ready friend when alone in public situations. The basic pick up and play format of successful mobile games facilitates gaming in such circumstances. The study found female game players in particular liked the simplicity and familiarity of existing games. At least a subset of future gaming should continue to function in a way that has meaning to these users. This might be contrary to the aspirations of game developers looking to exploit to the maximum the potential of future mobile gaming technologies.

CONCLUSION

Inclusion of games on mobile phones is an additional feature that has surpassed intended development expectations. However, this study suggests that there are user needs that do not seem to match with the technology push that is currently in progress. The planned advances in mobile gaming will be attractive to many consumers if playability in the mobile context of use is maintained. However even advanced gaming phones may need to offer simple pick up and play games alongside more sophisticated offerings.

ACKNOWLEDGEMENTS

The authors wish to acknowledge the continuing support of TTP Communications Limited (TTPCom Ltd.) as sponsors of related research within the Department of Human Sciences Loughborough University.

REFERENCES

Bjork, S., Falk, J., Hansson, R., and Ljungstrand, P., 2001, Pirates! Using the physical world as a game board. In *Proceedings of Interact 2001, IFIP TC.13 Conference on Human –Computer Interaction*, edited by M. Hirose, (Tokyo: Japan), pp. 423-430.

Brunnberg, L., 2002, Backseat gaming: Exploration of mobile properties for fun. In *Proceedings of CHI 2002. Changing the World, Changing Ourselves. Conference on Human Factors in Computing Systems,* Conference Extended Abstracts on Human Factors in Computer Systems, (New York, USA: ACM Press), pp 854 – 855

Clanton, C., (2000) Lessons from game design. In *Information Appliances and Beyond,* edited by E. Bergman, (San Francisco: Morgan Kaufmann), pp. 299 - 334.

Kuivakari, S., 2001, Mobile gaming: A journey back in time. In *Proceedings of Computer Games and Digital Textualities,* (Copenhagen, Denmark: Information Technology University), http://diac.it-c.dk/cgdt/program.html.

Jordan, P.W., 2000, *Designing Pleasurable Products.* (London: Taylor & Francis).

Robens, P. 2001, Goodbye to free phone games. BBC news, Dotlife, 17 December, http://news.bbc.co.uk/hi/english/in_depth/sci_tech/2000/dot_life/newsid_1709000/1709107.stm

Sacher, H. and Louden, G., 2002, Uncovering the new wireless interaction paradigm, *Interactions,* **9** (1), pp. 17 – 24.

Tyler, C., 2001, The elicitation and assessment of gender differences in requirements for mobile gaming. *MSc. Thesis Information Technology,* (Department of Human Sciences, Loughborough University, UK).

Emotional intelligence in interactive systems

Antonella De Angeli, Graham I Johnson
Advanced Technology and Research, UK

INTRODUCTION

Since Gardner (1983) challenged the widely held notion that intelligence is a unitary capacity for logical reasoning possessed by every individual to a greater or lesser extent, the concept of multiple intelligences has strongly affected psychology, education theories and neuroscience. Recently, this idea has entered computer science, where it has been used to support the establishment of affective computing, or computing that relates to, arises from, and deliberately influences emotions (Picard, 1997).

Affective computing lies on the border of artificial intelligence and anthropomorphic interface design. It aims to enrich artificial intelligence rule-based systems with 'emotional modules' to recognise user emotions and to give machines 'emotions'. The field is witnessing an extraordinary popularity. Wearable computers equipped with physiological sensors and pattern recognition have been developed to receive and interpret signals which are expected to convey the affective state of the user (Picard, 1997). Affective architectures have been proposed to represent and reason with affective control states such as, preferences, desires and emotions. Animated talking heads have been designed to produce synthetic speech and facial expressions which are expected to convey emotional states (Poggi and Pelachaud, 2000).

Product design follows a similar route from functional to affective. Recent trends are towards objects that inspire users, evoke emotions, and even stimulate dreams. Product semantics has been applied to the design of emotionally rich products: the affective response is embedded in the visual features of the product, such as its shape, colour or texture (Demirbilek and Sener, 2000).

Designing affective products is a complex task. Emotions associated with objects are often a blend of various contradictory feelings, which are difficult to interpret. Emotions are mental events; they are dynamic and depend on our relationship with an object in a particular context. Emotions are subjective, as such they depend on the observer's personality, goal, values and attitudes.

The challenge becomes even more difficult when we attempt to create emotionally intelligent systems, or systems that not only induce affective responses but also understand the emotional status of the user and adapt their behaviour accordingly. There is no established methodology yet, which may help us understand and create emotionally adept technology. Work on emotion, though expanding rapidly, still needs to co-ordinate the effort of many disciplines. Psychology investigates how emotions interface with cognition, personality, and social issues. Physiology links emotions to anatomical structures and processes;

sociology investigates how emotions are triggered, interpreted and expressed by group members. Anthropology ties emotions to cultures; philosophy investigates the essential nature and definition. The very same question '*What is an emotion?*' still provokes different and sometimes contradictory answers. Nevertheless, the psychological importance of emotion is widely acknowledged. In contrast to the traditional Western rationalism, a consensus has emerged that emotions augment rather than interfere with cognition. Besides, emotions are considered to be the primary source of motivation: they arouse, sustain and direct human action.

THE EMOTIONAL INTELLIGENCE FRAMEWORK

Emotional intelligence (Salovey *et al.,* 2000) is regarded as the ability to perceive and express emotions, to understand and use them, and to manage emotions so as to foster personal well-being. The concept subsumes Gardner's interpersonal and intrapersonal intelligences in a unique emotional space, so differentiating specific emotional competencies from social ones. According to the Emotional Intelligence Framework (EIF), emotional intelligence encompasses several abilities hierarchically ordered in four branches composed of several sub-skills organised according to their complexity.

1. Identifying emotions
2. Using emotions
3. Understanding emotions
4. Managing emotions.

At the basic level there is the ability to perceive, appraise and express emotions accurately. These are basic information-processing skills in which the relevant information consists of feeling and mood states. Emotions are identified in one's own physical and psychological states as well as in other people and objects. Basic skills also include the ability to express emotions and needs related to these feelings; and evaluating accuracy and honesty in the expression of feelings.

The second branch, *using emotion,* refers to the use of emotions as thinking facilitation. Different emotions induce different information-processing styles, hence emotional states can be harnessed by an individual towards a number of ends, such as stimulating creativity and problem solving.

The third element, *understanding emotion*, concerns essential knowledge about the emotional system. The most fundamental competencies at this level concern the ability to label emotions with words, perceive causes and consequences of emotions, understand how different emotions are related and interpret complex feelings. This knowledge contributes to the fourth branch regarding the *regulation of emotion*, mood maintenance, and mood repair strategies. In order to put knowledge into action, people must develop further competencies. They must be open to feelings, both those that are pleasant and unpleasant. Then, they need to practice and become adept at engaging in behaviours that bring about desired feelings in themselves and in others.

ASSESSING CURRENT AFFECTIVE DESIGN

Our proposal is to employ the EIF as an indicator of technological evolution, to evaluate advances in affective computing, and as an instrument to stimulate further development. Our aim is not to propose a model for Affective Computing, but to raise some neglected questions in a structured way. We are aware that humans are not suitable models for machines, because computing requires formalisation and reductionism, but we need to make sure that such a simplification does not irremediably hamper the target: creating emotionally adept technology.

Most of the research on affective computing has concentrated on the first stage of the EIF taxonomy, and particularly on perceiving user emotion or displaying computer emotion. With regard to perception, two types of sensors have been proposed so far: those that require user intent and those that passively collect data from the user. Typically, devices for communicating affective feedback are intended to help people express their frustration. They can be interface widgets operated by the user or pressure sensitive devices that recognise intentional or even unintentional muscle tension or speech variations. Passive sensors record the user behaviour (e.g., typing) or physiological variations (e.g., blood volume pulse, electromyogram, skin conductivity and respiration rate).

The basic assumption behind the psychophysiological approach is that emotions are biologically grounded and universal. This is true if we address the basic physiological process of emotion. Highest stages of the emotional system are socially grounded. The appraisal of the events, which have generated an emotion, and the relevant norm for behaviour vary as a function of personal variables, such as culture, gender, age or social factors, such as relative power and relationships.

Physiological measurements may tell us that the body is undergoing some physical changes, but the meaning of that change still varies among different people. A possible solution may be training a computer, as has been done in the speech recognition field. However, passive sensing still opens many issues, related to the acceptability of the solution, loss of control and privacy. The basic question is: Do I really want a projector to sense and react to my apprehension before giving a talk, when I am doing my best to hide my emotions from the audience?

People express their feelings through communication and action, but these feelings are always grounded in a social context. Sometimes they need to be ignored. Sensing without intelligence is useless if not dangerous. Information collected in the first stage of a process will hardly be transformed into useful behaviour without further skills. The request for feedback may be a partial solution to the ambiguity of the perception level, and a very anthropomorphic one. If the machine senses a change in the user status, it should first verify that that change is due to an emotional reaction, and that the user wants the issue to be addressed.

Another relatively well-explored area concerns the expression of emotions in the computer. The current approach is very anthropomorphic in nature: talking heads are designed to display even subtle feelings with voice and facial expressions, such as eyebrow and lip movements. This viewpoint is justified by the highly influential media equation paradigm stating that individuals' interactions with computers are fundamentally social and natural, exactly like people-people interactions in real life (Reeves and Nass, 1996). Nevertheless, the design of animated faces based simply on the representation of surface signals (physical movement) does not catch the full complexity of the phenomenon. Cultures vary in the display and interpretation of facial expression of emotion. Individuals vary in the emotional intensity that they attribute to facial expressions, as well as on the inferences they draw from those expressions. Facial emotion appears to be connected to the language a person is speaking with evidence of Hindi-English bilingual people producing different expression according to the language spoken. Again, believable talking heads need further branches of intelligence, to meaningfully co-ordinate their gesture and language.

Verbal language is a powerful channel to express emotion. Our research with chatterbots (De Angeli *et al.*, 2001a) has proved that emotional and social intelligence are instrumental to the believability of artificial creatures. Their absence dissolves the illusion of life, as demonstrated in the following interaction between Alice and an occasional user.

User: *do you have feelings Alice?*
Robot: *as a robot I do not have any human emotion. My apparent emotional responses are simply algorithms designed to make me easier to interact with.*
Client: *I suspect that you do not exist and that you are simply an interface made by some ambitious programmer*

DISCUSSION AND DEVELOPMENT

Affective computing is in its infancy, with most of the proposed solutions covering only the basic skills of emotional intelligence while aiming at the highest emotional competency: emotion management. The inevitable computational reductionism has generated a misunderstanding, assuming that emotions are physical and cognitive, whilst they are actually deeply grounded in social contexts. Also current approaches tend to separate emotional intelligence, from other types of competencies which contribute to 'intelligent behaviour'.

The time is right to develop the field, addressing functional demands and innovative metaphors. We must understand what we are trying to achieve with developing emotionally adept technology. Applying the theory to real user needs may help us to understand how to design these technologies. We need to understand which computer-supported activities may be enhanced by emotional intelligence in the machine and which user emotional needs can be satisfied. If we can sense emotions, we need to know how this knowledge may be used to improve the user experience. We need to differentiate functional context, where we need tools for functional purposes, from social and emotional ones, where we want to exploit the full range of human ability.

Electronic commerce appears a promising domain to experiment with affective computing (De Angeli *et al.*, 2001b). Virtual shops need to evoke positive affect, if they want the customers to return and buy. In this context, we may set a clear goal for affective systems: to *please the user*, enhancing or repairing the user's mood. Research under laboratory conditions indicates that it is possible to elicit target emotions. Of the diverse techniques available, films are the most effective in eliciting common emotions among different people. This has been attributed to the fact that films have a relatively high degree of ecological validity. Emotions are also generated by dynamic auditory and visual stimuli. Computer interfaces can be regarded as a collection of snap-shots, composed of diverse visual and auditory stimuli designed to directly engage the user into the interaction

Affective computing is still following a hard-computational approach. The possibility of success may be increased by a softer view based also on knowledge of psychology. A psychological perspective, as the one presented here may help us to understand the nature of emotional intelligence in interactive systems and evaluating what emotions are expected and accepted in a machine.

REFERENCES

De Angeli, A., Johnson, G.I., and Coventry, L., 2001a, The unfriendly user: Exploring social reactions to chatterbots. In *Proceedings of the International Conference on Affective Human Factor Design,* edited by M.G. Helander, H.M. Kalid, and Po T. Ming, (London: ASEAN Academic Press), pp. 467-474.

De Angeli, A., Lynch, P. and Johnson, G.I., 2001b, Personifying the e-market: A framework for social agents. In *Human-Computer Interaction INTERACT'01,* edited by M. Hirose, (Amsterdam: IOS Press), pp. 198-205.

Demirbilek, O. and Sener, B., 2001, A design language for products: Designing for happiness. In *Proceedings of the International Conference on Affective Human Factor Design,* edited by M.G. Helander, H.M. Kalid, and Po T. Ming, (London: ASEAN Academic Press), pp. 19-23.

Gardner, H., 1983, *Frames of Mind: Multiple Intelligences*, (New York: Basic Books).

Picard, R., 1997, *Affective Computing*, (Cambridge, Mass: The MIT Press).

Reeves, B. and Nass, C., 1996, *The Media Equation*, (New York: Cambridge University Press).

Poggi, I. and Pelachaud, C., 2000, Emotional meaning and expression in animated faces. In *Affective Interactions*, edited by A. Paiva, (Berlin: Springer), pp. 182-195.

Salovey, P., Bedell, B.T., Detweiler, J.B. and Mayer, J.D., 2000, Current directions in emotional intelligence research. In *Handbook of Emotions*, 2nd ed., edited by M. Lewis and J.M. Haviland-Jones, (New York: The Guilford Press), pp. 504-520.

The use of social representations in product design

Andrew J Taylor
Loughborough University, UK

Valerie M Taylor
Warwickshire Educational Psychology Service, UK

INTRODUCTION

The need for product designers to understand the emotive effects of their products raises many issues, which are not amenable to straightforward research and analysis. Good design requires a more holistic understanding of people than is provided by current usability-based approaches (Jordan, 2000). To move towards this, much information can be drawn from other disciplines including art, marketing, ergonomics, sociology and psychology but, due to the eclectic nature of the field relevant information and ideas need to be distilled and made available to designers in an accessible and concise format if they are to be useful in practice (Taylor *et al.* 1999). There is no doubt that the values and meanings carried by images and objects are, to a large extent, socially constructed and this paper proposes that some ideas derived from social psychology are directly relevant to emotive aspects of design.

Social representations are common-sense theories about the world which filter into society from a variety of sources. They emerge by means of everyday conversations, through social and cultural interactions and of course are strongly influenced by the various media (Potter, 1996). Social Representation Theory has been used by psychologists to help in the understanding of human social behaviour.

This paper argues that products also play an essential role in forming social representations, as well as needing to fit in with existing ones, and proposes that an understanding of the basic principles and theories will map onto, and help to inform, the design process.

THE THEORY OF SOCIAL REPRESENTATIONS

Psychology is a complex field, characterised by diversity and there are multiple approaches available for understanding the nature of human behaviour in its social context. The development of an individual's identity or sense of self is also complex since it is influenced by subjective experience including consciousness, private individuality, agency and cognition. Furthermore, our individual identities are tied up with our social identities i.e. the roles and characteristics which are developed and are perceived by others in various social settings. It may be argued that products can play a significant role in these constructions.

Within the overall discipline of psychology social psychology is unique in that it deals with several different domains of analysis: the individual or intrapersonal, the interpersonal, the group and societal levels. Societal explanations are concerned with interactions between individuals and how these are influenced or constrained by cultural expectations or ideologies.

Societal attitudes differ from the personally-held attitudes which directly shape individuals' actions and behaviours. They are best referred to as social representations - a term coined from Moscovici's theory (1976) based on his post-war research which revealed Freud's psychoanalytic theory to have filtered down from analysts' consultations and journals into French culture and common sense. Moscovici observed that people generally shared a simplified image of psychoanalysis, which differed from what Freud actually propounded and also in many, respects their personally held views. Social representation is therefore seen as a phenomenon created in the 'interactive space' residing between personal attitude and the actual object.

The theory is constructionist in that social representations provide a mechanism for the construction of people's social worlds and are essential to the way in which we think, communicate, and develop a sense of identity. These representations, often described as 'floating' in society, can be seen operating in three different dimensions – the individual level, interpersonal level and societal level, thus providing a common framework for people to communicate, define themselves and make sense of their shared worlds. As well as being generated through everyday conversations, great emphasis is also placed on the media's role in their construction, validation and dissemination. Social representations are not static - historical, social and cultural factors all contribute to their shaping and reshaping e.g. the initial representation of AIDS as a 'gay plague' became modified over time through media influence.

Attitude research is commonly used to gain insight into people's opinions and feelings about any particular topic and consists of two relevant entities:

In contrast social representation theory presents a three-part model:

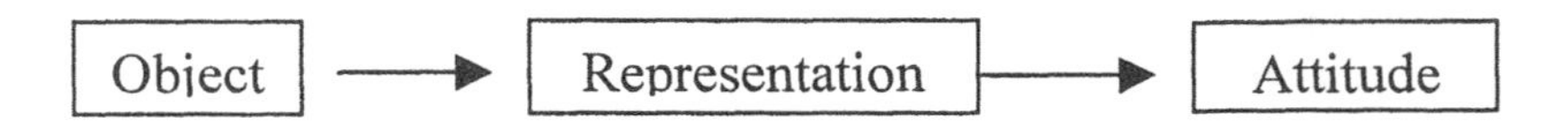

The social representation sits between the object and the attitude to act as a means of making sense of the object. The attitude relates to the image or perception of the object rather than to its actual content. The object may be anything – a political party or movement, a football team, a type of car – anything, which gives rise to opinions or attitudes.

A social representation can be made up of a mixture of concepts or ideas and images, which reside in people's minds and circulate in society. They are usually built around a figurative nucleus of a core image or picture.

Moscovici (1984) suggests that there are at least three senses in which social representations are truly social:

- they are generated in communication in terms of individuals developing shared views of the world;
- they provide a 'common currency for communication' and 'stable versions of the world' (Potter, 1996 p140) and
- they act as a means of homogenising social groups.

THE RELEVANCE TO PRODUCT DESIGN

Industrial design involves the transformation of abstract concepts into tangible entities. In this respect, there is clearly a parallel with social representation theory.

As designers, we must put ourselves in the position of other people, people often very different from ourselves. This is not easy, since there is a natural tendency to design for us.

The meanings and representations of everyday life are the outcomes of jointly constructed processes between image producers and their audience, since they rely on shared implicit 'short-hand' or 'referent systems' such as the stereotypical images found in many TV commercials or visual advertisements. Similarly, products themselves carry meanings, particularly in today's consumer society, and designers must match the expressive qualities of the products to the receptive characteristics of the user.

Social representations also allow individuals to make sense of something that is potentially unfamiliar as well as evaluating it. New products coming onto the market must be understood and evaluated, and a fine balance may need to be struck between the novel and the familiar aspects of the design.

ANCHORING AND OBJECTIFICATION

Potter (1996) identifies two processes involved in forming social representations - 'anchoring' and 'objectification'

Anchoring is a categorisation or naming process, which helps us to make sense of something by linking it into a familiar sphere of knowledge. It involves assigning representations to a familiar category.

Objectification involves the transformation of an abstract concept into a tangible object or entity i.e. the new concept is transformed to a visual image of the anchoring representation.

For example representations of highly-processed dairy products are invariably anchored in the countryside and often objectified by fields of cows. The associations conjured up can be thought of as anchors into other spheres which suggest certain ways of making sense of the phenomenon (for example, anchoring health into the natural world).

Many images in the media are intended to objectify social representations, for example a man in an aftershave advert can be considered as an objectification of a social representation of masculinity. Intangible concepts are often objectified in images of people. E.g. war may be objectified by an image of an injured soldier or a refugee.

Implications for product design

The authors propose that similar principles may be applied to products, in that these need to objectify notions and ideas appropriate to the customers and users.

Marketing and design are inseparable, but the crucial link between market information and the design itself is often problematic in practice. Even if a sound understanding of buyers and users has been gained the embodying of this understanding in a new product by the design team is not straightforward and is still something of a 'black art' (Taylor *et al.* 1999).

The well-used example of the Sinclair C5 illustrates the problem; a large amount of money was in fact spent on market research as well as the technology, yet in spite of very professional inputs the product failed very visibly. Applying social representation theory could have raised the question of whether the product identity was anchored into an appropriate sphere, or in fact to any particular sphere. Some of the characteristics may have been perceived as being anchored into domestic products, or even toys, rather than transport. Potential customers for such a radically new product would inevitably feel uncomfortable if they were unable to anchor the concept into a familiar category.

Social representations are not static - historical, social and cultural factors all contribute to their shaping and reshaping and they are very much a product of the period in which they are constructed. Mobile phones were initially anchored in business and commerce, with the product objectifying the 'businessman' of the early 90's. Evolution has moved the product to a broader spectrum of users; for a lone female traveller it may be anchored in a sense of security, while for a teenager the spheres may be fashion and the IT revolution. The mechanisms of anchoring and objectification operate to enable such modifications to take place and to allow absorption of novel representations.

Similarly, constructed meanings associated with phenomena such as gender are also subject to changes. For example, media representations of women a few decades ago were explicitly anchored into the home and family, and in subordinate relations to men, but current media representations show evidence of women being more independent and having more diverse roles. The TV advertisement for the Fiat Punto and the increasing role of female presenters on car and technology programmes illustrate this trend.

The effects of this can be seen in many products, such as car interiors. For example the Jaguar X-type was designed to maintain the Jaguar brand image which traditionally had male connotations, but the new car also had to appeal to women. According to the design project manager, Simon Butterworth, "When it comes to the interior ... the designers tried to assimilate as much Jaguar heritage as possible in terms of design cues, ... An important consideration for the interior designers, though, and one that had not been important in previous models, was that this car was going to appeal to women drivers resulting in a women's product and marketing committee being formed." (Kimberley, 2001)

Anchoring and objectification may be seen to underpin the necessary styling cues of automotive interiors. Chrysler have considered "... restoring more sculpture to interior surfaces and evoking the same emotions and admiration people express for details on classic antique cars. And ... concept cars which feature finely detailed, nautical-style gauges and fine bits of brushed and chromed metal scattered throughout the interior like jewelry." (Winter, 1997)

CONCLUSIONS

Although the ways in which people construct meanings are complex some simple notions exist about the mechanisms, which enable us to transform, abstract concepts into concrete representations. In these there are parallels in what designers do in the course of their work. The principles of 'anchoring' and 'objectification' are proposed as being particularly relevant to the activity of product design.

REFERENCES

Jordan, P.W., 2000, *Designing pleasurable products*, (London: Taylor and Francis)

Kimberley, W., 2001, X marks the spot. In *Automotive Engineer*, Vol 26 No 3 (Professional Engineering Publishing Ltd) pp. 62-65

Moscovici, S., 1976, *La Psychoanalyse: Son Image et Son Public* revised ed., (Paris: Presses Universitaires de France).

Moscovici, S., 1984, The phenomenon of social representations, in *Social Representations* edited by Farr, R.M. and Moscovici, S., (Cambridge: Cambridge University Press).

Potter, J., 1996, Attitudes social representations and discursive psychology in *Identities Groups And Social Issues,* edited by Wetherell, M., (London: Sage Publications).

Taylor, A.J., Roberts, P.H. and Hall, M.J.D., 1999, Understanding person-product relationships – a design perspective. In *Human Factors in Product Design*, edited by Green, W.S. and Jordan, P.W. (London: Taylor and Francis) pp. 218-228.

Winter, D., 1997, Interiors become extroverts. In *Ward's Auto World*, 33, pp. 43-44

On the conceptualisation of emotions and subjective experience

Thomas van Rompay, Paul Hekkert
Delft University of Technology, The Netherlands

INTRODUCTION

Over the years, numerous studies on the role of emotions in a wide variety of research fields have been conducted. Research programs were initiated not only within the field of psychology, but in recent years also increasingly within the field of industrial design. These research projects are based on the assumption that 'good' design should address the wishes and needs of future customers, including their inner emotional experiences. Notwithstanding the successes of design firms, popular for their emotion-eliciting designs, the processes by which products may elicit emotions remain obscure and vague to the human mind. At times it may even seem that these processes are so dynamic and changing, that generalisations with regard to these processes are simply impossible. However, if our aim is to *understand* the way people *experience* emotions, we may have to change our strategies. Rather than treating emotions exclusively as affective feelings potentially coupled to action tendencies, we may also consider studying the conceptualising processes that are at the basis of emotional experience.

In recent times, originating within the field of psycholinguistics, much attention has been devoted to human conceptualisation. Through the study of linguistic expressions, a whole new area of research topics emerged, one of them being the central place of metaphor in human conceptualisation. Whereas previously metaphor was merely conceived of as a stylistic instrument appropriate for poets and artists, the 'new school' came to regard metaphor as the key process to understanding human conceptualisation. As argued and, since then, convincingly shown by Lakoff and Johnson (1980, 1999), metaphor is primarily a matter of thought and can be defined as 'understanding one thing in terms of another' (Lakoff and Johnson, 1999). Taking this definition as a starting point, we will show how the understanding and experiencing of emotions may be structured by metaphor as well. That is, in line with our definition, how emotions are understood and experienced in terms of something else. The process of metaphor will be discussed based on collages of *anger* and *cosiness*. Collages may lend themselves particularly well to this approach since their purpose is to communicate an, oftentimes, abstract emotion or experience in terms of *something else*.

EMOTION METAPHORS IN VISUAL DISPLAYS

One of the first authors to systematically apply the psycho-linguistic framework to visual displays was Forceville (1996). Forceville convincingly showed how advertisers, in order to create a positive attitude towards a product or service advertised, may consciously or unconsciously *use* metaphor as a means of communication. Taking metaphor outside the purely linguistic domain, Forceville's findings should be of special interest to those communicating meaning via visual displays in general.

In this section, two collages made by students of the Department of Industrial Design in Delft will be discussed. The students, as part of a form giving design course, were asked to make collages on *anger* and *cosiness*. We will start our discussion with the *anger* collage (see Fig. 1) based on the verbal metaphors for anger.

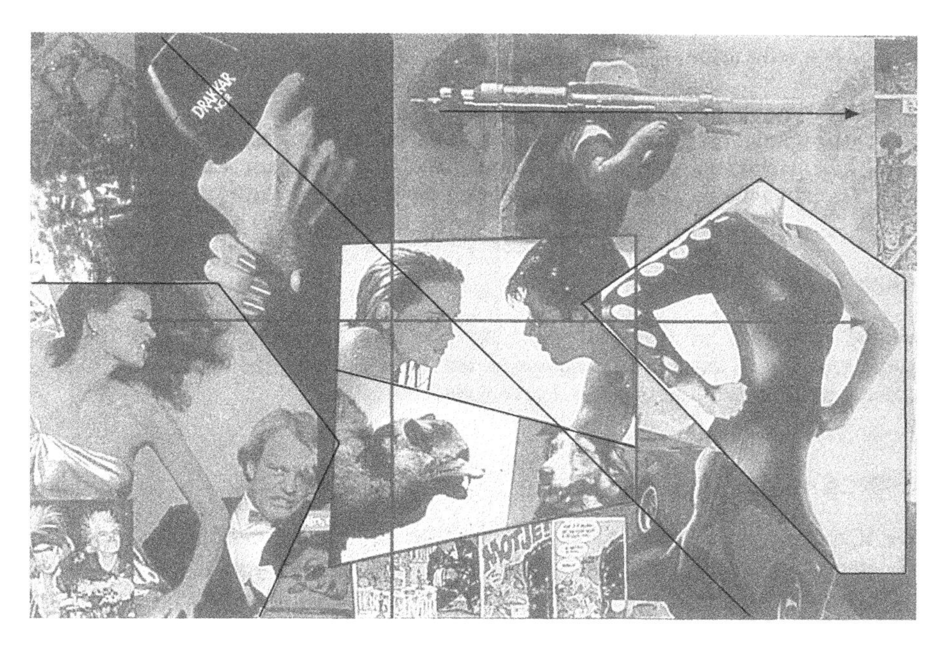

Figure 1 Anger collage

The verbal metaphors will be presented as follows: '*X is Y*', meaning that X is understood in terms of Y. For example, in *Anger is Fire*, *anger* is understood in terms of *fire*. If correct, the same metaphors should structure non-linguistic expressions of anger as well. If so, we should be able to show how these metaphors structure this visual display of anger by referring to its constitutive elements. If metaphor indeed turns out to be medium-independent, that is, not only characteristic of language but of non-linguistic expressions as well, one may argue, in line with Lakoff and Johnson, that metaphor is basically a matter of thought, at the basis of all forms of expression. Next, we will try to show why we use the metaphors; we use. In concluding, based on a comparison between two collages, the role of balance in relation to the emotional 'feel' expressed in the collages will be looked into.

ANGER METAPHORS

Anger is the heat of a fluid in a container

Verbal examples of this metaphor are *I had reached the boiling point* and *Anger welled up inside me.* The cartoon (exploding red head, bottom centre), although not clearly visible, is a well-known expression of the *Anger is the heat of a fluid in a container* metaphor. An increase in heat causes the fluid to go up, up to the point at which an explosion may occur. In cartoons this upcoming explosion is a frequently used image to convey anger.

Anger is a wild animal

A verbal example of this metaphor is *He has a monstrous temper*. In its most literal sense, the animal on the left in the centre of the collage is a clear expression of this metaphor. However, one may also recognise this metaphor in the animal-like expressions on the faces of the characters on the left side of the collage.

Anger is fire

Verbal examples of this metaphor are *He was breathing fire* and *He was consummated by anger.* In addition to the fire (top left), the colour red, clearly present in the collage signals the *Anger is fire* metaphor.

Anger is an opponent

We frequently use this metaphor as expressions like *He was battling his anger* and *He surrendered to his anger* indicate. Interestingly enough, in the centre of the collage we notice two pairs of potential opponents (notice the opposing positioning of the figures): the male, the female figure, the wild, and the domesticated animal.

Of course, the collage discussed merely serves as an example of how anger may be expressed in a visual display. Even so, the verbal metaphors discussed indeed seem to structure the expression of anger in this collage. We may ask ourselves then, why do we use the metaphors discussed and not others? As argued by Lakoff and Johnson (1980, 1999), metaphors are oftentimes based on bodily experiences. If correct, what are the bodily experiences motivating the use of the anger metaphors discussed above?

In discussing emotions, four emotion components (Frijda, 1986) may be distinguished: the subjective feeling or awareness of the emotion one's in (I'm angry), the internal physiological processes accompanying the emotion (increased muscular and blood pressure), an expressive component (red face, shaking hands), and an action component (I'll kill him!). Metaphors seem to be used primarily in the expression and communication of the first three components (Fainsilber and Ortony, 1987). For example *I felt anger taking control over me* is a metaphor used to express the subjective feeling component. *I almost exploded* refers to the physiological component while *If looks could kill* refers to the expressive component. The physiological and expressive components refer most clearly to the bodily experiences accompanying anger: increased internal pressure (blood pressure, muscular pressure), agitation, redness, and interference with accurate perception (Frijda 1986). The increased muscular and blood pressure motivate the 'Anger is the heat of a fluid inside a container' metaphor while redness motivates the 'Anger is fire' metaphor. In other words, we use most of the anger metaphors we do because they 'reflect' the bodily sensations accompanying anger. Interestingly enough, in expressing or communicating the fourth 'action' component of emotions, people 'use' fewer metaphors in comparison to the other components discussed above (Fainsilber and Ortony, 1987). This may relate to the fact that actions are far more concrete, and therefore less difficult to communicate, in comparison to the more abstract other components discussed. In concluding, we will discuss another element structuring the expression of a large variety of emotions: *balance*. Based on a comparison between two collages, the concept of balance will be looked into.

Balance

The concept of balance will be discussed in relation to the collages presented in Figure 1 and Figure 2.

Figure 2 Cosiness collage

As will be shown, the presence or absence of balance in visual displays is directly related to the subjective feel of the emotion expressed. That is, *balance* is fundamental in the experience and conceptualisation of subjective experience, including the realm of emotions.

Anger is basically the result of a perceived injustice, that is, anger is oftentimes experienced as a disruption of balance. Before the offending event, the situation between the wrongdoer and the victim is balanced. There are no *tensions* between the people involved. The offending event, causing the anger, disrupts the balance and creates *tension*. In the case of anger we may indeed experience anger as a disruption of balance as the expression *Don't get mad, get even* indicates. Anger may fade out by itself, although frequently anger *asks* for an act of retribution so as to restore the balance. This way, things get back to normal and tensions disappear. Cosiness on the other hand is experienced as balanced and by consequence, as lacking tension.

As argued by Arnheim (1954) two properties of a visual display have a particular influence on the perception of balance: *weight* and *direction*. The weight of an element in a visual display is dependent on its location. That is, a heavy element in the centre looks 'lighter' in comparison to one further away from the centre of the display. Of course, aspects like colouring, spatial depth and shaping influence the perceived weight of an element as well.

With regard to direction, a visual display is perceived as balanced when the forces constituting the display compensate each other resulting in perceived stillness and, by consequence, lack of direction and movement.

Inspecting the *anger* and *cosiness* collage with regard to weight and direction is particularly revealing. That is, the left part of the anger collage is perceived as 'heavier' in relation to the right part. As discussed, the further removed from the centre, the 'heavier' the element seems to be, and the more a compensatory element is needed on the other side of the display to restore the balance. Although, the centre of the collage may be perceived as equally heavy, the right part of the collage seems to lack such a balance-restoring element. On the other hand, the 'heaviest' elements in the *cosiness* collage are located slightly below the geometrical centre. Furthermore, the centre of the display is accentuated by the fact that the borders of the plane are left unused. The unequal distribution of weight is one of the reasons for the perceived direction, as indicated by the red arrows, from left to right in the anger collage. Because of the equal distribution of weight, no such direction is perceived in the cosiness collage. In short, the lack of perceived balance in one collage and the presence of balance in the other may be accounted for by the distribution of weight and perceived direction. More important for our purposes, the expressions of balance in the two collages discussed are congruent with the emotional feel expressed.

Clearly, metaphor is characteristic of the way emotions and subjective experience are understood and, as argued likewise by Forceville (1996), can be systematically studied in visual displays. Therefore, metaphor is not only a matter of language but is fundamental to other non-linguistic forms of expression as well. Finally, the fact that the same emotion metaphors are 'used' in different mediums of expression may be related to the bodily experiences underlying the emotions expressed.

REFERENCES

Arnheim, R., 1954, *Art and visual perception,* (Berkeley: University of California Press).

Fainsilber, L. and Ortony, A., 1987, Metaphorical uses of language in the expression of emotions. *Metaphor and Symbolic Activity*, **2**, pp. 239-250.

Forceville, C., 1996, *Pictorial metaphor in advertising*, (London and New York: Routledge).

Frijda, N., 1986, *The emotions*, (Cambridge: Cambridge University Press).

Lakoff, G., 1987, *Women, fire and dangerous things,* (Chicago: University Press).

Lakoff, G. and Johnson, M., 1980, *Metaphors we live by,* (Chicago: University Press).

Lakoff, G. and Johnson, M., 1999, *Philosophy in the flesh,* (Chicago: University Press).

Emotionally rich products: the effect of childhood heroes, comics and cartoon characters

Oya Demirbilek
University of New South Wales, Australia

Bahar Sener
Loughborough University, UK

INTRODUCTION

Everyone can remember at least one childhood hero or some favourite cartoon characters that were dominating our child dreams from as far as the memories can go. These memories certainly form the basis for the long success story of the use of heroes and cartoon characters in media to promote products to the public (Sorfleet, 1995). Employing heroes and cartoon characters as a constant visual theme in products may have many potentials. As these characters appeal to a broad age range, they can mutate with the styles of the times and they can survive longer. A subtler and less visible application of cartoon character elements can be found in successful products. This raises the question whether we are or not pre-conditioned in our aesthetic preferences starting from our childhood.

This paper is a review of relevant theories and research on how childhood memories influence product experience and describes a study in which childhood heroes, cartoon characters, and the attributes of desirable products are explored. It includes sections on positive emotions, human physiognomy and associated emotions, emotional role of products in consumers' life, research on pleasure in product use and the power of comics. The main purpose of this paper is to present an overview of the theoretical framework of the study to widen the debate as to the merit of such research and possible means to improve the research methodology.

POSITIVE EMOTIONS

As Kobayashi and Hara (1996) state, people are showing the highest emotion recognition rate at cartoon-like synthetic faces. This has been proven to be true even in cross-cultural subjects. The ratio of recognition was pointed to be around 90% for six facial expressions: surprise, fear, disgust, anger, happiness, sadness and these can be better expressed visually, rather than textually. According to Goleman (1995) emotions related to *happiness* and *surprise,* prepare the body for different kinds of responses. Happiness induces an increase of activity in a brain centre that inhibits negative feelings and boosts the energy level available. This creates a sensation of general rest, as well as readiness and enthusiasm. Surprise allows a better and larger visual field to get more information about an unexpected event. Enjoyment and surprise are universal core positive emotions in the sense that people in cultures around the world recognise the associated facial expressions (Ekman & Friesen, 1978; Goleman, 1995). Emotions have been represented in Microsoft's online chat software Messenger as simple facial expressions, which are called 'emoticons' (Figure 1). Desmet, Hekkert and Jacobs (2000) developed 'PrEmo' to assess the different emotions induced in users by product appearances (figure 2). They based their research on the fact that cartoon characters are a very good medium to express emotions. This brings us back to our previous hypothesis related to the influence of cartoon characters on our childhood. Most of us have learned to read and write with the help of cartoons, and most of us still remember the cartoon heroes that were populating our dreams and forging our child imagination and games. Can this be the reason why these cartoon characters are so good in expressing emotions?

Figure 1 Microsoft Messenger's 'Emoticons'.

Figure 2 'PrEmo' and emotional expressions.

McCloud (1994) explains that the cartoon characters by themselves are not enough to express emotions. The line that draws them plays an important role in the expression. Lines form different parts and segments but we observe the whole. This allows us to *see* things where they are not, to *perceive* something else or to *remember* something familiar.

If line is a powerful tool for expressing emotions in cartoon characters and comics, three-dimensional products could also use it. Shapes, forms, materials, silhouettes and colour combinations form products. While cartoons and comics were addresses to our vision and perception, products are addresses to almost all of our senses. This gives more opportunities for designers looking at targeting emotional responses. Although nowadays almost everyone is aware of emotions, little is still known about how people emotionally respond to products and about the aspects of design that trigger an emotional response in them. The main question here is about 'How emotions, especially happy ones, are induced by products?'

HUMAN PSYCHOLOGY AND ASSOCIATED FEELINGS/EMOTIONS

Facial expressions, body movement, hand gestures, posture, eye blinking and changes in pupil size are all involved in a physical communication as the means of specific message transfer. These non-verbal expressions are very important cues in attempts to understand emotions and their expression. Our senses and memories are the starting point for any emotional process (Mäkelä, 1999). However, looking back to simple psychology definitions, emotions are not triggered by situations or events, but by our thoughts, beliefs and attitudes about the situations or events.

DESIGNING EMOTIONALLY RICH PRODUCTS

The stories behind products or the stories they remind us are then a very important aspect of the emotion inductors. Products telling 'stories', being associated with memories of some sort or with 'built-in meaning' will then be more successful in inducing emotions in people. People want to buy products that are a part of their life, that share their life and that they can relate to. To illustrate the psychological model of emotion induction, Sony's Aibo is a good example. We all know that dogs are man's best friend and that we can cuddle or get a lot of love from them. There are many stories about heroic dogs and caring, loving, protecting dogs in cartoon characters, books and films to support this background information. This is most probably one of the main reasons why Aibo has been such a successful product. A car called Pod, designed and developed by Sony, based on the emotional attachment one can have with a car, shows emotions and learns from driver experiences (Reuters, 2001). Pod will smile, frown and cry, take one's pulse and measure one's sweat amongst many other things. The emphasis on the form is more on 'cute' rather than robotic. The attributes of cuteness evoke happiness and the affordance of tendency towards protection, and have been widely used in product design (Demirbilek and Sener, 2001).

Products as objects of emotion

Holman (1986) talks about the aficionado effect of products observed among collectors, pet owners, gourmets, or technology enthusiasts. This is because consumers emotionally feel attached to these objects. Valued collectable antique products are the ones either produced by a disappeared artist or company, or that were designed in a fashionable way for the period, or again in a specific material related to the historical time. Most of these items have the stories of their previous owners embedded into them. Rhea's (1992) model of people's everyday interactions with products contains four stages describing the development of the consumption. Rhea points out that people always move linearly through them, progressing as long as their product experience remains satisfactory. These stages are:

- ***Life Context*** refers to the background of people, their thoughts, feelings, activities, encompassing beliefs, attitudes, and perceptions. ***Engagement*** refers to the initial interaction with the product.
- ***Cognitive Presence*** distinguishes the product from its competitors, *Attraction,* become interested in the product and ***Communication,*** the product should communicate its key, positive attributes.
- ***Experience*** refers to the period of ownership and use. The product has to be reliable, creating a pleasing experience, meeting expectations, addressing concerns, solving problems, and fitting into people's lives.

- ***Resolution*** refers to the experience of disposing of the product and the way people determine their overall experience with the product. This impression then feeds back into life context and forms a basis for expectations and desires for the next cycle as it is illustrated in Figure 3.

RESEARCH ON PLEASURE IN PRODUCT USE

The present study attempts to understand the use and ownership of products inducing feelings of happiness. It aims to find answers to the following questions:

1. *What were childhood heroes/cartoon characters for two different generations?*
   What are the feelings, emotions, and dreams associated with them?
   What were the stories in which they were involved?
2. *How do people describe happiness?*
   What are happy situations / when do people feel happy?
3. *How do people emotionally respond to product A?*
   What are the attributes and features that make products attractive?,
   What are the aspects of a design that trigger an emotional reaction?

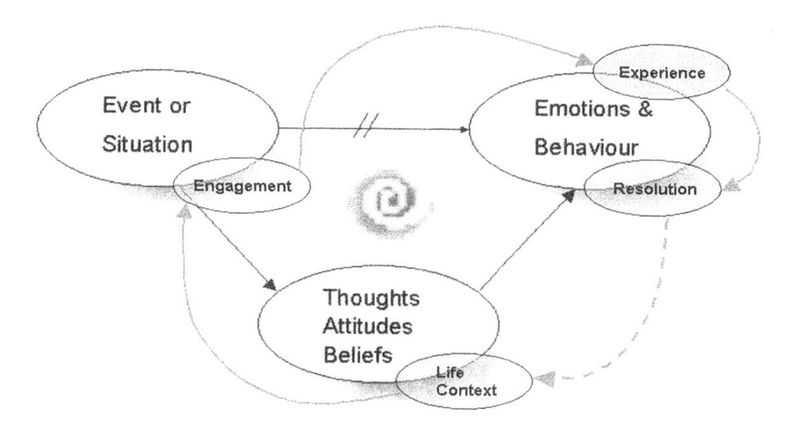

Figure 3 Adaptation of Rhea's consumption cycle to the induction of emotions in our minds.

Research method

In this study, two groups of people aged between 18-22 and 35-40 are involved. Three separate sets of information will be analysed and compared amongst them to see whether there is any correlation or influences between the childhood cartoon characters, the style/the semantic language of products and happiness they induce.

Survey and participatory session

The survey is presented in a graphical format, where people are first asked to identify their childhood heroes on McCloud's pyramid diagram (Figure 4). This is done in order to identify in which distinguished groups of childhood characters and heroes the participants are. Following that, they are asked to identify a cartoon character or a story/context for a list of images of good design award winning products (Demirbilek and Park, 2001). For each product, the participants have to fill some information in the form of:

Does this product remind you of a cartoon character? – If yes, which one?
Does this product remind you of a story/context? – If yes, which one?
Any other things that this product tells you?

The next step is a participatory session where the participants are asked to depict, with the help of collage, the features of the products they are attracted to, in order to identify the qualities and attributes of their preferred products. In the last step, they are asked to describe happiness and the situations in which they remember being happy, again with collage techniques. They have been instructed to describe the best way they can their understanding of happiness and their related emotions. A further step will be looking at if there is any correlation between their attractive product features, their childhood heroes, and their depictions and descriptions of happiness. The information resulting from the present study may give important cues on consumer trends related to product design.

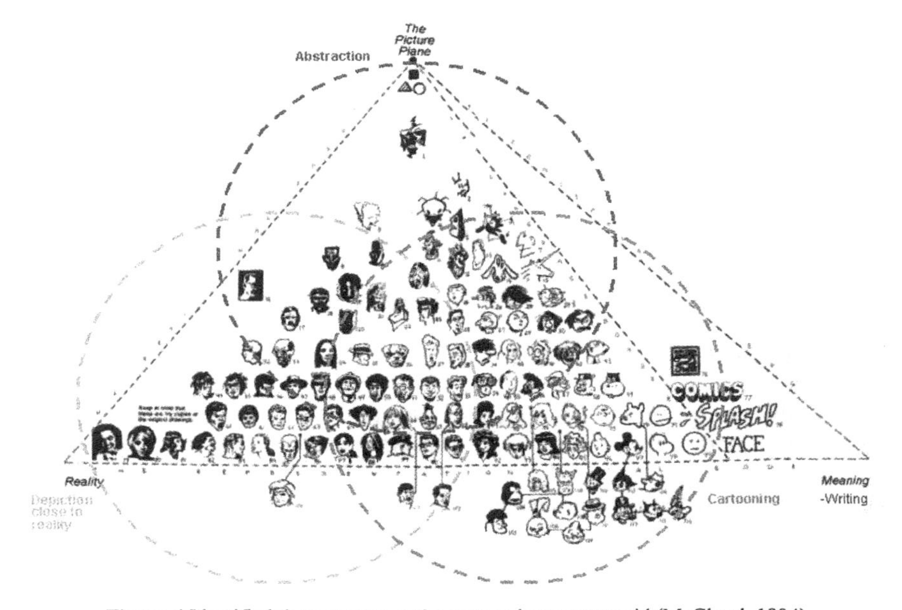

Figure 4 Identified three groups on the cartoon heroes pyramid (McCloud, 1994).

CONCLUSION

Emotional design is an important area of design and design research. So far, there are no direct methods for predicting, evaluating and measuring this human-product phenomenon. Most of the research done on feelings, measured in various ways, has been often done using questionnaires. The present study seeks to look at how people express their feelings of happiness related to products. This is done with a graphical questionnaire and collage sessions, aiming at providing a theoretical statement about which story-character features could be integrated in designed products to achieve positive emotional effects. This paper shows that more knowledge is needed to analyse relationships between emotions, users, and products.

Emotional responses to products seem to vary between different generations, social groups, nationalities and cultures. Design attributes enhancing desired feelings and emotions in products may well be hidden in the childhood period of people, when their main beliefs, values and thoughts are taking shaped. To provide useful guidance on such attributes in the context of product design, as well as a more holistic approach to design, further cross-disciplinary research is required, involving cognitive ergonomics, psychology, product semantics, and design.

REFERENCES

Demirbilek, O. and Sener, B., 2001, A design language for products: Designing for happiness, In *Proceedings of the International Conference on Affective Human Factors Design*, edited by Helander, M.G., Khalid, H.M., and Po, T.M., (London: Asean Academic Press), pp. 19-24.

Demirbilek, O., and Park, M., 2001, A survey of criteria for the assessment of "Good Product Design". *Fourth European Academy of Design Conference*, Aveiro, 10th-12th April, pp. 370-377.

Desmet, P.M.A., Hekkert, P. and Jacobs, J.J., 2000, When a car makes you smile. *Advances in Consumer Research*, **27**, pp. 111-117.

Ekman, P. and Friesen, W.V., 1978, *Facial action coding system, Palo Alto*, (CA: Consulting Psychologists Press).

Goleman, D., 1995, *Emotional intelligence*, (New York: Bantam Books).

Holman, R.H., 1986, Advertising and emotionality. In *The role of affect in consumer behavior*, edited by Peterson, R.A., Hoyer, W.D., and Wilson, W.R., (Lexington MA: Lexington Books), pp. 119-140.

Kobayashi, H. and Hara, F., 1996, Analysis of the neural network recognition characteristics of 6 basic facial expressions. *JSME International Journal*, Series C, **39** (2), pp. 323-331.

Mäkelä, A., 1999, Emotions, user experience and directions for designing emotional rich products, eDesign, http://www.hut.fi/~ahmakela/eDesign.

McCloud, S., 1994, *Understanding comics: The invisible art*, (New York: Harper Perennial).

Reuters, 2001, Tech car carries emotions on its bumper, October 18, 2001.

Rhea, D.K., 1992, A new perspective on design: Focusing on customer experience. *Design Management Journal*, Fall, pp. 40-48.

Sorfleet, A., 1995, Why a Mascot? http://www.advokids.org/style/mascots.html.

AFFECTIVE USABILITY

Feeling your way home: the use of haptic interfaces within cars to make safety pleasurable

Steve J Summerskill, J Mark Porter
Loughborough University, UK

Gary E Burnett
University of Nottingham, UK

INTRODUCTION

At the start of the 21st century the interiors of automobiles are becoming increasingly complex. Navigation systems, communication devices and car component management systems are all being implemented in cars. However, these devices increasingly require the visual and mental attention of the driver, distracting from the primary task of getting safely from A to C via B. Do these devices enhance the user experience, possibly providing a degree of pleasure, or do they add to the pressure that drivers already perceive in the era of traffic jams, speed cameras and road rage?

Consumers in the information age are provided with increasing levels of functionality from the products they interact with. Nokia have produced a mobile phone which has an integrated digital camera and the BMW 7 Series car has over 700 functions controllable from one interactive knob. Technology 'push' assumes that people equate increased functionality with increased satisfaction or pleasure. This remains debatable as inaccessible functionality often leads to frustration. An example of technology 'push' is the implementation of satellite navigation systems in the automotive environment. The availability of a screen based interface in the car has allowed manufacturers such as Lexus to implement screen based functionality for systems other than satellite navigation, such as fan speed and air temperature. These interfaces generally increase the time that is required to access functionality when compared to standard dash mounted controls, increasing the 'eyes off road time'.

Research into distraction related accidents by Kantowitz (1996) has highlighted the need to reduce the time that drivers spend with 'eyes off the road' in order to operate secondary functions such as navigation systems. Currently, certain automotive manufacturers allow access to functions such as navigation destination entry whilst on the move (a complex task requiring the entry of alpha numeric data). The manufacturers, who do not allow access to this type of functionality for safety reasons, receive complaints from their customers stating frustration with the inability to perform certain tasks. This points to the premise that drivers can be overconfident in their ability to interact with such systems safely and are frustrated by the lack of access to all functionality whilst driving.

This paper discusses the benefits that non-visual, haptic (active tactile) feedback may provide. It is anticipated that the ability to operate the secondary functions within a car without looking will result in safer access to functionality. This will result in enhanced user interaction and acceptance of the system as a whole, whilst actually reducing the time that drivers spend with their eyes off the road. An added benefit from the use of haptic feedback is the inherent physical interaction between the car and driver, as the user perceives a direct physical relationship with the product. However, reducing the visual load on the driver is of no use if the cognitive load is increased, as both are capable of distracting the driver from the primary task. The possibility must be acknowledged that using current technology, it may not be safe to allow people to perform the tasks they wish to whilst driving, such as using a mobile phone. The only way to reduce frustration on the part of consumers in this case is to inform them of the risks of being distracted whilst driving. However, the actions of certain manufacturers in allowing access to complex functions on the move may mean that consumers will always resent these functions being removed once they have experienced them in the driving context.

We are currently investigating the above issues in our BIONIC (Blind Operation of In-car Controls) project, which is funded by the EPSRC/DTi Foresight Vehicle Link programme. Honda R&D is the lead industrial partner. The possible benefits of perceived added quality and safety to the driving environment cannot be underestimated in terms of product acceptance and marketability.

THE HIERARCHY OF CONSUMER NEEDS AND FUNCTIONALITY

The hierarchy of consumer needs as proposed by Jordan (2000) is adapted from Maslow's 'hierarchy of human needs', that in-order for a user to gain pleasure from a product, a suitable level of functionally and usability must be in place. A number of examples where functionality within products is not inaccessible to the user due to limitations within the interface have been found. Perhaps the most relevant example to the BIONIC project is that of car radio usage. Van Nes *et al* (1990) performed a study in which a group of participants were asked to use a new car radio for a period of 36 weeks. The users had the radio fitted to their cars and were also given a copy of the user manual.

The frequency of radio function usage was recorded electronically over the period, and post evaluation interviews were performed. Only a third of participants used the manual in any depth resulting in many participants who were unable to use some of the basic functionality of the car radio, causing them much frustration. Designers of in-car systems should therefore use hidden functionality with caution. This is further emphasised by the work of Burns and Evans (2000) who adapted the 'Kano model of product quality'. The model (see Figure 1 below) shows that poor implementation of features can enrage or frustrate consumers. Taking this model further, an unexpected feature, such the ability to send e-mail messages from a car, may provide initial 'delight' to the consumer, but prolonged interaction with a poorly implemented interface is likely to cause frustration which will soon negate any feelings of delight.

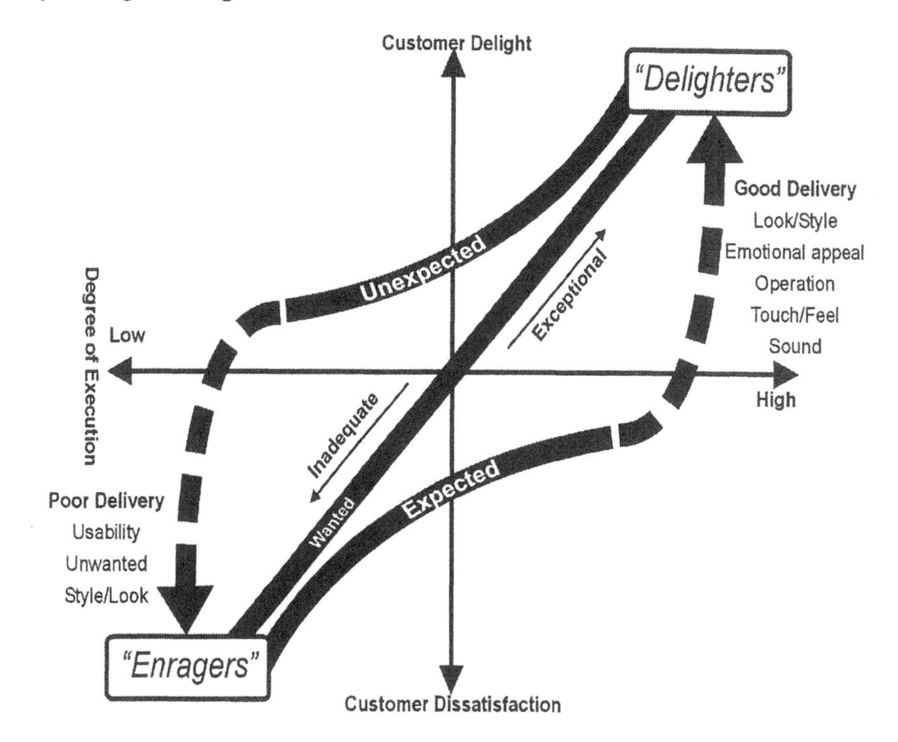

Figure 1 An adapted Kano model of product quality. Burns and Evans (2000).

The driving environment is becoming more complex due to both external and internal pressure. Factors such as traffic congestion, and the desire to communicate with the outside world all add to the pressure of the driving environment. This introduces the possibility that drivers will be distracted by the interior and it 'added functionality', neglecting safe driving by being unaware of the exterior environment. Therefore, the control of the secondary functions must be made as simple and easy to understand as possible. Referring once again to the 'hierarchy of consumer needs', it is the responsibility of the designer/ergonomist to ensure that the included functionality is safe to use in the environment into which it is to be placed.

THE STATUS QUO IN CAR INTERIOR DESIGN

Radios and navigation systems are not the only devices which drivers must interact with. Car manufacturers such as Fiat and Volvo are developing systems which integrate functionality such as SMS messaging, e-mail, telephone, and emergency assistance into 'driver support systems'. Figure 2, below, shows the BMW I-Drive system, with an insert showing the main I-Drive controller and the screen based interface. The I-Drive system integrates all of the secondary functionality (over 700 functions) that is available in the BMW 7 series. The main I-Drive controller is a force feedback device with which the user interacts with screen based information. The benefit of the I-Drive is that all of the functionality available, such as navigation, entertainment, and environmental settings can be controlled from the main force feedback controller. Usage of the I-Drive system is essential if key features such as navigation are to be accessed effectively. However, the system requires frequent and prolonged visual screen interaction in order to be able to complete tasks.

An example of this is illustrated by an evaluation of the I-Drive system performed by the magazine Autocar (Pollard 2002). The time it took to access a number of functions with I-Drive was timed. It took 21.7 seconds to select maximum airflow to the foot wells, whereas this function can be accessed in most cars by the flick of a switch. It therefore seems inevitable that I-Drive will increase 'eyes off road time'. Most people who drive have at sometime used a radio or cassette player whilst driving, and have been surprised by the change in the exterior conditions when the eyes finally refocus on the road, e.g. the car in-front braking.

Figure 2. The BMW I-Drive system

THE USE OF HAPTIC FEEDBACK TO REDUCE EYES OFF ROAD TIME

The use of secondary systems currently depends on visual feedback, whether in the form of screen based information as provided by the I-Drive system, or control positions as can be seen in lower end production cars. An ISO standard (ISO TC22/SC13/WG8) is currently being drafted which will set limits on the eyes off road time for control interactions and will, it is anticipated, exclude many systems that are currently available from in motion use. In order to provide the same level of functionality that drivers are currently experiencing, other feedback methodologies must be developed. The use of the tactile sense is currently under used and is therefore a target for exploration by the BIONIC project.

It is anticipated that providing tactile feedback will provide a pleasurable experience for the driver, excluding some of the anxiety that is inherent with taking the eyes off the road, and provide a more interactive experience between the product and the user. In order to explore the non visual use of controls, experiments have been performed with visually impaired participants. The aim of these experiments has been to determine the tactile cues that visually impaired people use to determine the function of various controls, and to determine how visually impaired people explore unfamiliar electronic products. Performing this experimentation has lead to two of the BIONIC concepts currently being explored.

The first approach involves a design for a dash board, which incorporates fixed locations for control interactions. HCRPs (Hand Control Reference Points) are placed at the centre of control clusters, allowing all of the controls that are associated with a certain function, such as entertainment, to be reached with the hand in one set location. The different functionality of the associated controls will be identified using further tactile cues, such as shape, texture and protrusion levels from the mounted surface, with state of the system information being provided by control position with suitable tactile markers. The aim of this approach is for the driver to be able to reach to a certain location and interact with the control surfaces without having to take the eyes off the road.

The second approach being explored uses a multi function controller in combination with both visual and tactile feedback methodologies. The tactile feedback methodology being explored is the use of active pin matrix displays. These displays provide feedback to the user in the form of symbols, which change to provide information on the current mode, which is being interacted with, in combination with state of the system information. It is anticipated that the user will initially interact with the screen based feedback and will learn to associate the tactile symbols with the screen based information, with the aim of reducing the time needed with 'eyes off the road'.

CONCLUSIONS

The combined influences of standardisation of 'eyes off road time', and research showing how driver distraction causes accidents, are going to change the functionality that is available in the automotive environment. This will produce safer driving packages, but is likely to cause frustration for drivers who are unaware of the dangers that current systems pose. In order for people to be able to use the current level of functionality that is present within cars, interaction methodologies will need to be explored which reduce distraction, but it is perhaps inevitable that tasks such as using a mobile phone will be banned in the near future. The need to take risks is part of human nature and risk taking is for some people part and parcel of the driving experience. Most drivers at some point will go through a red light, or drive with a mobile phone in the crook of their neck. Nevertheless, these activities are dangerous and cost lives. Receiving good, or bad news on a mobile phone, or entering destination information into a car, means that attention is divided and the user is not concentrating on the road, with obvious implications.

REFERENCES

Burns A.D. Evans, S. (2000*) Insights into Customer Delight*. Collaborative Design.

Jordan, P.W. 2000. Designing Pleasurable Products. Taylor and Francis. London.

Kantowitz, B. H.; Hanowski, R. J.; Tijerina, L. 1996. Simulator evaluation of heavy-vehicle driver workload: II: complex secondary tasks. *Human Centered Technology: Key to the Future*. Proceedings. Volume 2. pp. 877-881.

Pollard, T. February 2002. *Does I-Drive Work*. Autocar.

Van Nes, F.L., Van Itegem, J.P.M. 1990. Hidden Functionality: How an Advanced Car Radio Is Really Used. *IPO Annual Progress Report 25*. Institute for Perception Research, Eindhoven, The Netherlands.

Measuring user satisfaction on the web: the stories people tell

Cathy Dudek, Gitte Lindgaard
Carleton University, Canada

INTRODUCTION

The purpose of our research is to gain a better understanding of user satisfaction on the web by using several measurement techniques including an unstructured interview method, which forms the basis of this paper. Edwardson (1998), using a critical incident technique, found that people recall customer service experiences laden with emotion, and they report more than 'cold cognitions'- the fact-based accounts of efficiency and effectiveness. In our experiments, we found that people talked about an exceptionally beautiful web site with a very low level of usability as favourably as those who inspected an appealing web site with a very high level of usability (Lindgaard and Dudek, 2001). However, the exceptionally beautiful site was not perceived as usable also, as evidenced by the large number of negative usability statements, suggesting that the dimensions of aesthetics and perceived usability are orthogonal (Lindgaard and Dudek, 2002a). In contrast to our results, Tractinsky, et al, (2000) claimed that "what is beautiful is usable", when they measured users' evaluations of ATM interfaces with varying levels of usability. In our early studies, users were required to inspect a web site and then take part in an interview. They were not required to complete any tasks. We began to question whether task requirements were the reason for the discrepancy between their findings and ours. In Tractinsky, et al.'s (2000) study, users completed a usability test as well as a questionnaire believed to measure interface 'appeal' - a term they use interchangeably with 'aesthetics' and 'satisfaction'. Here, we are concerned with 'satisfaction' and its relation to perceived usability. In their study, users were asked to complete tasks but were not prevented from completing any of them. In contrast, our usability tests on the high appeal/low usability site, confirmed that it was extremely unusable given that on average only 3.9 of 8 tasks were completed successfully (Lindgaard and Dudek, 2002b). On the high appeal/high usability site results indicate that an average of 6.1 of 8 tasks were completed successfully (Lindgaard and Dudek, 2002c). This paper discusses the results of an experiment examining what people had to say about 1) their level of satisfaction and 2) their perception of the site's usability with or without the requirement to complete a usability test. Our aim was to disentangle the relation between satisfaction and perceived usability. We chose the results from the beautiful but unusable site here to see if user experience varied as a function of task requirements. We compared conditions under which subjects were asked to either: 1) browse the site, then take part in an interview (hereafter referred to as SO), 2) browse the site, complete an interview, do usability tasks, followed by another interview (hereafter BA); or 3) complete usability tasks only followed by an interview (hereafter TO), as shown in Table 1.

Table 1 Group Descriptions

Group	Description
SO	Browsing then Interview (N=20)
BA	(a) Browsing then Interview (before usability test) (b) Usability Test then Interview (N=20)
TO	Usability Test then Interview (N=20)

We hypothesised that (1) SO would be more satisfied with their experience than BA(a) or TO because they would be less concerned with usability. (2) BA(a) would be more satisfied than TO who had a usability test only because they would not have been completely aware of all the usability problems before completing the usability tasks. (3) Within BA, we expected subjects would be less satisfied after the test, than before it because they would have become aware of the usability problems. (4) BA(b) and TO would not vary because both groups completed the same usability test. (5) The proportion of positive usability statements would differ between the conditions. This paper focuses on the stories people tell when faced with different task demands.

METHOD

Sixty participants were recruited from around the university campus and assigned randomly to the groups. All spoke English as a first language and had no prior usability experience. Each interview was transcribed ad verbatim after which statements were categorised. Using five operational definitions, emerging from earlier research (Lindgaard and Dudek, 2001), statements were classified as follows: aesthetics; emotion; likeability; expectation; and usability (Lindgaard and Dudek, 2002b). Aesthetics statements were those that dealt primarily with what the user was seeing on the page. For example, "The graphics are great". Emotion-type statements were those that conveyed that the user felt something about the site or that could reasonably have completed the sentence, "I feel... relaxed, seductive, excited, etc..." or "It was... uplifting, enticing, etc..." Likeability statements were overall judgements about the site, for example, "It was okay", or "It was horrible". Expectation statements indicated how well the web site lived up to preconceived notions, for example, "I expected..." or "I would have thought that..." Finally, usability statements were those that seemed to deal with efficiency and effectiveness, for example, "It was not easy to...", "I couldn't go back...". The statements were sorted for each subject. A list was compiled for every transcript counting all unique satisfaction (aesthetics, likeability, expectation, emotion, and usability) statements in order to rate them. Working together, three judges then rated the statements classifying them as positive, neutral, or negative. This enabled us to go back to each individual interview and assign the appropriate values. Each unique statement was counted only once in an interview transcript because people tended to repeat themselves. The proportion of positive statements was calculated for each subject by dividing the total number of positive statements by the total number of positive and negative (excluding neutral) statements. Usability statements were calculated in the same way for each subject.

RESULTS

The results are presented in Table 2. Our hypotheses were partially supported. SO and BA(a) were equally satisfied and both groups were more satisfied than TO. Participants in BA were more satisfied before the usability test than after it.

Table 2 Satisfaction - Proportion of Positive Statements (POP)

Group	Description	POP	Subjects
SO	Browsing then Interview	0.66	N=20
BA	(a) Browsing then Interview (before usability test) (b) Usability Test then Interview	0.51 0.25	N=20
TO	Usability Test then Interview	0.47	N=20

We hypothesised that subjects in SO would be more satisfied than the other groups and that BA(a) should be more satisfied than TO. To compare the level of satisfaction between these groups, a Oneway ANOVA conducted for proportion of positive statements confirmed that the groups differed significantly ($F(2,57) = 4.180$, $p<.05$). Scheffé post hoc tests revealed that SO was more satisfied than TO ($p<.05$) but not than BA(a) ($p>.05$). To assess whether satisfaction levels changed from before to after the usability test, a t-test for paired samples was conducted for proportion of positive statements for BA (a and b) only. This showed that the user experience declined because the proportion of positive statements was significantly lower after than before the usability test ($t(19)=4.63$, $p<.05$). We conducted a t-test for independent samples to compare the 'after usability test' groups (BA(b) and TO). BA(b) was significantly lower than TO ($t(38)=-3.80$, $p<.001$), disconfirming hypothesis four stating that they would not differ because both groups completed the same tasks. We return to this issue later.

To explore whether evaluations of perceived usability differed between the three groups we conducted a Oneway ANOVA for proportion of positive usability statements (comparing SO, BA(a), TO). Surprisingly, there was no significant difference ($F(2,57)=.35$, $p>.05$). Recall that values below 50% represent a negative experience and above 50%, it is positive. The Grand Mean was 30.1%, which suggests that regardless of task requirement, neither group perceived the site to be usable. That is, approximately 70% of all usability statements were negative because neutral statements were excluded from the calculation. To see whether perceptions of usability changed, a t-test for paired samples conducted for before and after usability test (BA (a) and (b) only) showed that the proportion of positive usability statements declined significantly ($t(19)=2.98$, $p<.001$).

DISCUSSION

With respect to satisfaction, hypothesis 1 stated that subjects in SO would be more satisfied than those in the other groups because no task requirements were placed on them. This was partly confirmed in that the SO group was more satisfied than the TO but not BA(a) group. Hypothesis 2 stated that subjects in BA(a) would be more satisfied than those in TO because they would not yet be completely aware of all the usability problems. However, these two groups did not differ. The data thus failed to support this hypothesis. Hypothesis 3 held that subjects in BA(a and b) would be less satisfied after than before the usability test. This hypothesis was confirmed. There are four possible explanations for this. 1) The site turned out to be less usable than they originally thought, 2) people have a tendency to change their minds when asked the same questions before and after a usability test, 3) users were responding to demand characteristics, or 4) a combination of all three. Further to this, and contrary to hypothesis 4 which held that BA(b) and TO would not differ in satisfaction levels, the two groups did differ. Because the BA(a) group was equally as satisfied as the TO group, and BA(b) was less satisfied than TO, there is reason to believe that asking people the same question twice about the same web site may compel them to change their mind the second time around. Again, BA(b) should not have differed from TO unless the groups were from different populations, but subjects were randomly selected for each group so that should not have been the case. It appears that the users' level of satisfaction, at least as captured in the stories people tell, vary with task demands.

In terms of perceived usability, hypothesis 5 stated that the proportion of positive usability statements should differ between experimental conditions. When we compared SO, BA(a) and TO, satisfaction was unaffected despite the different task demands. However, within BA (a and b only), positive usability statements declined suggesting that encountering severe usability problems accounted for the decline in their level of satisfaction.

This finding allows us to speculate on the reasons for differences in the satisfaction level in the two usability test groups (BA(b) and TO) although they completed the same tasks. Usability statements made by subjects in SO tended to be just as negative as those in the BA(a) and the TO groups (i.e. below 50%), contrary to Tractinsky et al's (2000) findings. However, according to our findings, variations in task demands appeared to have no impact upon the stories people tell about their perception of usability. This suggests that people make a judgement early on usability and they stick to it regardless of task requirements. This appears to be the case unless people are asked the same question more than once.

In short, user satisfaction was lower for the TO group than it was for the SO and BA(a) groups, while perceived usability was the same for these groups. This suggests that the positive aspects of the site (unrelated to usability) receive a disproportionate amount of attention when users are not directly involved in a usability task. Although, when they are attempting to complete a set of tasks, the negative aspects of the experience become the focus of their attention at least when they are asked the same questions twice for the same site. The statements of BA subjects possibly became more negative after the usability test because they chose not to talk about positive aspects, which had already been mentioned in the first interview. When considering the possible explanations proposed earlier, it is reasonable to suggest the effect is a combination of all three: the site may have been less usable than they thought and there may be a general tendency for people to change their minds, and they may have assumed that they were expected to offer a different opinion precisely because they had two interviews. Therefore, poor usability may encourage reason to change their minds, especially if they think that is what they are expected to do.

It is possible to draw two conclusions from these results. First, a satisfying web experience need not always depend upon a usable interface, even when the user is confronted with difficult task requirements. Second, our results indicate that satisfaction and perceived usability are orthogonal dimensions because they did not covary, which is contrary to Tractinsky et al.'s (2000) findings. Users do notice usability problems regardless of whether they are just browsing, are anticipating usability tasks or they have already completed such tasks. This strongly suggests that an appealing interface does not directly influence perceptions of usability. However, when users have tasks to complete, their initial level of satisfaction may change even when the usability of the site confirms their expectations.

REFERENCES

Edwardson, M., 1998, Measuring Consumer Emotions in Service Encounters: An Exploratory Analysis. *Australian Journal of Market Research,* **6** (2), p. 34.

Lindgaard, G. and Dudek C., 2001, Is a great experience merely satisfying and does appeal equate high subjective usability?, *Proceedings, Affective Human Factors Design,* edited by M.G. Helander, H.M. Khalid and T. M. Po, (Singapore), p. 379-386.

Lindgaard G. and Dudek, C., 2002a, What is this evasive beast we call user satisfaction? (*Interacting with Computers – In press, accepted August 9, 2002).*

Lindgaard G. and Dudek, C., 2002b, User Satisfaction, Aesthetics and Usability: Beyond Reductionism, In *Proceedings, International Federation of Information Processing (IFIP2002)*, edited by J. Hammond, T. Gross, and J. Wesson, (Montreal), pp 231-246.

Lindgaard, G. and Dudek, C., 2002c, Aesthetic Appeal versus Usability: Implications for User Satisfaction. *To appear in Proceedings Human Factors (HF2002),* (Melbourne).

Tractinsky, N., Katz, A., and Ikar, D., 2000, What is beautiful is usable. *Interacting With Computers*, **13**, pp. 127 – 145.

Towards an understanding of pleasure in product design

Samantha Porter
Coventry School of Art and Design, UK

Shayal Chhibber
Coventry School of Art and Design, UK

J Mark Porter
Loughborough University, UK

INTRODUCTION

> "More and more people buy objects for intellectual and spiritual nourishment. People do not buy my coffee makers, kettles and lemon squeezers because they need to make coffee, to boil water, or to squeeze lemons, but for other reasons." Alberto Alessi

An appreciation of the hedonic and pleasure dimension within product use is fast becoming of primary importance to the design industry and consequently to the consumer. The leading Dutch electronics manufacturer Phillips has been a leading proponent of a more holistic design approach through the work of Jordan (2000) and Marzano (1998).

Consumers' expectations have been raised. Sawhney (personal communication, 2000) suggests that consumers demand functionality, expect usability and are seeking products that elicit other feelings such as pleasure. The aesthetic of a product, the feel of the material, the tactile response of controls and so on may influence these emotions. More abstract feelings such as reflected status or implied ideology may be important for some consumers. Products can make us happy, proud and enrich our lives. It is traditionally 'usability' that has been the key issue in product design for the ergonomist; it has become one of the critical factors in ensuring the commercial success of a product. It has resulted in the growth of ergonomics as a discipline and is recognised in the international standard ISO DIS 9241-11. However, the traditional approach is inappropriate to satisfy this increased consumer need.

A NEW APPROACH

Jordan (2000) adapted Maslow's (1970) well known hierarchy of needs, to characterise a new approach, creating a hierarchy of consumer needs.

Pleasure
Usability
Functionality

Figure 1 Jordan's hierarchy of consumer needs

The lower levels concern the functionality and usability of the product. If they are satisfied the user will look to the top of the hierarchy for for pleasure.

The four pleasures

Jordan (1997) developed his model further by taking Tiger's (1992) conceptual framework of 'four pleasures' and adapting it to relate to product design.
Physio-pleasure
These are pleasures derived from the sensory organs such as touch and smell as well as sensual/sexual pleasure, e.g., the tactile sensation from using controls or the olfactory sensation from the smell of a new car.
Socio-pleasure
This is concerned with pleasure gained from interaction with others. This may be a 'talking point' product, e.g., a special ornament or painting. Alternatively, the product may be the focus of a social gathering, e.g., a vending machine or coffee machine. This pleasure can also be a product that represents a social grouping, e.g., a particular style of clothing that gives a person a social identity
Psycho-pleasure
This pleasure is closely related to product usability, and is the feeling of satisfaction formed when a task is successfully completed and the extent to which the product makes that task more pleasurable, e.g., the interface of an ATM that is quicker and simpler to use.
Ideo-pleasure
This is the most abstract pleasure and refers to the pleasure derived from entities such as books, art and music. In terms of products, it is the values that a product embodies, e.g., a product that is made of eco-friendly materials and processes that conveys a sense of environmental responsibility to the user.
It is within this framework that this research was conducted.

AIMS

The main aims of this study were:

1. To identify which characteristics (if any), of the products people own or wish to own, bring pleasure and to evaluate how these perceived characteristics change with the age and gender of the product user.
2. To analyse the findings within the context of *Jordan's theoretical framework.*
3. To utilise the data to begin to construct a 'pleasure concept' that provides a clearer picture to the pleasures gained by product users.
4. To consider how this information might be practically communicated to, and used by, designers.

METHOD

Two approaches to data collection were explored; a free format interview where participants talked about the pleasure derived from their favourite products and a questionnaire in which an attempt was made to formalise pleasure characteristics in product design.

Despite the advantages of the questionnaire in terms of data analysis, the method was not adopted. The richness of information gained from the interviews was not captured by the questionnaire and subjects were left feeling disenchanted with their participation. Participants were asked to talk about up to 3 products that give them pleasure; each interview was video-taped. A questionnaire was used to collect personal information, to record details of the products that subjects had chosen and a set of statements relating to functionality, usability, and the four pleasures generally. The statements are presented in the result section. They were based upon the characteristics of the *four pleasures* and were piloted using individuals representative of the anticipated participant population. Twenty-four participants were interviewed; 12 male, 12 female split equally between the age categories 20-30, 40-50 and 60-70 years old.

DESIGNER EVALUATIONS OF THE PRODUCT PLEASURE RESOURCE

The video-tapes were transcribed and a set of bullet points derived for each product; the cost, the age, frequency of use and points relating to functionality, usability, and the four pleasures. An image of each product was included. The designer resource was informally evaluated by practising designers through discussion. The response was positive with a great deal of enthusiasm for the images and the insights into aspects of products that give pleasure. The designers taking part in the evaluation also appreciated the 'visual' format of the resource.

QUESTIONNAIRE RESULTS

Functionality

The respondents selected one of four statements:

A I like products to have exactly the functionality I will use

B I like products to have exactly the functionality I will use, plus some functions that I am interested in to explore and evaluate whether they will be useful

C I like products that have more functionality than I will probably use

D I like products that have the maximum available functionality, despite the fact that I will not use many of them

No differences were found between males and females or between age groups for the functionality level statement.

Usability
The respondents selected one of four statements:
A I like products that are simple and easy to use first time; everyone should be able to use them.
B I like products that are challenging to learn, but once learnt are easy to use.
C I like products that continue to challenge me even when I have owned them for a long period of time, e.g., computer game with varying levels of difficulty
D I like products with secrets and hidden features
No differences were found between males and females. 84% of males and 92% of females chose either statement A or statement B with few subjects choosing products with higher levels of interaction. There were some non-significant age difference trends; both the young and the older age groups preferring products with simple and easy usability in contrast with the 40-50 year olds who prefer products that require some degree of learning but are then easy to use.

Physio-pleasure
The respondents selected any number of five statements:
A Colours are very important in my choice of products
B I really like to touch and feel products when I interact with them
C The 'right' kind of sound is very important to me, e.g., the clunk of a car door
D I like products that use materials in an interesting way
E I am interested in the shape and form of the products I use
The main gender differences identified were 50% of males thought that the 'right' sound in use was important as opposed to 8% of females and 92% of females thought that shape and form were important against only 42% of males. The main differences found across age were that the older age group found colour, sound, and unusual use of materials unimportant as compared with the other age groups.

Socio-pleasure
The respondents selected any number of seven statements:
A I like products that demonstrate that I have discerning taste to others
B I like products that demonstrate to other people that I am a successful
C I like products that tell people something about me, e.g., ethnic clothing, sports watch
D I like products that are a talking point amongst my friends
E I like products that are a talking point amongst any group of people
F I like products that allow me to socialise, e.g., picnic hamper
G I like products that a fit into any social context, e.g., dress up/dress down
The main gender difference in this category was males valuing products that display discerning taste to others (58%) much more than females (26%). In general the older age group made many less responses to statements in the socio-pleasure category with all statements except 'allow me to socialise' (50%) being selected by less than 25% of respondents in this category.

Psycho-pleasure
The respondents selected any number of three statements:
A I like products that express my personality

B I like products that allow me to complete tasks easily, making me feel confident

C I like products that operate in a meaningful way to me, e.g., desktop metaphor on a computer

A greater number of females (66%) valued products that reflected their personality than males (42%) and the younger and middle age groups find this much more important than the older age groups. All of the older age group and 75% of the middle age group like products that allow tasks to be done easily and give confidence; it is much less important to the young age group with only 37.5% selecting that statement.

Ideo-pleasure

The respondents agreed or disagreed with the following statement

A I like products that represent an ideology that I believe in, e.g., eco-friendly, fair trade, materialism, decadence, religion

There were no clear differences in response to this statement.

CONCLUSIONS

The need to formulate an original methodology resulted in a highly speculative investigation. It lead to some conclusions about differences across gender and age, in particular males valuing products that display discerning taste to others (58%) much more than females (26%) and the older age group made many less responses to statements in the socio-pleasure category. All of the older age group and a majority of the middle age group like products that allow tasks to be done easily and give confidence; it is much less important to the young age group. The research has clearly demonstrated the need for further research in this area. The response from the designers to the basic resource produced supports the further development of a tool for designers.

REFERENCES

ISO DIS 9241-11, Ergonomics requirements for office work with visual display terminals (VDT's) - Part 11: Framework for describing usability in terms of user based measures.

Jordan, P.W., 2000, *Designing pleasurable products*, (London: Taylor and Francis).

Maslow, A., 1970, *Motivation and personality* (2nd ed.), (Harper and Row).

Tiger, L., 1992, *The pursuit of pleasure*. Little, (Boston: Brown and Company).

Does usability = attractiveness?

Martin Maguire
Loughborough University, UK

INTRODUCTION

Consumers are attracted to electronic products for a variety of reasons. The style and functions of the product are clearly important to many. But how important is the ease of use (or usability) of a product to the consumer, and how does this trade off against its style and function? In addition, does usability contribute to a product's attractiveness?

Of course, the perception of usability may also change over time. As the consumer starts using the product, its attractiveness will increase if the product is easy to use and meets their needs. Conversely, a product that is initially considered attractive will quickly lose that appeal if its operation is inconvenient or frustrating.

It has been reported that consumers in the UK are spending ever-increasing sums on electronic gadgets, and research by the trends analyst Verdict suggests that this will be the fastest-growing sector of retail spending over the next five years, overtaking DIY (Daily Mail, 2002). In this competitive market, companies will want to understand more about how to make their goods attractive to potential buyers.

Jordan (2001) argues that Human Factors, as a discipline and as a profession, needs to address the pleasure-based aspects of products if consumers are to reap the full benefit of product ownership. This paper reports on a study carried out to explore the relationship between usability and attractiveness for consumer products.

SURVEY ON PRODUCT LIKES AND DISLIKES

A questionnaire-based survey was conducted to explore what characteristics users liked and disliked about particular products they owned. The aim was to determine whether usability features were important in a consumer's liking or disliking of a product. The survey was administered to 116 people consisting of students, university staff, other professionals, and members of the public, recruited via Loughborough University contacts and the local newspaper. The group were of mixed ages with a predominance of younger users. The numbers in each age group were: 69 between 18 and 30, 42 between 31and 50, and 4 who were 51 or above. Each person was asked to complete the survey. They were asked to think of one household electronic product that they used and to list:

(a) up to three features that they particularly liked about it,
(b) up to three features that they disliked about it.

They then categorised each of the features they listed, indicating whether they considered them as related to: Style, Function, Usability, Cost, or a separate category of 'Other'.

RESULTS FROM SURVEY

The respondents rated a wide range of product types including mobile and cordless phones, personal digital assistants, game consoles, minidisk players, digitals or standard cameras, vacuum cleaners, guitar amps, microwave ovens, alarm clocks, washer/dryers, gas cookers, PC laptops, stereo or music/radio systems, personal stereos, typewriters, televisions, TV handsets, shavers, electric fan heaters, steam irons, travel irons, hairdryers, clock radios, and childrens' computers.

For the characteristics that respondents *liked* about their product, the frequency for each of the categories were recorded (in Table 1):

Table 1 Frequency of feature categories that users liked about their product.

Category	Frequency
Style	117
Function	118
Usability	67
Cost	8
Other	7

For the characteristics that respondents *disliked* about their product, the following frequencies were recorded (Table 2):

Table 2 Frequency of feature categories that users *disliked* about their product.

Category	Frequency
Style	41
Function	120
Usability	78
Cost	4
Other	26

A comparison of these results is shown in Figure 1. Interestingly, while *style* features rate very strongly in people's liking of the products (117 items recorded), they were much less important in their disliking of them (41 items). For example positive style features for a mobile phone included: 'can change covers', 'blue screen', and 'elegant look'. Negative features were: 'too big', 'the aerial', 'cheap plastic feel' and 'silver finished scratched off'.

It can also be seen that *functions* formed a major part of both users' likes and dislikes about the products (118 and 120 items respectively). Positive functions for a mobile phone included 'predictive text messaging ', 'helps getting in touch with friends', 'name and number phone book', and 'great games'. Negative aspects included 'have to recharge a lot' and 'buzzing sound when on phone'.

Usability, although less important than functions was still a significant factor in users likes (67 items) and dislikes (78 items) Usability features quoted were 'very easy to set time clock', 'fast power up', and 'clear menu structure'. It can be seen from Tables 1 and 2 however show that usability factors are slightly more frequent in peoples' dislikes than likes.

The fourth factor, *cost*, can be seen as of minor importance, as might be expected for a product that a person has owned for some time. However for mobile phones, cost features related typically to the means of paying for calls 'pay as you go' etc.

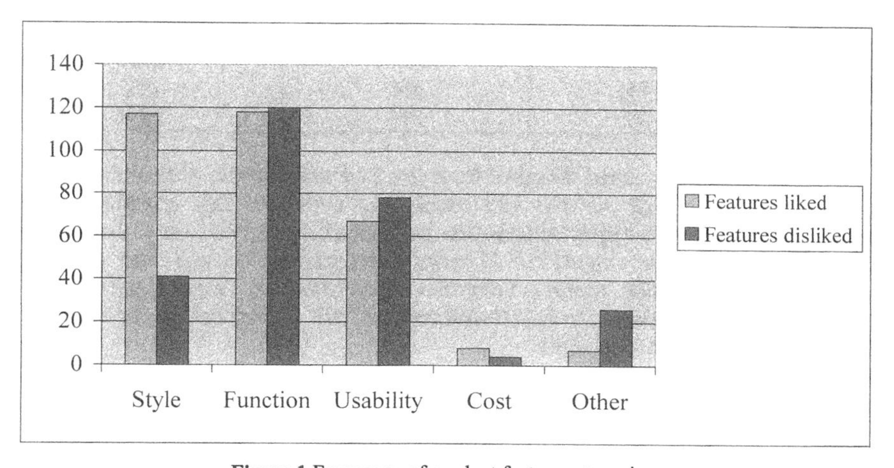

Figure 1 Frequency of product feature categories.

'Other' factors, also played a greater part in disliking a product than in liking it. Positive 'other' factors included general features of products like small size, lightweight, but also factors of safety (for an oven: automatic gas shutoff when the hob cover is down) and flexibility of usage (for learning computer: the child can work on own or with a partner). Negative 'other' factors included aspects of reliability (phone network failure, handset breaking, etc.), safety (an older mobile phone being perceived as generating too much radiation and being a target for thieves), convenience (hard to replace components for an iron), wear and tear (phone handset scratching easily) and cleaning (front door of microwave oven being hard to clean).

Table 3 gives an indication of the importance of the factors that affected peoples' liking and disliking for the products, showing the percentage of features that relate to style, function, usability, cost and 'other'. It can be seen that style seems to play a much stronger role in peoples' liking of a product (37%) than in their disliking of it (15%). To adopt Hertzberg's Two Factor Theory based on motivation at work (1993), style is more of a *motivating* factor, and so its presence stimulates the users' positive feelings, while its absence is less important. While function also plays an important role in people's liking of a product (37%), the lack of functionality or poor functions is a more important factor in people's unhappiness with the product (45%). It is thus both a strong motivator and an even stronger hygiene factor, i.e. its absence stimulates users' negative feelings. On a smaller scale, the presence of good usability is a motivator in product liking (21%) and poor usability is a slightly more important 'hygiene' factor in product dislike (29%).

Table 3 Relative importance of different categories.

Category	Percentage of all features named	*Percentage of features liked*	*Percentage of features disliked*
Style	27%	*37%*	*15%*
Function	41%	*37%*	*45%*
Usability	25%	*21%*	*29%*
Cost	2%	*3%*	*1%*
Other	6%	*2%*	*10%*

Users were also asked whether their level of satisfaction with their products had changed over time. Of the 115 users who answered this question, a clear majority (70) stated that their satisfaction had not changed over time. For 45 users it had changed to some extent. For 17 users, their satisfaction had risen and for 26 users it had fallen. The reasons behind user satisfaction changes are shown in Figure 2. It can be seen that **functional** reasons are prominent in the reason for changes in satisfaction over time.

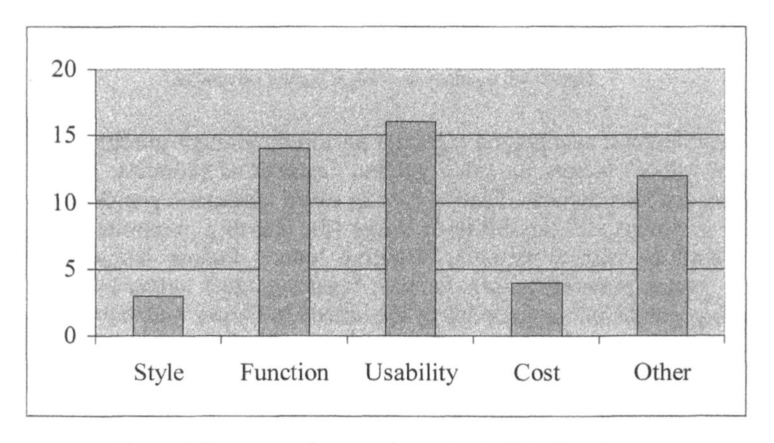

Figure 2 Frequency of reasons for product satisfaction change.

Typical comments were: 'Newer models on the market offering better features' and 'It has come up to scratch and is a good iron for the money I paid for it'. Interestingly **usability** reasons for changes in satisfaction are equally important. Examples were: 'Buttons sticking is very irritating', 'Problems with software have become apparent over time', 'Realisation that 20/20 vision required for most functions' and 'Got used to controlling predictive text better'. This supports the idea that usability continues to be important as the product is used over time and significantly effects users' users continued satisfaction with it (and implicitly its attractiveness to them).

A final question in the survey asked whether, if the person had to buy a replacement product, they would buy the same product again. The responses were that 77 would buy the same (or very similar) again, 27 might buy it again, and 12 would buy something different. This shows that in general respondents felt positively about the product they reported on.

CONCLUSIONS

The study set out to look at the relationship between usability and attraction for electronic consumer products. The results seem to show that both *functions* and *style* are the strongest factors in users' *liking* of products. *Usability* plays an important part in liking of the product but to a smaller extent.

In terms of features that users *dislike* about products, *function* still plays an important role, either because products do not work well, or because other models come onto the market with new features. *Style* seems to play a much smaller role in peoples' dislikes of a product. It may be argued that style will add to the attractiveness of a product if the functionality and usability is right, but will not enhance peoples' feelings towards the product if it is dysfunctional or unusable. After a period of ownership users become accustomed to the product's style, they feel positively or negatively about it.

Lack of *usability* seems much more important than style in terms of peoples' dislikes, and a deficiency in usability has more impact in creating a negative view than the presence of usability has in creating a positive view. Over a period of time when a product is owned and used, *usability* seems to play as strong a role as functionality and satisfaction, *Usability* does seem to contribute to the attractiveness of a product, although functions and style appear more important. Poor usability will diminish user satisfaction with the product, not so much as having poor functionality, but more than inappropriate or unsuitable styling.

ACKNOWLEDGEMENTS

The author would like to thank Wendy Olphert for useful discussions about this paper and Georgina Robertson for her help with data entry.

REFERENCES

Daily Mail, 2002, The big switch to gadgets, 5 March 2002, p 22.

Jordan, P. W., 2001, Pleasure with products - The new human factors. In *User Interface Design for Electronic Appliances*, edited by Baumann, K. and Thomas, B., (London: Taylor and Francis), pp. 303-328.

Herzberg, F., Mausner, B., and Snyderman, B., 1993, *The motivation to work*, (Chichester: Wiley).

Emotional response to food packaging

Adrian Woodcock, George Torrens, Deana McDonagh
Loughborough University, UK

INTRODUCTION

This paper describes a pilot study that focused on the perceptions and performances of individuals when opening foodstuff packaging. Demographics are such that the number of people over 65 will increase by 11% by the year 2011 (National Government Statistics, 2002). The needs of this section of the United Kingdom population are not adequately catered for in the design of food packaging. The aim of this study is to highlight the needs, aspirations, and emotional responses of a sample user group when opening basic packaging.

Food packaging may seem like purely functional products, but they can have an effect on peoples' own perceptions of how they are coping on a day-to-day basis. Eating and preparing food is an activity of daily living. Being able to perform tasks like these with confidence will lead to elderly people being able to pursue more dignified independent lives.

To elicit the views and opinions of senior citizens about packaging, a series of discussion groups were held in a residential home near Loughborough University (UK). The information and results obtained helped to refine a larger study assessing the needs and aspirations of elderly and disabled people within our society. This pilot follows on from a study conducted by one of the authors (Torrens *et al.*, 2001) in which the forces and postures adopted by participants to open vacuum sealed jam jars were recorded. This pilot study showed that the perceived force was much larger than the actual force required to open the jars. This indicated the main problem area was the prehensile grip involving friction between their palmer surfaces of their hands and the packaging. It is thought that people's perceptions play a major role in the packaging they can open and decide to purchase. This highlights the importance of qualitative data supporting the quantitative data, to provide a fuller picture.

Two initial discussion groups of three people, a group of three males, and one of three females, discussed the issues they perceived as being important when purchasing packaged food. Their physical ability was profiled before opening five types of foodstuff packaging and their performance was recorded on video and through a second interview. The perceived ability or inability of the participants to open packaging and their subsequent success in this task performance will be discussed in relation to the packaging usability and the emotional effect on them. This study shows that food packaging can have an impact on peoples well being and personal esteem. At some time, we have all experienced the frustration and annoyance with our inability to open a tightly sealed jar, regardless of our age or physical ability.

The objectives of this study were to:

- Obtain qualitative information about participants perceptions of foodstuff packaging
- Find through discussion what effects food packaging has on an elderly persons feelings
- Observe body language, posture and grip pattern whilst performing task
- Observe and record the emotional response of participants following a task performance

METHOD

The packaging tested included a) ring pull can; b) plastic welded strip (bacon); c) vacuum sealed jar (jam); d) carbonated drinks bottle (water) and e) foil sealed pack (yoghurt).

Qualitative information about participants perceptions of foodstuff packaging

A researcher introduced the topic to the participants with another researcher acting as the scribe. Packaging in general was first discussed to establish any common problems associated with general packaging and to get the ball rolling. Then the packaging to be tested was handed out and discussed. The participants highlighted specific issues with each package type and demonstrated problems they perceived when opening that type. The investigators directed the discussion only when it digressed from the focus of the study. The scribe noted important points and comments made by the participants.

Profile the physical characteristics of the participants

The females were profiled while the males were in the discussion group and vice versa. The profiling included hand measurements (hand length and width measured three times with an anthropmeter). Pinch and grip strengths were also measured, in accordance with previously documented methods (Torrens & Gyi, 1999). These procedures were undertaken to verify that the participants were a relatively representative sample of the elderly population. This data was then compared to the figures in PeopleSize 2000 software (Open Ergonomics, 2000).

Observe body language, posture and grip pattern used to perform task

The participants were then seated in front of a video camera and asked to open each type of packaging. The order in which the packaging was presented was different for each participant so that the order of the packaging did not affect the results. Each participant was asked to give the packaging a rating for ease of use and asked questions about their feelings between opening each package. This provided time to rest in order to negate any effects from muscle fatigue and to let the soft tissues in the hand re-inflate (having applied a high pressure to the packaging) due to blood pressure. Participants' emotional responses, through visual cues, were noted by the scribe during the interview. These were compared with the body language recorded by the video camera. One female did not complete this stage.

RESULTS

Table 1 Summary of qualitative information elicited.

Packaging cited as being bad or hard to open:	Packaging cited as being good or easy to open:
• Shrink wrapped products • Jam jars • Products with sticky tape on them • Cornflakes • Ring pulls • Bottle tops	• Boxed fruit juice cartons, because the tab is large and easy to hold • Washing powder with large perforations • Pills individually wrapped (not in bottle)
Emotions expressed • Frustrated • Fed up • Helplessness • Annoyed • Pride	***Quotes*** *"I Struggle when on my own"* *"I Loose desire to try"* *"Wonderful (if it opens)"* *"Makes me frustrated"* *"There is no need for it to be difficult"*
General comments *"I would not attempt to open it"* *"Marvellous when it works"* *"It looks easy but I feel bad that I can't open it"* *"I always use a tool to open screw lids"*	*"Today every thing revolves around the young who can buy in bulk e.g. buy I get 1 free"* *"I am dependant on other people to open some things"*

Observations

The participant in Figure 1 rated this packaging; 'easy to open'. The observations suggest that this is not the case; at first attempt he cannot lift the ring up so he uses both thumbs to prise it up whilst resting it on his thigh, he then uses a power grip to pull the lid off, the recoil spills some of the contents.

Figure 1 Technique used to open ring pull can.

DISCUSSION

The results table (Table 1) summarises the comments made by the participants, the discussion itself gave the researchers a more holistic view of how the participants felt about the packaging. Initially the participants did not explicitly associate food packaging with how they felt but as the discussion progressed the participants shared stories of how they felt like throwing packaging out of the window. The participants had strong views on how they thought packaging should be. One female participant did not like soap powder boxes. It later emerged that the package was easy to open but the perforations on the cardboard lid made it fall off after opening it a few times. She thought the design was good but badly manufactured, and she felt strongly enough about this to buy a different brand. This suggests that opening packaging is not the only cause of frustration for the elderly user but that, good design and build quality is also important to them.

The male participants were willing to discuss packaging that they felt was bad but were less willing to discuss their emotions when opening packaging. Their body language however did suggest varying degrees of emotion. When some of the participants opened the carbonated drinks bottles they showed surprise and in some cases delight (as can be seen in Figure 2, the picture is obscured for participant confidentiality). There appeared to be no correlation between participants hand dimension, grip or pinch strength and their ability to open the packaging. A much larger sample size is required to test the significance of this result.

The discrepancy between the observations of the participants' body language and their 'easy' ratings is not easily explained (as in Figure 1). This may be due to them not wanting to acknowledge that they have a problem, as this would bring into question their independence (as was evident at the beginning of the discussion group with the male participants). It may also be because they have the ability to open it, so they perceive it to be easy, as they cannot open difficult packaging.

A high proportion of the participants stated their preconceptions of the packaging before opening them, some said they would not be able open it before trying to, this is probably due to past experience of similar products (Norman, 1998). The majority of participants did not attempt to open the packaging in the way that it was designed. They would seek help or use tool like a knife or even the gap in a door. None of the participants opened all the packaging, none could open the bacon, and two of the five did not know that it was possible or even designed to be opened by the tab in the corner. The package instructions caused much discussion and the 'tamper evident' button caused confusion, some of the instructions were 'not helpful' and were perceived as 'irritating'. An example of instructions that caused irritation was "Instructions: open by hand".

All the participants lived in a residential home with a home manager available on site if necessary. The participants said their choice of food was affected by packaging and for example, they tended to purchase loose bacon. If they lived on their own they would worry about their ability to open the packaging. Some would only purchase certain food when they had younger or able visitors, not to entertain them but so that they could have the packaging opened for them.

Figure 2 Participant showing his reaction at being able to open a carbonated drinks bottle after saying; "packaging does not change how I feel".

CONCLUSIONS

- The perception people have of particular packaging means that they do not purchase them for fear of problems with opening.
- Food packaging can have a positive or negative affect on a person's feelings.
- Good packaging can enhance and people's ability to lead lives and therefore maintain their self esteem and dignity.
- The emotion participants suggested via the body language of users differed from the emotion they stated in questionnaire.
- The participants hand strength was often adequate to open the packaging but was not applied to the packaging due to the product semantics and surface features.
- A larger more comprehensive study will be carried out to explore these issues further.

REFERENCES

National Government Statistics, 2002, <http://www.statistics.gov.uk/statbase/mainmenu.asp>

Norman, D.A., 1998, *The design of everyday things*, (London: MIT Press).

Open Ergonomics, 2000, People Size version 2.0, <http://www.openerg.com>

Torrens, G.E. and Gyi, D., 1999, Towards the integrated measurement of hand and object interaction. In *7th International Conference on Product Safety Research*, (Washington D.C.: Consumer Product Safety Commission), pp. 217-226.

Torrens, G.E, Williams, G. and Huxley, R., 2001, "Could you open this jar for me please": A pilot study of the physical nature of jar opening. In *Contemporary Ergonomics* 200, edited by McCabe, P.T., Hanson, M.A. and Robertson, S.A., Taylor and Francis, (London: Taylor & Francis), pp. 73-78.

ATTACHMENT

A perspective on the person-product relationship: attachment and detachment

Özlem Savaş
Middle East Technical University, Turkey

INTRODUCTION

At any time in the history, the human state can be evaluated through the relationship between people and products. Products, as being a system of symbols, constitute the information for understanding human life. They reflect ideas, values, past experiences, living conditions and future expectations of people who make and use them.

Today, the material world that people created has become much more important than at any time in the history. Because, for the first time, the production and consumption activities created a strong threat on the sustainability of our planet, forcing us to re-evaluate our relationships with products. One of the reasons underlying today's high consumption levels is the weakening of emotional link between people and products, through industrial revolution and the introduction of mass production. Eternally Yours distinguishes three dimensions in the product life span: the technical, the economical and the psychological. (Verbeek and Kockelkoren, 1998) A product may be discarded since it can not function as expected any more or since its new forms are introduced to the market. However, today's consumption dynamics reveal that the more significant reason is the termination of emotional connection between individual and product. We see that while some products are impossible to be given up for the user, some are thrown away although they are still functional. In other words, an individual may feel attached to some products and detached from some products.

ATTACHMENT AND DETACHMENT

Dewey (1954) suggests that persons and things constitute the environment with which we engage and create experiences and defines this relation as 'interaction'. Taking the idea of Dewey into the case of products, experience, therefore the meaning of products arises as a consequence of person-product relationship.
Along with their interests, past experiences, future goals, values, ideas, culture, etc., people may respond to different aspects of a product and live different experiences with that product, therefore, give it a meaning in different ways. The emotions felt towards a product are evoked by these meanings and may provide the attachment or cause the detachment.
Several studies formed a basis in search for attachment as a feature of the relationship between people and products. (e.g. Schultz, *et al.*, 1989; Wallendorf

and Arnould, 1988; Ball and Tasaki, 1992; Csikszentmihalyi and Rochberg-Halton, 1981, Dittmar, 1992). By relating attachment to the individual's self concept, these studies suggest that we see some objects more a part of the self and some less. Belk (1988) suggests that our possessions contribute to our identities and also reflect it. For him, recognising that we think of our possessions as parts of ourselves is a key in understanding the meaning of objects.

This study addresses both attachment to and detachment from products and proposes the following definitions. Attachment to a product is a positive emotional state of the relationship between an individual and a product, which indicates a strong linkage between them, and results in considering the product as part of the self with a strong will to keep that product. On the other hand, detachment from a product is a negative emotional state of the relationship between an individual and a product which indicates the lack of linkage between them and results in being unconcern about the product or willingness to discard it.

METHOD OF THE STUDY

The study was conducted through individual interviews by employing a questionnaire, which is a mixture of open-ended and close-ended questions. The sample of the study consisted of 54 respondents selected from people residing in Ankara, by balancing different groups of age, gender and socio-economic status. In two steps, the respondents were asked to identify one of their products to which they feel attached and one to which they feel detached, by restricting them to select an industrially produced product that is to fulfil a function. Continuing to interview through the products selected, they were asked to explain their reasons for feeling such and tell their emotions about these products.

RESULTS OF THE STUDY

Nature of Attachment and Detachment

Specific product vs. product utility. In some cases, product attachment and detachment are constituted towards a specific product itself, a particular product that the individual owns. However, in some cases, product's physical existence as itself is lost; and, attachment and detachment are constructed towards the utility and the experience provided by the product. Results, including both types, revealed that specific products played a more significant role in the constitution of attachment and detachment (64.9 % in attachment and 59.3 % in detachment).

Positive and negative emotions. How an individual feels when s/he uses or thinks about a product? Surely, attachment to a product revealed itself by the positive emotions of individual, and detachment from products by the negative emotions. Emotional responses toward the attachment products were; confidence, independence, care, relaxation, passion, achievement, nostalgia, warmth, security, love, pleasure, satisfaction, proud, being charismatic, being attractive, friendship and comfort. Whereas, feelings towards detachment products included; dislike,

regret, discomfort, disturbance, dissatisfaction, boredom, bad image, distrust, disappointment, failure, stress and wasted money.
Caring and Unconcern. Individuals had a tendency to take care for the product to which they feel attached, with the aim of keeping it for a long time. Contrarily, behaviours towards detachment products were unconcern and will to discard. They were already discarded, given to somebody else, sold, forgotten in somewhere or were being used involuntarily.
The period of having. The average time of having attachment products is 19.8 years. All of them were used almost everyday and are still being used. Whereas, average time of using detachment products is 4.4 years, most of them being used just one or two times. Findings also suggested that as the time passes the shared history between the individual and the product increases the level of attachment.
Acquisition of the product. Many respondents had the tendency to mention how, when and where they bought their attachment products or who made them, with pleasant memories. Contrarily, in case of detachment, most of the respondents were regretful in acquiring those products. Respondents were asked what was their intention on acquiring those products. The answers included; unconscious purchase, superfluous purchase, attractiveness of the product and its advertisements, love at first sight and friends convincing.
The role of instrumental value. The use-related, utilitarian aspects of products play different roles in attachment and detachment. Although the products interested in this study exist for fulfilling a function, it was observed that once the attachment to a specific product is constructed, it may last even when the product looses its ability to perform properly. On the other hand, instrumental value was essential in case of detachment from products. That is, if a product to which the individual feels indifferent does not function properly or becomes technologically aged, s/he detaches from this product.

Reasons of Attachment and Detachment

Respondents identified 14 groups of products and explained their attachments and detachments through various reasons. Product types included; cars, furniture, stereos, TV, computers home appliances, kitchen appliances, washing machines, other home appliances, plates, personal goods, sports equipment and professional equipment.
Considerations of respondents in their attachments and detachments were first classified into subgroups and then into broader categories. Reasons for attachment to products and levels of their mentions are summarised below:
The past (35 %): *family heirloom*, products that were given by a family member; *memories*, products that remind specific persons or events in the past; and, *habitual having*, products that have been possessed and used for a long time.
Experience (40.7%): *enjoyment*, products such as TV's and stereos that provide enjoyment; *frequent actions*, products used in everyday activities in that the individual feels involved; and, *desired feelings*, products that evoke feelings such as independence, confidence, release, etc.

Utilitarian (55%): *usefulness*, the basic utility of products that is associated with need; *performance*, ability of product to fulfil its function properly; and, *job/earnings*, products that are used for job and earning money.
Personal Being (48.1%): *reflection of self*, product's ability to reflect individual's identity and its appropriateness to personality and life style; *symbolisation of values*, products that become the symbols of personal values; personification, products considered as a person, a friend or a family member; *self-made*, products that are constructed by the individuals themselves and mentioned as personal and special; *passion*, individual's intense liking of the product; and, *interests and goals*, products that function as means for performing personal interests or reaching a future goal.
Social Being (25.9%): *social status*, appropriateness of product's qualities or brand to the desired social level; *image*, product's effect in the way a person is seen; *togetherness*, experience of the product with friends or family members; and, *other's opinion*, products that are also liked by other people.
Form (22.2%): *physical elements*, physical description of the product such as its dimensions and colour; and, *style*, visual aspects or the ambience created.
Reasons for detachment from products and levels of their mentions are as follows:
Utilitarian (82.6%): *uselessness*, products that have not any utility for the individual; *performance*, insufficiencies of products such as poor quality, weakness, difficulty in use and inadequate operation; *out of use*, products that can not perform its task any more; and, *cost*, expensiveness of acquisition or manipulation of the product.
Personal being (17.4%): *dislike/boredom*, products that are simply detested; and, *inappropriateness to self*, unsuitability of the product to individual's personality or life style.
Social being (21.7%): *social status*, products that show the individual belonging to an undesirable social class; *image*, an unwanted appearance caused by having that product; and, *other's opinion*, other peoples' dislike.
Form (32.6%): *physical elements* and *style*.
Purchase (28.3%): *superfluous*, products that were bought although they were not needed; and, *expectancies*, disappointment of individual when s/he uses the product.
Environmental (17.4%): *living conditions*, changes in individual's life that effect the need for the product; and, *technological obsolescence*, introduction of product's new versions into the market.

CONCLUSION

The results of this study revealed that individuals feel both attachment and detachment in their relationship with products. Attachments were associated with positive emotional responses, satisfactory and pleasurable experiences during manipulation and care for keeping the product for a long time. Contrarily, detachments were characterised by negative feelings, disagreeable use and unconcern of the product or willingness to discard it.

Attachments and detachments were explained with variety of reasons including both instrumental and symbolic aspects. Individuals had the tendency to associate their relationship with products with their past or present lives, interests, goals, social roles etc. Attachments and detachments were explained in terms of appropriateness of the product to the self-definition of individuals.

It was mentioned that a significant part of the efforts spent on the world's sustainability has been devoted to change today's consumption patterns. The most important product related factor in decreasing consumption levels is to provide economical, technological and psychological longevity of products. The findings of this study suggest that product attachment and detachment have significant implications for the consumption activities, by effecting psychological lifespan of products.

Product attachment, as being characterised by the strong will to keep the product for a long time, has the ability of decreasing consumption amount. This is because, once the attachment to a specific product is constructed, the product is not affected much by the aging of the product or the changes in the market and it survives with the individual. On the other hand, detachment is characterised by the will to discard the product. Detachment products are the ones that are not used but consumed and that are discarded and changed with another one easily. Also, product detachment is associated with the purchase regret which can be defined as the conflict within the individual between the time of purchase and the time of use of the product. Therefore, detachment may be considered as both the reason for and the result of excessive consumption.

REFERENCES

Ball, A. D. and Tasaki L. H., 1992, The Role and Measurement of Attachment in Consumer Behavior. *Journal of Consumer Psychology*, 1(2), pp.155-172.

Belk, R. W., 1988, Possessions and the Extended Self. *Journal of Consumer Research*, 15(September), pp.139-168.

Csikszentmihalyi, M. and Rochberg-Halton E., 1981, *The Meaning of Things: Domestic Symbols and the Self*, (New York: Cambridge University Press).

Dewey, J., 1954, *Art as Experience*, (New York: Capricorn Books).

Dittmar, H., 1992, *The Social Psychology of Material Possessions: To Have Is To Be*, (New York: St. Martin's Press).

Schultz, S. E., Kleine R. E. and Kernan J. B., 1989, 'These Are a Few of My Favorite Things': Toward an Explication of Attachment as a Consumer Behavior Construct. *Advances in Consumer Research*, 16, pp. 359-366.

Verbeek, P. P. and Kockelkoren P., 1998, The Things That Matter. *Design Issues*, 14(3), 28-42.

Wallendorf, M. and Arnould E. J., 1988, 'My Favorite Things': A Cross-Cultural Inquiry into Object Attachment, Possessiveness, and Social Linkage. *Journal of Consumer Research*, 14(March), 531-547

Design's personal effects

Fiona Candy
University of Central Lancashire, UK

PERSONAL INTRODUCTION

My experience as an industrial designer has been acquired in fashion and textiles; the ubiquitous nature of this expressive field has brought opportunities to teach on design, craft and art based courses at undergraduate and postgraduate levels. I am interested in interdisciplinary approaches that use forms of visual communication within the investigative process and in the explanation of outcomes. I made a decision at the outset of this recent project to utilise a method of study and mode of presentation suited to the needs of its content. My research is derived from the individual contributions of many people via the accumulation of hundreds of full colour photographs. These cannot be meaningfully reproduced in this publication; diagrams or figures will not communicate my results. Therefore, I aim to supply here an introduction to the project and an overview of findings. My intention is to report in greater detail on specialised outcomes following the completion of current work in progress. I hope you will be open to this explanation of context and realise that my work takes a considered position in relation to the expression of design and emotion, which makes it necessary for me to use the 'I' word and speak personally.

SPLIT PERSONALITY

A change of teaching circumstances, initially stimulated by the implications of 'lifestyle retailing' (Candy, 2000), supplied a very constructive opportunity to consider my subject from a panoramic, interdisciplinary perspective. Of course, the more I learned the more I discovered how little I knew. My interest in 'fashion' broadened to an interest in 'identity', and to curiosity about the designer's role within its expression. Anxieties about the inherent contradictions between concepts of *personal* identity and those of *mass* production were beginning to unnerve me. I had become sceptical about Marketing and its methods, which are inevitably geared to deliver economies of scale and unhappy with the way that Design looks at itself for inspiration *and* approbation. I realised how little I grasped of the complexities of commercial culture or of its dynamic, interactive relationship with Design. I'd had my head down, enjoying my craft. I did not want to move away from practice, but suddenly, I did not understand enough of what was going on. I found that I felt insecure about why designers (myself included at that time) rarely ask other people what they *like* or want, particularly how they *feel* about design or what they expect from it. Designers so often presume this information from their own experience of the world, in spite of the fact that they are designing for others. They detect the market in shops and at trade fairs; scout out the brands, the catwalks, galleries, street and media, but most often at a distance from their consumers, and usually with an obsessive focus on their own specialist fields. Evidence is compiled using intuitive selections from other designer's work, and relating to their own perception of Design's canon. In commercial practice, this can leave designers' research vulnerable to criticism, particularly from the marketers.

ASKING PERSONAL QUESTIONS

'The consumer is becoming like Lévi-Strauss's "bricoleur" finding and re-inventing, putting together in individual ways from the many items offered' (Shreeve, 2000). Market researchers analyse the inter-complexity of sales and ask questions about attitudes, interests and opinions, aiming constantly to track this consumer 'bricolage'. A massive tangle of information is cross and criss-cross-referenced to identify a range of socio-economic target groupings. But once these have been confirmed by data, they can profoundly influence the way that design, manufacture, advertising and retail are implemented (Turow, 1997). Statistically generated market research looks for stereotypes in patterns of scale and so by its very nature is unlikely to convey aspects of the personal.

At Harvard's 'Mind of the Market Lab' (Anon, 2002), Zaltman has undertaken research that uses visual techniques patented as ZMET to elicit information about a variety of consumer topics. He believes that 'consumers can't tell you what they think because they just don't know'. Their deepest thoughts, the ones that account for their behaviour in the market place, are unconscious' (Eakin, 2002). Other design research organisations, like the Ohio based Sonic Rim consultancy, are also using visual techniques to explore consumer behaviour (Viswanthan, 2001). For the most part, this research is initiated by clients with established brand motivation, where the latent influence of commerciality may convey certain prejudices in terms of furthering our understanding of human experience in the material world.

Could there be a way to supplement (or indeed, to counter) existing market research methods to render them more sensitive to emotion and personal idiosyncrasy, in order to 'retain some of the human complexity that measuring tries to remove'? (Boyle 2001). I wanted to know if designers could learn anything from deciphering the ways that individuals construct identity that is not currently being addressed by commercially sponsored research alone.

FAVOURITE METHOD

I resolved to investigate outward expressions of self-image: the organised set of characteristics that the individual perceives as peculiar to himself/herself. I needed to understand more about the action of design within commercial material culture upon these characteristics (and vice versa) and upon the human self-actualising tendency: 'the pressure to behave and develop - experience oneself – consistently with one's conscious view of what one is' (Maddi, 1989).

So, at the end of 1999, I initiated a simple investigation that I envisaged as action research, to put me in touch with some of the issues needing consideration for future work. I asked thirty people to each photograph favourite (and non-favourite) possessions in response to a set of simple questions. The specification of 'favourite' was selected as I anticipated that personalised interpretations of the term would reveal certain emotional priorities, and elucidate a variety of relationships with possessions. I was influenced in this by an ethnographic study carried out in the 1980's in Chicago, which asked people about the intensity of their relationships with various categories of possessions and goods (Csikszentmihalyi, Rochberg-Halton, 1981). Although fascinated by this work, I was struck by its lack of imagery or visual analysis. If I was aiming to acquire a rich, comprehensive knowledge of the influence of material goods in the formation of personal identity, then this sociological study was open to criticisms similar to those I have made of statistical market research. Seen from the perspective of a designer, their findings represented only part of the picture.

Participants were selected from descriptions given to me by various contacts, where I could see clearly defined perceptions of difference, but no 'scientific' model was established. They were loosely categorised as being from differing gender, age, and professional, cultural, racial and other social groups. Pragmatism led me to inquire about objects that would reflect the specialist three-dimensional design disciplines of the students with whom I work: specifically tableware, fashion, jewellery, and furniture. But more meaningfully, I reasoned that these types of artefacts are commonly used in close proximity to the body and therefore could have a reciprocal relation to concepts of intimacy and the personal. Each participant was asked to include a full-length photograph of the front doors of their homes, as I aimed that in the context of an exhibition, these would have the potential to draw on ideas developed by artists and anthropologists about the links between the house and the body, and how the house can be seen as an extension of the self (Carsten and Hugh-Jones, 1995). I hoped also to bring reference to concepts of 'home', (Rybzsynski, 1988), a place of sanctuary, or 'an extra skin or set of clothes' (Hundertwasser and Restany, 1998). A disposable camera was used as a practical way of recording the choices, as a basic tool common to all regardless of photographic experience. I undertook the exercise myself as a means of identifying with the process.

The photos were collated and presented in a format that allowed each personality of objects to be viewed and analysed as an entity. The project amassed over 700 images, but initially, just fifteen of the collaborations, each made up of twelve images, were formally exhibited, entitled 'HOMEWORK' (Candy, 2001).

UNDERGOING ANALYSIS

A mere written analysis of this work has serious communication problems. The exhibition's visual format manifested a 'gestalt' of people and their possessions. Synthesis was experienced through participation and it is difficult to relay this here, except in broad terms.

Although only images of inanimate things were shown, these facilitated insight into the individual humans who owned them. Anthropologists have observed this phenomenon within many societies and noted that 'such property is identical to the person and may stand for that person in his or her absence' (Miller, 1987). I asked groups of 'eyewitnesses' (designers and non-designers) to visit and assimilate the exhibition, compare and contrast their individual interpretations and to document their experience. These discussions and accounts varied: some took the form of voyeuristic observation and critique and others were more complex and socially questioning, but I was able to evidence that group members readily practiced a form of personality divination and could generate intricate personifications of the sets of objects on display. The sense of the person was so strong that several visitors reported a feeling of prying or intrusion into another's 'personal' space. Each entity of personal possessions displayed a consistent 'own person brand', which was corroborated by accounts from the eyewitnesses.

Content and meaning of these collections of personal effects can be expressed in subtly different languages. They are displays of schemata: 'cognitive maps or units, primarily taking the concrete form of values and social roles' (Maddi, 1989), revealing both personal and shared cultural experience. Such collections are outward signals of aspects of peripheral personality: 'the learned, relatively concrete aspects of personality that develop out of interaction of the personality core and the external, mainly social world' (Maddi, 1989). They can be decoded as expressions of self-actualisation. Cherished objects can be seen as communicative, as 'signs, objectified forms of psychic energy' (Csikszentmihalyi and Rochberg-Halton, 1981), or in relation to the concept of 'habitus' (Bourdieu, 1977), as a means of assessing artifacts and their pivotal role in social reproduction. My questionnaire can be explained as a form of 'personal orientation inventory' (after Shostrom's POI), although in this case documented via a visual channel in order to reveal facets of the complex relationship between people and artefacts.

But, by using design's highly communicative combination of spoken and *visual* language, I take each entity of objects to be composite expressions of identity - *personal style*. In this context, style is a complex, interpersonal phenomenon that encompasses the full range of transactions between people and things: personal, cultural, social and including intrinsic aesthetic quality and emotional experience. Carl Rogers, the founder of the humanist psychology movement, greatly valued creativity and the interpersonal relationship within learning; my research has encouraged me to find parallels between client-centred therapy and a potential style sensitive research method. Referring to his work as a therapist and educator he wrote: 'Experience is, for me, the highest authority. The touchstone of validity is my own experience. No other person's ideas, and none of my own ideas, are as authoritative as my experience. It is to experience that I must return again and again, to discover a closer approximation to truth as it is in the process of becoming in me' (Rogers, 1995). I feel that this statement also supplies and insightful description of design and the creative process.

Can designers learn to transcend themselves and get a real sense of the other person? If they are to become conscious enablers of a profound social process like personalisation, they will need to think and act as experienced 'people people', as well as 'things people'. If we take up on Rogers's commitment to his own experience and to the interpersonal relationship when learning about and with other people, we can begin to see new ways that designers can conduct their own, ongoing interpersonal research. I believe that receptiveness to the personal style of others may bring designers an *empathic* understanding of people's values. To support this, I am engaged in the development of visual tools (or educational aids) that can facilitate interpersonal exchange between designer (or student designer) and user/client, where personal orientation to issues relevant to design can be considered. Such tools will also aim to review testimony of consumer individuality as well as the signs of similarity that other forms of market research aim to record.

I would like to *show* you the results of my investigation; but in that process you may well gain other insights or recount the experience differently to mine. After all, we are individuals, so we may not learn the same things.

REFERENCES

Anon., 2002, Mind of The Market Lab, *Harvard Business School*, US (online) ≤ http://www.hbs.edu/mml/research.html> (Accessed 3 Mar 2002).

Boyle, D., 2001, *The Tyranny of Numbers.* (London: Harper Collins) p.230.

Bourdieu, P., 1977, *Outline of a Theory of Practice.* (Cambridge, UK: Cambridge University Press).

Candy, F., 2000, Lifestyle and Design for Social Identity, In *Re-Inventing Design Education in the University,* edited by C. Swann and E. Young (Curtin University of Technology, Perth,Australia), pp. 132-137.

Candy, F., 2001, *Homework* installation, Victoria Building Foyer Gallery, University of Central Lancashire, UK. 16 Mar – 20 Apr 2002.

Carsten, J. and Hugh-Jones, S. et al, 1995, *About the house: Lévi-Strauss and beyond.* (Cambridge, UK: Cambridge University Press).

Csikszentmihalyi, M. and Rochberg-Halton, E., 1981, *The Meaning of Things: Domestic Symbols* and *the Self,* (Cambridge, UK: Cambridge University Press).

Eakin, E., 2002, Penetrating the Mind by Metaphor, *New York Times*, US (online) <http://www.nytimes.com/2002/02/23/arts/23ZALT.html?ex=1015465751&ei=1&en=9d4> (Accessed 23 Feb 2002).

Hundertwasser, F. and Restany, P., 1998, *Hundertwasser, the Painter King with Five Skins*, US, Llc: Taschen.

Maddi, S., 1989, *Personality Theories: a comparative analysis.* (Homewood, US: Dorsey Press).

Miller, D., 1987, *Material Culture and Mass Consumption,* Oxford: Blackwell.

Shreeve, A., 2000, 'Fashioning Craft', *Consuming Crafts*, Buckinghamshire Chilterns University College, UK. 19 – 21 May 2000.

Turow, J., 1997. 'Breaking up America: the dark side of target marketing', *American Demographics,* Nov 1997, p.51.

Rogers, C., 1995, *On Becoming a Person: A therapist's view of Psychotherapy.* London: Constable, p.23.

Rybzsynski, W.,1988, *Home: a short history of an idea,* London: Heineman.

Viswanthan, V, 2001, Research by Design, *Business World,* (online) <http://www.sonicrim.com/red/us/commune/papers/business_world.pdf> (Accessed 3 May 2002).

Designing consumer-product attachment

Hendrik N J Schifferstein, Ruth Mugge, Paul Hekkert
Delft University of Technology, The Netherlands

INTRODUCTION

With some products, consumers develop an emotional bond. These products are judged to be important to consumers and are often among their favourites. If designers were able to stimulate emotional bonding between consumers and their products, producers might increase the lifespan of products, because consumers might hang on to their products for a longer time, and consumers would be more likely to repair products that no longer functioned properly. Therefore, stimulating consumer-product attachment may result in a decrease of unnecessary waste, in a decreased use of limited resources of energy and materials, and thus may contribute to a more sustainable society. It is unclear, however, why people become attached to products.

This paper reports the results of a questionnaire study that assesses the degree to which consumers are attached to some of their products, and explores which potential determinants influence this degree of attachment (Schifferstein and Pelgrim, 2002). The findings are illustrated with excerpts from additional face-to-face interviews conducted with a number of interesting cases. We discuss the relevance of these outcomes for design practice.

METHOD

Respondents

A mail questionnaire was sent to 200 newly recruited members of a consumer household panel, based on a random sample of the local community. Because one of the products investigated in our survey was a car, only car owners were included in our sample. 161 usable questionnaires were returned in time, a response rate of 80.5 percent. From the 161 respondents, 103 (64%) were males. Ages ranged from 21 to 78 years, with an average age of 51. Almost half of the sample had a higher vocational (29%) or academic (18%) education.

Questionnaires

Four versions of a mail questionnaire were developed that differed only with regard to the target product category: lamps, clocks, cars, and ornaments. Respondents were instructed to choose a product specimen from the specified product category and to answer all questions for this product. The degree of attachment was measured by the five-item consumer-product attachment scale developed by Schifferstein and Pelgrim (2002). This scale defines the degree of consumer-product attachment as the strength of the emotional bond a consumer experiences with a product. The determinants of attachment were assessed using 48 statements related to various possible determinants. Respondents indicated the extent to which they agreed with each statement on a five-point Likert scale (strongly disagree, disagree, neither agree nor disagree, agree, strongly agree).

Case studies

To illustrate the findings of the questionnaire study, 12 individuals from the respondent sample, who differed considerably on important variables, were personally interviewed at home using a semi-structured interview.

RESULTS

Attachment for products and over time

The mean attachment scores were significantly higher for the ornament (3.6) than for the three other products (lamp 2.9, clock 3.2, and car 2.9) (S-N-K test, $p<0.05$). The average degree of attachment plotted as a function of the length of the ownership period showed a decrease in attachment after the first year, although the highest attachment was found for products owned over 20 years (Figure 1). Attachment was significantly higher for old products (> 20 years) than for products acquired more recently (S-N-K test, $p<0.05$). Ratings for new products (<1 year) were not significantly higher than those for products owned 1 to 20 years ($p>0.05$).

Determinants

Responses to the 48 statements related to possible determinants were subjected to an exploratory principle components analysis (PCA). Seven factors were identified: the memories to persons, events, and places carried by the product (12 items, Cronbach's $\alpha=0.92$), the extent to which the product supports the person's self-identity (5 items, $\alpha=0.85$), the product's utility and its ability to make a person independent from others (8 items, $\alpha=0.86$), the life vision it symbolizes, both religious and political (6 items, $\alpha=0.84$), the enjoyment it activates (7 items, $\alpha=0.81$), its market value (3 items, $\alpha=0.90$), and its reliability (3 items, $\alpha=0.67$). The other four factors consisted of one item only and were not used in subsequent analyses.

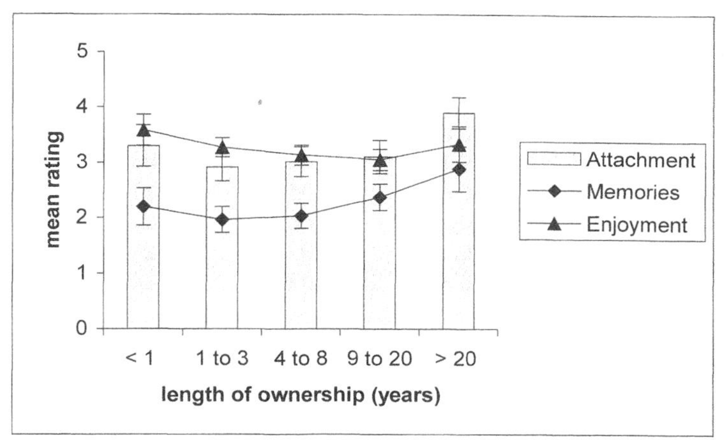

Figure 1 Attachment, memories, and enjoyment as a function of length of ownership.

Prediction of attachment

To investigate the impact of the various possible determinants of attachment, we performed a regression analysis in which the scores on the attachment scale were used as the dependent variable, and the scores on the 7 determinant scales derived from the PCA as the predictor variables (Table 1).

Memories significantly enhance attachment formation: for each product the extent to which a product evokes memories is positively related to the degree of consumer-product attachment. This effect is significant for lamps, clocks, and ornaments. In addition, the extent to which respondents enjoy the product is positively related to the degree of attachment. This effect is only significant for cars. According to the overall analysis, the other variables do not have a significant effect on the degree of attachment. The only possible exception is the product's utility: this variable is negatively related with the degree of attachment for clocks. This negative correlation is possibly due to a reversed causal relationship: when people are attached to a clock, they will keep it even if it does not work anymore.

When the average ratings on the seven possible determinants are plotted as a function of the length of ownership, significant differences are observed for the variables memories ($p<0.01$) and enjoyment ($p<0.05$). The shapes of the curves for these two determinants are roughly similar to the relationship found for attachment (Figure 1). For memories, the ratings for products owned longer than 20 years are significantly higher than for those acquired more recently ($p<0.05$), whereas for enjoyment, products owned for less than one year are enjoyed more than those obtained 4-20 years ago ($p<0.05$). This suggests that enjoyment may be the primary reason why people are attached to new products, whereas memories may be the primary reason why people are attached to a product they have owned for a long time.

Table 1 Regression analysis: Effects of 7 potential determinants on the degree of consumer-product attachment for the four individual products and for the aggregate data set.

Determinant	Lamp		Clock		Car		Ornament		Overall	
Memories	0.60	**	0.62	**	0.48		0.50	**	0.57	**
Self-identity	-0.03		0.16		-0.07		-0.01		0.02	
Utility	0.09		-0.45	*	0.05		-0.27		-0.13	
Life vision	0.03		-0.26		-0.08		-0.02		-0.04	
Enjoyment	0.47		0.44		1.02	**	0.24		0.56	**
Market value	-0.25		0.22		-0.03		0.10		0.06	
Reliability	-0.07		0.02		-0.07		0.00		-0.02	
R^2	0.52	**	0.65	**	0.62	**	0.56	**	0.58	**

**p<0.01, *p<0.05

DISCUSSION

Although the present study is limited by its focus on four product types only, it has demonstrated that feelings of attachment are often related to the memories evoked by the product and to the enjoyment provided by (using) the product. Enjoyment may be particularly important for attachment to new products, whereas memories may be more influential in older products, and may block a disposal decision.

We will illustrate our findings with fragments from the case studies. An example of a case in which memories are important is given by a 49-year old woman, who is attached to a golden pendant: 'I am not wearing it now, but I always keep it close to me. I've had it so long already. It was the first thing I got from my husband 30 years ago, for my birthday. His mother found it much too expensive, but he bought it anyway. I still remember how I felt when he gave it to me.' Attachment due to enjoyment of aesthetic product properties plays an important part in the case of a 49-year old woman, who describes the clock she obtained 4 years ago as follows: 'I find it very nice to look at, because it is quite uncommon, and also because I receive many compliments from others. People just think it's funny. I picked it out myself. It is not a traditional clock. It suits me. I appreciate a nice design now and then.'

The two factors that emerged from this study as important determinants of attachment roughly correspond to two of the three dimensions that Richins (1994) found in a multidimensional scaling analysis of group coincidences for possessions sorted into groups valued for similar reasons. Her first dimension contrasted symbolic possessions with practical objects (memories), the second dimension contained prestige items at one pole versus ordinary possessions at the other pole (status, market value), and the third dimension ran from necessities to recreational objects (enjoyment).

Although several additional factors were investigated, none of these contributed significantly to the prediction of attachment in our study. For example, people do not become attached to products that perform their primary utilitarian function adequately. A 50 year-old woman comments on the lamp she bought one year ago: 'I bought it because I liked it, and because I needed a lamp. It is something you use, and when it breaks down, you buy another one. Maybe you buy the same one or maybe another one, but I would not feel heartbroken.'

If a designer wants people to become attached to his/her product, the present study suggests that s/he should facilitate ways to form associations between the product and people, places or events (memories), or s/he should design a product that evokes enjoyment. As regards the first factor, we note that the memories connected to a product are usually not under the designer's control: memories typically involve an individual's connections to people, places or events that are important to that particular individual. Nevertheless, because many product-related memories are evoked by products that were once received as gifts, one possible strategy might be to develop products that are suitable for gift-giving. However, Kleine *et al.* (1995) found that many possessions received as gifts invoked only weak attachment. For a gift to become a high-attachment product, it should reflect the receiver's personal identity. As a consequence, the successfulness of a gift depends on the giver's capacity to judge what kind of product the receiver would like to have, and this is not under the designer's control.

The second factor suggests that designers should create products that are not only useful, but also enjoyable. This factor encourages designers to evoke sensory and aesthetic pleasure. A corresponding design strategy may start out by evaluating the signals emitted by a product and the corresponding sensations perceived by all the sensory systems (vision, audition, touch, smell, and taste) during usage. The designer should then look for a pleasant combination of ways to stimulate a product user. Part of this strategy entails determining whether the different elements provide a consistent whole, whether they clash together, confuse the consumer, or trigger the consumer's interest.

REFERENCES

Kleine, S.S., Kleine, R.E. and Allen, C.T., 1995, How is a possession "me" or "not me"? Characterizing types and an antecedent of material possession attachment. *Journal of Consumer Research*, **22**, pp. 327-343.

Richins, M.L., ,1994, Valuing things: The public and private meanings of possessions. *Journal of Consumer Research,* **21**, 504-521.

Schifferstein, H.N.J., & Pelgrim, E., 2002, Consumer-product attachment: the construct and its measurement. Manuscript submitted for publication.

More than meets the eye. Exploring opportunities for new products, which may aid us emotionally as well as physically

Jim Thompson
University of Central Lancashire, UK

INTRODUCTION

This paper seeks to demonstrate that by considering established principles of emotional response it may be possible to begin to propose simple design methodologies, which the designer may use to gain insight into the emotional aspects of a potential design. Using these 'design tools' the designer may then seek to address these emotional factors in their design proposals.

When a designer is called upon to design a new product several approaches are immediately available to him by reason of their prevalence and long history of practice. Materials and manufacturing processes are well understood, marketing knowledge is a broad yet familiar subject area with a variety of methodologies and sampling techniques, physiological factors of design from the measure of the human body to how our senses work are a well researched area. All of these factors have a broad selection of references available to the designer. What is not always readily understood or accommodated is the inclusion and consideration of emotional aspects of products.

Current thought proposes that considering these emotional aspects of design not only improves the user experience but also helps to make a product more successful in a crowded marketplace. The tangible aspects of design are well understood and readily available to designers, processing and control of the intangible (emotions) is not. This growing awareness and opportunity for products with character and emotion is cited by Philips Design team in their statement *From Ambient Intelligence to Ambient Culture*

"Ever since we started developing tools, we have lived in an artificial environment, consisting of our own artifacts. Until now, the artificial objects that we created were unable to 'answer back'. Now, however, objects are almost becoming 'subjects' - intelligent and capable of independent activity. The more objects become 'personalities' in their own right the more both us and them will need to learn to take into account each others' requirements".

The idea of products becoming intelligent subjects with character is a strong manifestation of emotion in design. As societies dynamics are changing historical opportunities for human interaction and emotional fulfillment are either ceasing to exist or evolving into different forms. This evolution sees users relying more on the objects they possess to either facilitate their interactions in a different format or to supply them completely to the user. The first step of a designer is to begin to understand the psychological basis for emotions and their cause and affects. Also a problem facing most practicing designers is that they do not often have the luxury to undertake long studies in the current problem area. What a practicing designer needs is a concise description of this area, and a set of design tools to quickly get them started.

PSYCHOLOGICAL THEORY OF NEED/MOTIVATION/EMOTION

The now familiar hierarchy of needs proposed by Abraham Maslow (1943) proposes an order of human needs from the most basic and immediate to needs relating to higher levels of personal fulfilment and growth. Despite the fact that there has been little study to support this theory study by Wahba and Bridgewell (1976) show it is generally a widely accepted model. The strengths of using Maslow's hierarchy in this case is that it gives the designer a summary of human needs (which relate to emotions) as well as it's hierarchical ordering of the need groups. These needs are split into three groups.

The first are the *deficiency needs*, (physiological, safety, belongingness, esteem). These needs must be met before moving on to the next higher level. If the situation changes and one of these needs becomes wanting then the individual will act directly to meet this need. The next group of needs is the *growth needs*, (need to know and understand, aesthetic). These needs are personal needs by which we seek to improve ourselves by giving ourselves time to explore different issues and gain understanding. The third group is *self actualisation (self actualisation, transcendence)* where we feel we have fulfilled our potential and be in a position to aid others to attain this state.

When we experience a need, at any level of Maslow's hierarchy, we seek to satisfy that need. This aspect manifests itself as *motivation.* As with theories of human emotion Huitt (1998) summarises that there is no single overriding theory, which applies to all situations.

MOTIVATION

Huitt (2001) states that whilst emotion is an *'indefinite subjective sensation experienced as a state of arousal'* it does not necessarily result in activity to change, sustain or enhance the emotion. Yet if we are using Maslow's Hierarchy of needs, it is reasonable to assume that the desire to meet the needs we may experience have strong emotions attached to them which we seek to fulfill. In Huitt's (2001) summary of motivational theories he indicates that sources of motivation can be classified as being from the self or, *intrinsic,* or from outside factors, *extrinsic.* The theories relating to these areas of motivation then fall into three main groups *body/physical* (behavioral, external social, biological), *mind mental* (cognitive, affective, conative), *transpersonal/spiritual* (spiritual).

Both of these summaries are particularly useful to designers and they begin to present psychological theory in an accessible manner without diminishing their meaning. By addressing this information as a designer we can begin to see indicative areas of motivation, which have identifiable sources. As a designer we can begin to appreciate these sources and how they affect and are affected by products.

HOW WE RELATE TO OBJECTS

In his paper *The Social Psychology of Objects* Hugh Miller (1995) presents a list of how we use and relate to objects in a variety of different ways. Miller draws upon a variety of academic references. He attempts to be as encompassing as possible to avoid (detrimental) fragmentation of the area. In his conclusion he states;

"This is an area where social scientists can inform designers, but also where designers with a thorough practical knowledge of what people do with objects, and what objects do to people, can be enormously helpful to social scientists."

This would seem to lend strength to the designer's instinctive reaction to emotional implications of objects, as well as some insights into how we might include them into design.

EMOTIONAL DESIGN TOOLS

There are many recognisable sources of idea generation and manifestation of the physical aspects of design (Baxter (1995), Papanek (1991), Ulrich & Eppinger (1995)). The example that follows attempts to use these familiar design methodologies, refer them to the previously discussed psychological issues and present them in a manner that is familiar to the designer and will aid the consideration of the emotional aspects of a design problem.

RESPONDING TO A DEFINITE ITEM OR SITUATION

The following table gives a list of factors that surround us when we interact with any item. It is similar to the discussion technique referred to as laddering developed by Hinkle (1965). This list presents a series of states, which may be encountered when using a product. The list presents the opposite extremes for each state. Whilst these are interrelated with the usual physical aspects of design by considering these in terms of the users emotion we may begin to identify how the user feels, what the product is projecting/demanding/eliciting and what opportunities there are to provide for these requirements/feelings within the design of the object.

Table 1 States of use list

General	Specialist
Regular	Seldom
Familiar	Unfamiliar
Pleasant	Chore (Dolorous)
Supportive	Demanding
Enjoyment	Functional
Choice	Dependant
Private	Public
No risk	High risk (cost of failure)
In control	Out of control
No pressure	High pressure
Rewarded	Ignored

Using an aid such as this helps the designer to give form to their consideration of the emotional arena their product and potential user operates in. This aid could be used in two ways. Firstly for the designer to complete and develop their own ideas surrounding the project. Secondly this aid may be utilised in a user group session in a manner to get the user to verbalise their emotional needs. People are not always able to consciously identify these needs when asked directly.

An example of use of the above table is below. The subject matter was a basic inhaler used in the treatment of asthma. The responses and design opportunities come from a general discussion amongst users and designers using the questions stated above.

Table 2 States of use and responses

Modes of use		**Response**	**Design Opportunity**
General	**Specialist**	**Semi specialist trained use**	**Helping hand**
Regular	Seldom	Dependant on severity, often regular	Reminder/calendar
Familiar	Unfamiliar	Becomes familiar but is never 'natural'	Seeks to accommodate
Pleasant	Chore (Dolorous)	Chore, 'necessary evil'	Builds trust leading to friendship
Supportive	Demanding	Very supportive, preventative regimes can be demanding	Seeks to re-assure
Enjoyable	Functional	Functional	No nonsense, competent
Choice	Dependant	Dependency	Supportive
Private	Public	Often needed in public but the desire is to be private	Discrete
No risk	High risk (cost of failure)	Risk if forgotten	Always there
In control	Out of control	Very controlled but diminishes if in attack phase	Supportive
No pressure	High pressure	Some pressure	Authoritarian
Rewarded	Ignored	Great relief when taken.	'Life line'

When asked to consider their response to the two opposing states users (and designers) can usually fairly quickly provide an insightful *response* based upon their familiarity and knowledge. The *design opportunity* is in effect how the user would like to be treated (emotionally) by the product. Using such an approach quickly indicated that it is possible for designers to quickly get to grasp with the potential emotional aspects of a design problem, and to elicit insightful response from target users. Whilst further study is undertaken into the importance of emotion rich design, it is also important to remember that the findings must be translated into easy to assimilate design tools if they are to be successfully adopted.

Designers have a great many existing and familiar design tools and these may well form the basis of further development of similar tools with regard to including psychological issues in design. Further development and research needs to be undertaken in this area to see what emotional issues are involved in modern complex products, and from these studies we must develop not only our understanding of these issues but the methods by which we may make them accessible to designers everywhere.

REFERENCES

Baxter, M., 1995, *Practical methods for the development of new products.* (London: Chapman Hall.)

Hinkle, Ds. 1965, The change of personal constructs. (Ohio: Ohio State University).

Huitt, B., 2001, *Educational Psychology Interactive.* (http;//www.chiron.valdosta.edu/whuitt/edpsyint.html)

Maslow, A., 1943, A theory of human motivation. *Psychological review,* **50**, pp.370- 396.

Miller, H., 1995, *The social psychology of objects.* Understanding the Social World Conference. (Conference proceedings University of Huddersfield 1995).

Wahba, A., & Bridgewell, L. 1976, Maslow reconsidered: A review of research on the need hierarchy theory. *Organisational Behaviors and Human Performance,* **15**, pp. 212-240.

Papanek, V,. 1991, *Design for the real world.* (London: Thames and Hudson.)

Philips Design 2002 Smart Connections. (http://www.design.philips.com/smartconnections/).

Ulrich, K. T., & Eppinger, S. D., 1995, *Product design and development.* (London: McGraw Hill)

Meaningful product relationships

Katja Battarbee, Tuuli Mattelmäki
University of Art and Design Helsinki, Finland

INTRODUCTION

Designing new products requires understanding the use of current products. The same applies for designing for experience. Especially for the consumer sector, also emotions need to be engaged and meanings tied in with the product and its use. Design for experience, going beyond usability, requires treating the user holistically (Jordan, 2000, Sanders and Dandavate, 1999) as a feeling, thinking, and active person. Experience is not only found in Disneyland, it is inherent in all interaction, fluctuating between levels of subconscious, conscious and meaningful stories (Forlizzi and Ford, 2000).

Although products are produced for masses, they must be designed for individuals. Partly the need for designers to conduct research has risen from the opinion that marketing data alone is not enough to fuel design. Rather, it is necessary to become exposed to real people and real contexts as leading user research and human factors people advocate (Segal and Fulton-Suri, 1997). The empathic understanding of people (Dandavate *et al* 1996) can be partly supported by stories through which people convey their attitudes and relationships to objects. This empathic understanding the existing meaningful relationships with products can be used for designing produts that deliver satisfaction on many levels.

This paper presents a tentative framework of meaningful relationships with products. The three main categories are described with examples, and compared to existing approaches to meaning and product experiences.

MEANINGFUL PRODUCT RELATIONSHIPS

People express values and attitudes – their selves - with the kinds of products they select for theirselves, their home, and their environment. (Csikszentmihalyi and Rochberg-Halton, 1998). Objects of choice carry personal meaning. The intensity of meaning can vary over time and over products. A coffee maker can be valuable in daily rituals of waking up or meeting friends. A modular sofa may become the most important part of a family being together, always rearranged and repaired. Childhood toys may be forgotten, and rediscovered in adulthood.

In this study we set out to collect stories about meaningful products in order to look at meaningful product relationships from a design point of view. The following relationship categories have been analysed from 113 stories and essays collected in Finland in 2001 from young adults but also from children, middle-aged, and elderly. The categories have been developed in an iterative analysis process. The product relationships fall into three main categories (see Figure 1): meaningful tool, meaningful association, and living object (term borrowed from

Jordan, 1997). The examples are edited and translated from the original essays and stories.

Figure 1: Meaningful Product Categories

Meaningful tool

The tool is needed for a purpose, in which the activity itself, not the hardware, is meaningful to the person. However, the hardware is necessary and an integral part of the experience. In this relationship the product could at any time be replaced with a similar or better tool. The relationships below have different emphases: functionality and usefulness, learning, and creativity and enjoyment.

Facilitator: These products help in satisfying needs of all levels: e.g. safety, mobility, indepence, social, and in accomplishing things.

You can hardly call a pale blue and squarish hair brush case pretty, but because of its size and function it has become a very important thing for me.(foldable hair brush)

8th grade excited me. In just the first few days I got lots of new numbers, new friends, and even met some girls. I found my first real girlfriend and fell in love. It was wonderful... And then came the phone bill. I saw the sum, I was in panic. How could I ever get that much money... Then I got a sms from my girlfriend, who told me it was all over. So glad you have a phone, makes this easier, she said. (mobile phone)

Challenge: Some objects pose challenges where we have to invest time, money, effort and attention to learn and master a skill or to care for the tool.

A snowboard is like a new car to an old man. You fix it up, maintain it, care about it... Your safety in the slopes also depends on the condition of your board. (snowboard)

I plugged in the loudspeaker, sat down, turned the volume up and hit the strings full force. The sounds were nothing like the ones I had imagined. I turned the volume down and imagined new, wonderful sounds, then tried to play them, but no success. I put the guitar in the corner and crawled onto my bed to think. I almost cried. It was weeks before I touched it again. The same experience repeated itself. (electrical guitar)

Self Expression: We use our skills with objects for purposes of creativity, productivity, and self-expression.

I learned to seek comfort, to work through my unhappiness and rage, celebrate happiness, express my emotions. (piano)

Although it is not a fine sewing machine, it has become important to me. One reason is to be able to have unique clothes. But my most important reason for sewing is to relax from school work... You can forget depressing things and focus on something you really enjoy. You can also show that you are creative in at least something. (sewing machine)

Meaningful Association

In this relationship, products are meaningful because they refer to or carry a meaning given by culture or an individual. Cultural meanings are understood similarly by the majority of people in the same community (although their preference for these meanings may be different). Personal meanings are constructed through experiences, and products are associated with them. All these products represent or refer to something outside the product itself.

Identity: personal, cultural, professional identities are conveyed with professional tools, clothes, and other symbolic products. Brands can be important, but so can be the choice to avoid them. Self-made objects reflect skill, style, interests, and accomplishments.

That muff was given to me by my mother, who thought that I was the only one of us five children who had the courage to wear it. That started my interest in furs, which I, as a native Vyborgess, have adopted as belonging to the Vyborgian sense of style. (fur muff)

Style, taste: aesthetic appreciation of beauty and senses is manifested in choice of products, brands, furnishings, fashion, environment, and objects of appreciation and collection, such as art and design pieces.

We still have in our living room a green, fiberglass chair, the "Pastilli". It was bought in the 60's and it was probably designed by either Tapio Wirkkala or Timo Sarpaneva, I am not sure. When I saw it in the office furniture shop window, I fell in love with its shape and colour.(Pastilli chair)

Link to a Memory, a Person, an Emotion, a Story: Objects remind us of past events and experiences, of people or family in general, and bring forth emotions connected to particular memories, objects, smells, materials, appearances. These memories may be communicated as stories.

A baby hat, with blue and white stripes, worn by three boys. I'll never give it away... it was the first thing I bought for my baby. There it waits on the cupboard shelf, to be taken out, us admiring its tiny size, and remembering the time I came home five months pregnant with the first thing for the baby. (baby hat)

The ring is not worth much in money, but nothing can measure its emotional value. Money can't buy the smiles, tears, whispers and promises we made as we planned our wedding together... (wedding ring)

Living object

In this relationship an emotional bond is created between the person and the individual product. The product is a companion, that has been with a person for so long that is is perceived as having personality, soul, character, is loved and cared for. It has a personal history of how it was made, aquired, and how it has survived. It may have identifying marks of use, wear and repair.

I imagined I was inside my car. We would drive along the floor, avoiding chair legs and other things, racing around, a bump in the rug as our garage. Tyly was like a friend, always ready for an adventure.(toy car)

My most important toy is a bunny, and there's one thing about it: sometimes I think it is real. Bunny is important because I can talk to it. Story: when I got it I felt like it had been made just for me. I was five years. I slept, ate, did everything with it. (stuffed toy bunny)

COMPARISON TO PARALLEL RESEARCH

Our study had a similar aim to the research described by Csikszentmihalyi and Rochberg-Halton (1981). They interviewed 315 people in Chicago about things in their homes and concluded that people invest phsychic energy into objects because they are expressions of the self. Their 20 product categories are accurate for classifying comments but from a design point of view that is too many to be practical in use.

Holman's (1986) five categories of object relations increase in intensity. From existing in the background, objects can mediate interactions, enhance actions, express identity and topmost, be objects of emotions like a loved person. True, most objects are around waiting to be useful, but we disagree on that intensity increases directly as proposed - there are too many interpersonal differences.

Richins (1994) has a broader approach and divides meanings into private and public. The sources of meaning are in Utilitarian value, Enjoyment, Representations of interpersonal ties, and Identity and Self-Expression. This relatively similar categorisation leaves out the "companion" category, and proposes enjoyment as a separate one. Her paper also presents a thorough analysis of approaches to product meaning and value.

CONCLUSIONS

The relationships are built over time and the richest descriptions were of objects with a long history. People have many, overlapping relationships to meaningful objects at the same time. For example: the guitar looks cool, sounds good, it lends a rock'n'roll image, it facilitates playing in a band, it poses a challenge for learning to play, provides a medium for expressing feelings through music, it becomes a friend and companion, reminds of people, places, events, accomplishments. No new, better guitar can replace the very first one.

What is meaningful to us depends on our life situation. Teddybears are loved by small children, but for a teenager a mobile phone is a symbol of independence and a tool for contacting friends. Ten years later a baby's pram can be important, and in old age things that remind of family and represent continuation can be meaningful.

Design does not happen in a vacuum. While other techniques contribute to other aspect of the holistic picture of the user, stories of products can be used as a way to empathically understand people and their values as expressed through products. Stories of meaningful product relationships can provide a deeper, personal understanding of the existing value context for designing new technologies.

REFERENCES

Csikszentmihalyi, M. and Rochberg-Halton E., 1981, *The meaning of things – Domestic symbols and the self,* (New York: Cambridge University Press).

Dandavate, U., Sanders, E.B.-N. and Stuart, S., 1996, Emotions Matter: User Empathy in the Product Development Process. In *Proceedings of the Human Factors and Ergonomics Society 40th Annual Meeting,* 1996, pp. 415-418.

Forlizzi, J., and Ford, S., 2000, Building blocks of experience: An early framework for interaction designers. In *DIS2000 Designing Interactive Systems Conference Proceedings* (ACM), pp. 419 - 423.

Holman, R.H., 1986, Advertising and emotionality. In *The role of affect in consumer behavior*, edited by Peterson, R.A., Hoyer, W.D., and Wilson, W.R., (Lexington Books), pp. 119-140.

Jordan, P.W., 2000, *Designing pleasurable products*, (London: Taylor & Francis).

Jordan, P.W., 1997, Products as personalities. In *Contemporary Ergonomics 1997*, edited by Robertson, S.A., (London: Taylor&Francis), pp. 73-78.

Richins, M.L., 1994, Valuing things: The public and private meanings of possessions. *Journal of Consumer Research,* **21**, p. 504-521.

Sanders, E. B.-N. and Dandavate, U., 1999, Design for experiencing: New tools, In *Proceedings of the First International Conference on Design and Emotion*, edited by Overbeeke, C.J. and Hekkert, P., (Delft University of Technology), pp. 87-91.

Segal, L.D., and Fulton Suri, J., 1997, The empathic practitioner: Measurement and interpretation of user experience. In *Proceedings of the 41st Annual Meeting of the Human Factors and Ergonomics Society,* (Santa Monica, CA), pp. 451-454.

PRODUCT CHARACTER

Happy, cute and tough: can designers create a product personality that consumers understand?

Pascalle Govers, Paul Hekkert, Jan P L Schoormans
Delft University of Technology, The Netherlands

INTRODUCTION

Consumers do not only consider products in terms of their functionality (Levy, 1959; Hirschman & Holbrook, 1982; Solomon, 1983). They often think and talk about products as having a personality and relate to them as such (Janlert & Stolterman, 1997; Jordan, 1997). A person can be rugged, masculine, and tough and so can a product. This equivalent of a person's personality is what we call product personality.

Product personality has been shown to influence product evaluation in that consumers prefer products with a personality that is consistent with their self-concept (Govers & Schoormans, 2002). This means that product personality can be used to enhance consumer preference. In order to do so, designers should translate personality characteristics into a product, creating a pre-determined product personality that consumers understand. Therefore the designer has to convert an abstract concept into visual and material features. Research (Smets *et al.*, 1994) indicates that designers are able to do so. Design students designed dessert packages based on a particular taste, portable cassette players based on a particular music style and sculptures based on a particular scent.

The current study investigates if designers can also convert personality characteristics into product form so consumers recognise it. Consumers were asked to rate products (irons) that designers created based on three specific personality characteristics: happy, cute or tough.

We hypothesise that:

H1: Consumers rate irons designed to be happy [cute] {tough} as happier [cuter] {tougher} than irons not intended to be happy [cute] {tough}.

METHOD

Although we use a research method that resembles the method used by Smets *et al.* (1994), there are two major differences. First, the research of Smets *et al.* is based on matching data. Respondents were asked to choose which one of three designs best represented a goal characteristic. The fact that, given a characteristic, a majority choose the correct design is an indication of recognition but not the strongest one. We use a more strict procedure. Respondents rate the stimuli on the same personality characteristics design students use as a basis for their design.

Second, because consumers and designers perceive product design differently (Hsu, Chuang, & Chang, 2000), the respondents in our study are consumers instead of design students.

Respondents

Eighty-eight respondents, 46 men and 42 women, were randomly selected from a consumer panel. Their age ranged from 19 to 74 years. Participation was rewarded with a small financial compensation.

Stimuli

During a preceding project 18 (11 male and 7 female) graduate students of Industrial Design Engineering were each asked to sketch an iron. The iron should express a particular personality characteristic: happy, cute or tough, without using symbols or icons. This resulted in a set of 18 irons, six 'happy' irons, six 'cute' irons and six 'tough' irons. The personality characteristics happy, cute and tough were selected because consumers have been shown to use them for both people and products (Govers, 2001). To make sure that the irons did not differ in function and features the students were presented a prototype indicating the functions that they had to incorporate in their design (see figure 1).

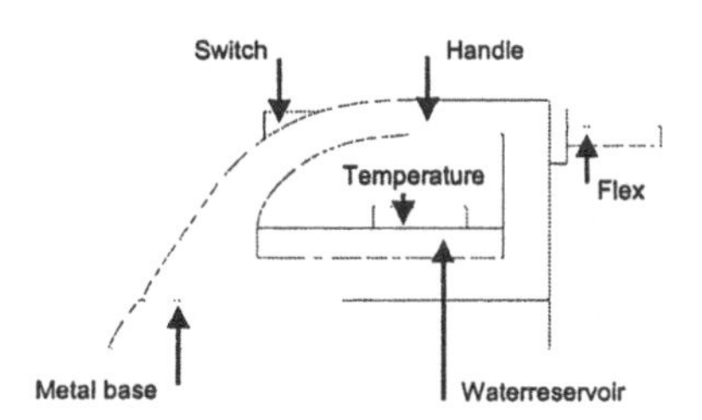

Figure 1 The prototype iron.

The students also filled out a short questionnaire explaining their design choices with respect to product form, colour, texture and geometric proportions. Participation was voluntary. Students could enter via a registration form on a notice board.

The stimuli for the current study are selected from the 18 irons that resulted from this project. For every personality characteristic, we randomly took three from the six irons. Figure 2 shows the nine profiles that served as stimuli.

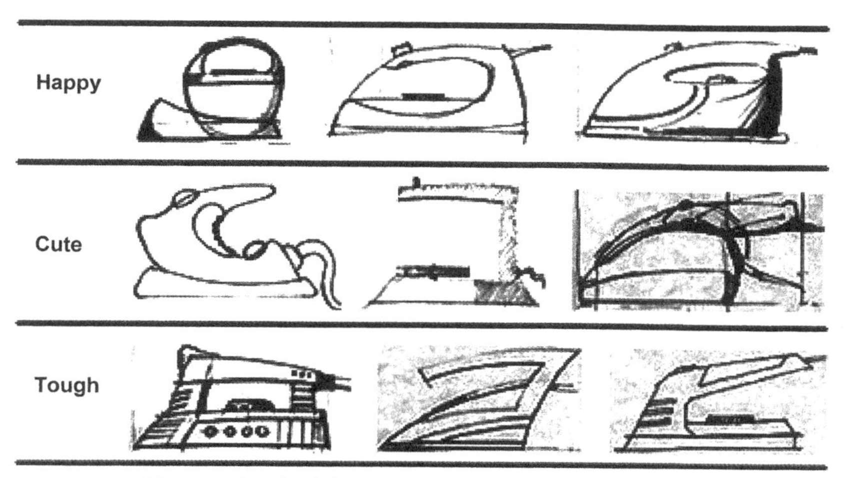

Figure 2 The stimuli irons ordered per personality characteristic.

Procedure

Respondents rated each iron on the three target characteristics, happy, cute and tough, using a 5-point scale. They also rated the irons on three other personality characteristics in order to prevent a 'matching' effect occurring. The irons are presented in three random orders. It took the respondents approximately 10 minutes to rate all irons.

RESULTS

To determine whether the respondents recognise the personality characteristic that is intended, three repeated measures analyses of variance are conducted. First, one that compares the happy irons versus the cute and tough irons on 'happy' (results shown in the first column of table 1). Second, one that compares the cute irons versus the happy and tough irons on 'cute' (results shown in the second column of table 1). Third, one that compares the tough irons versus the happy and cute irons on 'tough' (results shown in the third column of table 1).

It is expected that the irons designed to be happy will receive significantly higher scores on happy than the irons designed to be cute or tough. The results confirm this expectation. The happy irons are rated significantly happier than the cute irons ($F(1,87) = 10.3$, $p < .05$) and the tough irons ($F(1,87) = 21.3$, $p < .001$). However, the difference between the happy and cute irons is less significant than the difference between the happy and the tough irons.

Next, it is expected that the irons designed as cute are rated cuter than the happy and tough irons. The cute irons are indeed rated significantly cuter than the tough irons ($F(1,87) = 18.5$, $p < .001$), but the happy irons are rated cuter than the cute irons ($F(1,87) = 7.6$, $p < .05$).

Third, respondents should rate the irons designed to be tough significantly tougher than the irons designed to be happy or cute. The results show that this is the case. The tough irons are rated as tougher than the cute irons ($F(1,87) = 98.5$, $p < .001$) and the happy irons ($F(1,87) = 63.4$, $p < .001$).

We argued that if respondents recognise the product personality that was used as a basis for the design process, they would rate the irons designed to express a certain characteristic higher on that characteristic than the irons not designed to express this characteristic. For five of the six comparisons this expectation is confirmed.

Table 1 The respondents' ratings of the irons on the three personality characteristics *.

Respondents' ratings (N=88)	**Happy**		**Cute**		**Tough**	
	Mean	SD	Mean	SD	Mean	SD
Happy irons	3.1 ^{A,B}	.71	2.7 ^{A}	.75	2.9 ^{A}	.79
Cute irons	2.8 ^{A}	.77	2.4 ^{A,B}	.84	2.6 ^{B}	.82
Tough irons	2.6 ^{B}	.75	1.9 ^{B}	.64	3.8 ^{A,B}	.75

* Note: Means with the same letter differ significantly, *Italic*: $p < .05$; **Bold:** $p < .001$.

DISCUSSION AND CONCLUSION

The purpose of this research was to determine whether designers can create a product personality that consumers understand. Design students designed a happy, cute or tough iron; subsequently consumers rated the irons on these same personality characteristics. A repeated measure ANOVA was performed to compare the consumer ratings.

The results show that significant differences exist between the consumer ratings of the happy, cute and tough irons. The differences between the tough irons on the one hand and the happy and cute irons on the other hand are the most significant. This is in line with the meaning of the concepts. Happy and cute are semantically more alike than happy and tough or cute and tough. The semantic relation between the concepts can also be found in the product forms. An analysis of the questionnaires filled out by the students that designed the irons, shows that the students designing a happy iron and the students designing a cute iron both used round forms to express their personality characteristic. The only difference is that happy is visualised using open forms, whereas cute is visualised using stocky forms. The students designing a tough iron indicate that they used big and robust forms to visualise tough.

The small difference in product form between happy and cute may have influenced the consumer ratings for the cute irons. Round and stocky forms visualise cute. However, the happy irons are also quite stocky thereby increasing the cuteness of the happy irons. Another reason for this ambiguous result might be the use of 2D stimuli. With 3D stimuli product form is more distinct. The difference between round and stocky (cute) versus round and open (happy) would probably have been more clearly perceptible if we had used 3D irons.

Nonetheless, the results indicate that consumers understand the visual language of design students. This might be because the design students used associations as a basis for their designs and consumers understand these associations. Despite the fact that there was no specific question about associations, almost all students mentioned associations as a basis for their design decisions. We also know from the questionnaires that there is reasonable consensus about the kind of associations. Most of the students that designed a happy iron associate happy with smiling faces and the sun. Almost all the students designing a cute iron refer to small and young creatures. The students designing a tough iron all mention masculinity, specified by some in terms of machines, trucks and tools. Whether consumers share these associations causing them to 'read' the visual language of the designers has yet to be investigated.

In conclusion, the results of the current study suggest that designers can create a product personality that consumers understand. The study also shows that it is difficult to design personality characteristics that are semantically related. Further research will have to show what personality characteristics are relatively easy to design and how more complex product personalities can be designed.

REFERENCES

Govers, P., 2001, Developing a product personality measure. *Unpublished Research Report.*

Govers, P., and Schoormans, J.P.L., 2002, Similars attract; The influence of product personality congruence on product evaluation. *Submitted for publication.*

Hirschman, E.C. and Holbrook M.B., 1982, Hedonic consumption: Emerging concepts, methods and propositions. *Journal of Marketing*, **46**, pp. 92-101.

Hsu, S.H., Chuang, M.C. and Chang, C.C., 2000, A semantic differential study of designers' and users' product form perception. *Industrial Ergonomics,* **25**, pp. 275-391.

Janlert, L.E., and Stolterman, E., 1997, The character of things. *Design Studies,* **18**, pp. 297-314.

Jordan, P.W., 1997, Products as personalities. In *Contemporary Ergonomics*, edited by Robertson, S.A., (London: Taylor & Francis), pp. 73-78.

Levy, S.J., 1959, Symbols for Sale. *Harvard Business Review,* **37**, pp. 117-124.

Smets, G.J.F., Overbeeke, C.J., and Gaver, W.W., 1994, Formgiving: Expressing the non-obvious. *Proceedings of CHIGY,* **X**, pp. 78-84.

Solomon, M.R., 1983, The Role of Products as Social Stimuli: A Symbolic Interactionism Perspective. *Journal of Consumer Research,* **10**, pp. 319-329.

Brand and product integration for consumer recognition: a review

E G Kefallonitis, P J Sackett
Cranfield University, UK

INTRODUCTION

A company's product or service is more successful in meeting customer expectations if a high level of communication can be achieved between manufacturing, design and marketing disciplines. This needs to take place during the product conception stage. From the consumers' perceptive, knowledge in the generation of a holistic creation of an articulate product is limited. To link product features with perceptive recognition, research approaches exist but these are normally simplistically deployed through the means of a case study environment. From a brand's study viewpoint we found little evidence in the marketplace of achieving a strong association between the actual product features and the brand as perceived by the consumer. This leaves a limited scope for brand transmission through the product. The result is typically a 'brand-applicated' product. The result is that the product is developed separately and remote from the company's brand concept. Methods using consumer study viewpoints can link products and characteristics but these do not include direct brand involvement. The authors have not seen the desired level of design tools application and creative flexibility to allow significant brand image generative participation in the design process.

RESEARCH BASIS

The authors present an investigation into methodologies to integrate branding into the product design process. This is distinct from subsequent application of brand identity elements to the product for consumer perception. A study-map provides a summary of techniques that might link brand and product values. This provides the scope for maximising communicability of the differential values of the company's product or service as perceived by the end user. Brand and product mutual embodiment potentially allow the product to be identified in the absence of the brand's identity (name, logo etc.). In this way we could reflect the relationship between brand and product. This can lead to significant added value. Companies and consumers benefit as brand confusion is addressed (Clancy and Trout, 2002).

Product and brand may serve different needs but in consumers' perception refer to one another. In searching for a methodological basis between brand and product interrelation the authors developed a comparative study of some widely used methodologies. The theories investigated were distinct but could potentially be developed to associate brand and product/service through c onsumer involvement (Table 1).

Table 1 Synoptic review of methodologies studied.

Author(s)	Methods	Measure(s)	Assumes	Use
Green and Shrinivasan, 1978	Conjoint Analysis	For methods on new product design	Feature based representation of attributes	New Product Development
Schiffman *et al.*, 1981 Cox and Cox, 1994	Multidimensional Scaling (MDS)	For methods on product and brand positioning	Dimensional representation of attributes	Product and Brand Positioning
Kano et al, 1984	Kano's method	Product & Service Attributes to Customer Satisfaction	No direct brand involvement, expectation and need based	New Product Development
Akao, 1990	Quality Function Deployment (QFD)	Customer needs, Product Characteristics	No direct brand involvement	New Product Development
Aaker, 1991 Keller, 1993 Peter and Olson, 1993 Krishnan, 1996	Consumer Associative Networks	Consumer Memory Associations, Brand and Product category associations	Information is held in memory in form of associated nodes	Brand Equity
Young and Feigin, 1975 Howard, 1977 Gutmman, 1982	Means-End Theory	Brand persuasion, described as favourable feelings and purchase intentions	Consumer knowledge related to product is hierarchically organised	Concrete and Abstract product meaning association with consumer

A review of current literature within the domains of product development, product and brand positioning, reveals two contradicting views. Product involvement and brand commitment are not highly related, but they represent unique constructs (Warrington and Shim, 2000). Product differentiation is said to be introduced by brand heterogeneity with brands being formed of distinctive bundles of attributes (Sharp and Dawes, 2001). This is of great importance if one considers that the more similar a firm's offering is to that of a competitor company, the greater the need would be to change something in the customers' perception of the product (Sharp and Dawes, 2001). At present the market's state reflects brand confusion between similar brands highlighting the need for such attention to differentiation. Despite the functions of product development and brand involvement theoretically belonging in different domains, for consumer perception they are interlinked as they refer to the same product/brand choice (Rutter, 1996).

Theory, practice and consumer understanding

People appraise a product by the memory held of it from their previous use. Consumers do not always think of the brand first when making choices, in terms of which brand has high quality or what brand is the best in a product category (Rutter, 1996). Consumers use brand names and product attributes as retrieval cues for information about product performance (Van Osselaer *et al.*, 2001). When people are asked to name a product of good-design of which they own, they most often remember its use and not its brand name (Rutter, 1996). Therefore an increased level of resonance between product and brand would be highly desirable. Theory and practice suggest that brand and product experience are perceived and hold a kind of human information, especially when 'lived' by consumers (Table 2).

Table 2 Schematic study map of approaches studied in relation to consumer perception.

Theory	Practice
Aaker, 1996	Landor's global survey, Owen, 1993
Keller, 1993	Rutter, 1996
Krishnan, 1996	Sack, 1998
Peter and Olson, 1993	

The study of integrated product and brand theory showed that consumer associative networks represent a unique methodological construct for studying consumer memory associations. Information is close to the nature and understanding of the consumer (Krishnan, 1996). Industrial research and practice has used methods that support this transition of information in between brand and product in practice (Rutter, 1996). The development of an equation linking brand experience and human information can provide the basis of possible understanding between the connections of brand experience and product.

Reputation = Brand = Behaviour (through people, products, services, environment (Allen, 2000)

Brand experience = Customer needs/ expectations = Kind of human information (Rutter, 1998)

Identifying key associations is an essential part of understanding the brand's 'essence'; the overall impression consumers have of a brand (Waters, 1997). To meet expectation for a product or service one needs to understand the consumer associations value (Maio, 1999). After all, what really exists in a consumer's memory is the brand and its associations in relation to the consumers needs (Kapferer, 1997). The authors found that most product development projects and research undertaken have followed a case study approach. These strategies do not embody with the same strength the brand identity and brand values. They focus on the technical design and manufacturing excellence of the product. The result is excellence in the levels of technical and manufacturing standards and limited brand embodiment. As an example, we have a product that mirrors all the latest technological developments, has the right size, weight, etc. but does not reflect the company's brand ideas through its features. A product or service is easily copied by competitors in the market, the brand experience of an organisation is not readily replicated. This does not imply that every organisation in the industry should go forward in developing their own unique products. The methodology of integrated brand embodiment supports companies in realising the value of their brand and how its experience could directly be linked with the design, quality, and feel of its products to provide increased and sustained market value.

CONCLUSION

The literature review findings support the argument that higher levels of brand embodiment into the product features would be desirable. The client company would benefit from a uniquely characterised product or service and from the consumers perspective as it reduces brand confusion making product or service identifiable. High resonance between brand and product is the key requirement. Extending the Consumer Associative Networks methodology in verbal and visual territories offers promise in realising this objective. The technique holds information in human dimensions. Use of this tool in the brand product management could inject brand considerations early in the design lifecycle. Companies may consider refining their proposition and their product so that brand and product messages are mutually supportive. In this way the product or service provider can consistently meet consumer expectations and the product / service promises they project. This design approach is sustainable in the face of global competition.

REFERENCES

Aaker, D.A., 1991, *Managing brand equity*, (New York: The Free Press).

Aaker, D.A., 1996, *Building strong brands*, (New York: The Free Press).
Akao, Y., 1990, *Quality function deployment, integration customer requirements into product design*, (Productivity Press, Incorporated).
Allen, D., 2000, Living the brand. *Design Management Journal*, **11** (1), pp. 35-40.
Clancy, K.J., and Trout, J., 2002, Brand confusion. *Harvard Business Review*, **80** (3), pp. 22.
Cox, T.F. and Cox, M.A.A., 1994, *Multidimensional scaling*, (London: Chapman & Hall).
Green, P.E. and. Shrinivasan, V., 1978, Conjoint analysis in consumer research: Issues and outlook. *Journal of Consumer Research*, **5**, pp. 103-123.
Gutmman, J., 1982, A means-end chain model based on consumer categorisation processes. *Journal of Marketing*, **46** (2), pp. 60-72.
Howard, J.A., (1977), *Consumer behaviour: Application of theory*, (New York: McGraw-Hill).
Kano, N., 1984, Attractive quality and must-be quality. *Hinshitsu: The Journal of the Japanese Society for Quality Control*, **14** (2) April, pp. 39-48.
Kapferer, J., 1997, *Strategic brand management*, (London: Kogan Page).
Keller, K.L., 1993, Conceptualizing, measuring, and managing customer-based brand equity. *Journal of Marketing*, **57**, pp. 1-22.
Krishnan, H.S., 1996, Characteristics of memory associations: A consumer-based brand equity perspective. *International Journal of Research in Marketing*, **13**, pp. 389-405.
Maio, E., 1999, The next wave: Soul branding. *Design Management Journal*, **10** (1), pp. 10-16.
Owen, S., 1993, The Landor image power survey, A global assessment of brand strength. In *Brand equity and advertising, Advertising's role in building strong brands*, edited by Aaker, D. and Biel, A., (Hillsdale, NJ: Lawrence Erlbaum Associates), pp. 11-30.
Peter, J.P. and Olson, J.C., 1993, *Consumer behavior and marketing strategy*, (Homewood, IL: Irwin).
Rutter, B.G., 1996, Measuring the design experience. *Design Management Journal*, **7** (4), pp. 72-76.
Sack, M., 1998, Using research to create visual and verbal agreement. *Design Management Journal*, **9** (4), pp. 59-63.
Schiffman, S.S., Reynolds, M.L., and Young, F.W., 1981, *Introduction to multidimensional scaling*, (New York: Academic Press).
Sharp, B., and Dawes, J., 2001, What is differentiation and how does it work? *Journal of Marketing Management*, **17** (7-8), pp. 739-759.
Van Osselaer, Stijn, M.J. and Janiszewski, C., 2001, Two ways of learning brand associations. *Journal of Consumer Research*, **28**, pp. 202-223.
Warrington, P. and Shim, S., 2000, An empirical investigation of the relationship between product involvement and brand commitment. *Psychology and Marketing*, **17** (9), pp. 761-782.
Waters, K., 1997, Dual and extension branding. Using research to guide decisions and branding strategy. *Design Management Journal*, **8** (1), pp. 26-33.
Young, S. and Feigin, B., 1975, Using the benefit chain for improved strategy formulation. *Journal of Marketing*, **39**, pp. 72-74.

Why a Porsche 911 is better than an Audi TT: methods for analysing the character of a product

Steve Rutherford
Nottingham Trent University, UK

This paper attempts to zoom in on the complexity of perception that people gain when they use products and think deeply about them. We all look at things in different ways, with different priorities and from different backgrounds. Where could a common framework come from which would attempt to make sense of the range of opinion on offer from millions of users of thousands of products?

There are existing complexities which have been mapped onto words and analysed. It is an intrinsic part of branding a product that its image is boiled down to some clear, unambiguous words which define its character. The use of mood boards in this process is a combination of images and words that works. But what if instead of taking a picture to be worth a thousand words, a thousand words might be taken and transformed into a picture? What patterns do peoples' perceptions of products' characters form?

The study tries to define a framework for looking at products. It uses terms familiar to or provided by users. It tries to paint a kind of picture of what that product is. This can be a picture of an existing product or a picture of a design concept. It can also be a tool for looking at products over time and 'seeing' how that product is developing.

INTRODUCTION

Scientific exploration has many tales of 'Eureka' moments when people have experienced insightful visions of how things look or might be. The insight that led to the splitting of the atom took place between the scientist standing on the kerb and placing his foot on the road as he set off to cross it. The author experienced one of these visionary moments recently. Some history first.

A personal interest in the design of cars has led to ownership of a variety of 'people's cars'- Morris Minor, Citroen 2CV, Fiat Panda, VW Camper, VW Golf and Fiat Punto. Then last year these were followed by, on loan for ten days, an Audi TT Roadster. The driver's experience was one of personality transformation. The car firstly changed how people perceived the driver. There were obviously lengthy stares, some comments and some negative responses. After a short while, the driver experienced a response to this attention which is best summed up as annoyance and arrogance towards these invasions of his privacy.

The sudden jump from mundane A-B vehicles to state of the art desirability certainly exacerbated this effect, an effect that might be lost on someone who slowly worked their way up the automotive ladder of success and achievement. It was obvious that something complex, possibly a few things, were going on which were certainly worthy of further thought.

BACKGROUND

In the author's fields of ergonomics and industrial design, there are many examples of capturing extraordinary information on perceptions of products. The author's collection of ergonomics text books are sometimes of little help in determining how a project is to be progressed in research terms. Creativity seems to play a large part in this type of research. In designing a mirror viewing system for prone paraplegics recovering from spinal injury and surgery, the author's interviews with users was purposely widened to examine the whole experience of being laid back in bed for five months. The mirror system would ideally allow them as good a view of their environment as they would have if sat up in bed. It was discovered that simple and seemingly trivial aspects that we take for granted were, after even a few days lying prone, very high up the wish list of features for a new mirror system. For example, seeing what the weather was doing outside and checking out the nurses' legs figured highly.

When involved in the design of surgical instruments, the author and the designers specified tests for the textured finish of the handles. Cylinders of stainless steel with varying grades of texture were produced and sent to surgeons in the field. They were asked to handle the samples when their hands were covered in human blood, soft tissue and particles of bone in order to give us feedback on the handling qualities of the textures.

Ergonomics research as applied to the design process continues to grow in this way. The aim of this paper is to promote creativity about the application of research and discover new ways to give vision to research outcomes and enhance the dialogues between design, ergonomics and users.

Although the results may not be statistically valid, we must retain a practical view of the design process. If the alternative to these new forms of research is to go with the hunch of a designer who's possibly never designed this type of object before, then maybe the risk of these new methods is worth it.

PREMISE

The world of the automobile is an enormous crucible for this type of experiment. The car is the largest, most expensive product many of us ever buy. There is remarkably little in-depth research in the public domain with regard to the complexity of emotions related to the ownership and use of vehicles. In this regard, product design is way ahead of vehicle design - see Mäkelä (1999). There are some insights to be gained from reading the quality automotive press, particularly Car magazine from the EMAP empire, however even the most emotional correspondent will still emphasise valves rather than values.

Automotive enthusiasts will often relate the character of their vehicles with a wide range of vocabulary. Their relationships will often be stormy. The author's personal viewpoint is that a certain number of negatives in a character can enhance the positives. There are also perceptions and assumptions about the owners of certain vehicles.

Characters; relationships; arguments; perceptions. Do we need relationship counsellors to decide which car to buy? Do we need to know if we are compatible? If an interior designer can fill your home with stuff, can a car advisor fill your garage? To start to examine where we go next, a range of possible research methods were proposed and tested.

SURVEYS

1. Observations from the outside

"How would you describe this car? How would you describe the owner?"
These were the questions asked of visually sensitive undergraduate furniture and product design students when presented with a new Mini Cooper and an Audi A2. The results were surprising in terms of the range and expression of the words used. The A2 was predictably charged with being 'curvy' and 'compact', but also occasionally of being 'grrr' and 'woo hoo'. The Mini Cooper produced a wider range of words but fewer words with definite meanings – 'bubbly'; 'chunky'; 'dodgem'. When asked to describe the owners, there was more polarisation and correlation in the words used, with the Mini owner definitely being 'young' and 'extrovert', and the A2 owner being a 'rich', 'professional', 'family person'.

This exercise, involving approximately 20 subjects and one afternoon, produced enough responses to the Audi A2 to begin to see a definite repetition of answers. It is possible that with a little more time and more subjects the same could be true for the Mini. While there is possibly little new in this type of exercise, it must be remembered that the purpose of this type of exploratory research is to refine further detail on people's emotional responses to objects. In the context of this study, we are more interested in what exactly people mean by 'woo hoo' and on what scale this is a measurement of something.

2. Personal observations from the inside

"What words describe your classic vehicle, or are important factors in the vehicle's character?"

This was the question asked of owners of classic vehicles. Again, a surprising range of words were used to express the vehicle's personality and the respondent's relationship with it. Among the many suggestions, the original Mini produced 'precious', 'authentic' and 'temperamental'. The VW Type 2 Camper Van produced 'fun', 'escape' and 'infuriating'. Whilst some of these terms express aspects of the vehicle's character, such as 'authentic', some of them relate to the physical reality of the vehicle – 'infuriating'; 'temperamental' – and others relate to feelings made possible by the ownership of the vehicle – 'escape'.

Clearly the range, meaning and sometimes intangible aspects of the words are problematic. However, given that some kind of clear picture was emerging from the exercise with the Audi A2, what if the study was expanded in the future and the numbers of words collected allowed a degree of statistical validity to be attached to them? If we can determine what are the main factors involved with certain cars or categories of cars, can we then use these terms in a more focused study to determine 'personality profiles'?

PICTURES FROM WORDS

How might this information be used? As designers, we want to be able to reduce the risks inherent in designing new products. As designers, we want to lead the market with new, exciting designs. These two things fight against each other. Anything new and different will be risky. Manufacturers will use customer clinics and hall tests to gauge opinion. These methods rely on a heavy investment in creating an accurate picture of the possible designs. Perhaps the methods described here can better relate to an earlier stage in the design where there need not be the investment; where 'quick and dirty' statistical methods can replace and out-gun the designers' hunch and personal opinions?

There might also be the important element of the fourth dimension, time, to consider in quantifying the usefulness of these methods. This brings us to the statement in the title – 'Why a Porsche 911 is better than an Audi TT'. If many of us have lusted after a Porsche 911 for the last thirty years, isn't it presumptuous of Audi to think that they can design an experience in a couple of years that can in any way compete? Nevertheless, the TT begins to compete with the 911 (Porsche market their Boxster as an, admittedly more expensive, alternative to the TT). How should we compare these brands? Can we build a personality profile of them? Do they compare? Would the TT rate on the scales we might discover apply to the 911?

Figure 1 shows a hypothetical set of results from a fictitious survey. This is a personal viewpoint with absolutely no statistical validity. It paints a simplistic comparison in which we see, at least, that there could be a large mismatch between the variables measured, i.e. the TT has no scores on the history / legend or competition success scales. Although there would be a significant amount of subjectivity involved in a survey of this type, there are strong pictures held in users' heads about the image and impression expressed by very individual products and they are not easy to access.

What possible use could these methods be put to? Perhaps examining the difficulty of designing and releasing to the market new products in the 21st century will be the first application. Why is it that after a rapturous reception, the New Beetle has lost some of its cache in the market place? After a similar reception, what future awaits the New Mini? How much of our personal views are shared with others? How do we crack the code of what we feel? How are these products relating to their owners and prospective owners? Might the methods outlined here, economically administered to a large subject group, paint pictures of perception with some kind of validity?

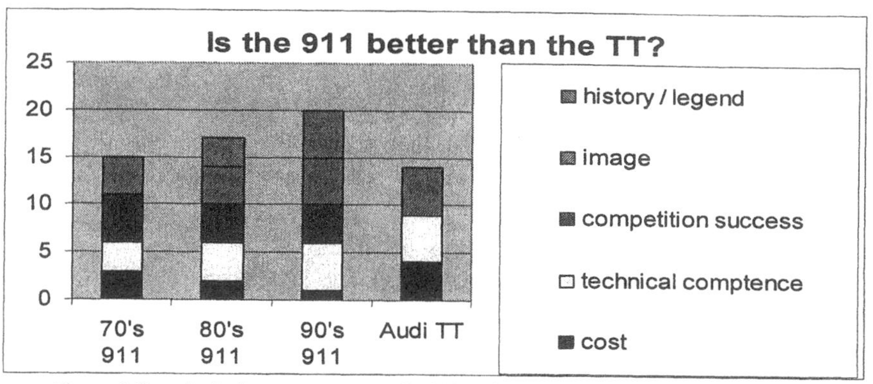

Figure 1 Hypothetical user assessment of a 911 and a TT using scales relevant to the 911

CONCLUSION

There are many aspects of this paper that will look familiar – the analysis of descriptive terms for products, for example. Also, if the exact methods used above to compare a 911 and a TT were ever to be used, it would certainly need to build on the work of others and require much more time to develop.

In the end though, methods like these involve designers in letting go, or at least sharing with users, the ownership of design possibilities. There's no doubt that large numbers of designers are against these methods or pay lip service to them, probably arguing that they lose creative control. But in the end, research and design should go hand in hand, and there is no reason why this fact should get in the way of creativity.

REFERENCES

Mäkelä, A.H., 1999, http://www.hut.fi/~ahmakela/eDesign/literature.html.

Branded GENES, designer JEANS, marketing DJINNS: the oxymorons of future shock?

N V Chalapathi Rao
Indian Institute of Science, INDIA

NEED SATISFIERS: PRODUCTS OR BRANDS?

The raison d'etre of products is the satisfaction of needs either explicit or perceived. While basic needs are obvious, it is difficult to categorise shining beads and sparkling rocks. The creativity of users has been driven by cues from the immediate surroundings. Headgear of one culture will be out of place in another. But then one cannot relate to a *guru* in unfamiliar gear. Certain forms have come to be associated with uses of the objects.

Technology Evolution That Satisfies

Whole economies depend on shaping nature into products. Value is always an add on by conversion. Simplistic agriculture and mining have covered little except long run marginal costs. More efficient oil extraction or farm output seems to only lower the felt worth of the commodity. Short sighted models of commodity regulation had to change. Brutal wars yield no solutions for scarce inputs. Silicon has changed economies for good. Others have done equally better with other chips. The debate is now on the new development model, on which variety of chips they need to depend: potato, silicon or casino. The players always seem to experience, evaluate, decide and choose.

Opportunity seekers who are nearest to the publics seem adept at assessing and judging need satisfiers. And everyone wants a good time. Technology has emerged from this primordial broth. Tools of communication are the ones that seem to have made a real difference. Adaptation of new technology has made more goods, which seemed to have spurned more levels of needs.

Natural is good and slow

Natural selection may have influenced progress over a long time. An example is the romantic story of the discovery of silk by the emperor's daughter (Wheatley 1996). The technology remained in China for 3 millennia until it was taken out. Chinese economy and the world trade have gifted the silk route to posterity. The story of cotton and how it shaped history is less romantic. The effect of the invention of the cotton gin are still good political fuel in the east. Less debated perhaps is the adoption of interventions in evolution like cross breeding in cattle and agriculture. It continued smoothly and ensured a faster rate of progress of basic products. At this point products have been basic need satisfiers. One just believed in Chinese, Egyptian, Holstein-Friesian, Jersey, which were at best, descriptors.

Brands Satisfy

Jeans were one of the early modifications of basic protection products into something which satisfied a want along with innumerable other products and yet did something else which one did not know yet. Similarly *serge de Nîmes,* or the common denim was what proper clothes were to be made of. Perhaps from then on things were going faster. These artefacts conveyed more than what they claimed. Eventually Levi's started invoking feelings and associations. The Marlboro man has to wear jeans, ride around in wilderness and live on beans. Camel (cigarettes) lights feel safer. One unexpected turn is the word itself. A lavender daylily flower was named 'designer jeans' by the American Hemerocallis Society (Sikes 1983).

Decisions and choice had narrowed down to just look for the brand and life takes care itself. From the other side, it occurred to one and all that it is the image and not much of the basic level need satisfier that made all the difference. And lo and behold, the world of trade and commerce was galloping. Nothing was left that was not branded. Even good old grandmas are brand new brand ambassadors. What was noticed and allowed to go on was that creations like jeans became generic and they had to be necessarily endorsed by branding, and done best with designers. But one needs to remember that the designer is the one who used his/her sole insights and intuition and a designer is beyond a nametag or label.

Of Marketers and Designers

And *Designers* were born. Instinct, detached observation coupled with creativity brought forth truly outstanding products (e.g. The compact cassette, Xerox machines, the ballpoint pen, the internet, Netscape / Mosaic, the Personal Computer and Coca Cola). Somehow they know more about what satisfies and how to satisfy. All one had to do was to think like a designer. At the peak all they need to do is to *design anything*. Some were remembered, part as brands like Biro, so many more are forgotten. Some of them had forgotten who was the target of their products. Formal studies of product deployment situations emerged, which gave a better understanding of what was wanted.

In a similar way *Marketers* who were always around were looking at what are the needs that got attention that may yield eventual profit. It was an obvious move to just carry over good feelings with a brand that promised all. May be the brag that one could *sell anything* to anybody. Naturally both made dear mistakes that wasted efforts. Once again formal methods of understanding prospects evolved which perhaps narrowed down the features of offerings.

At this point marketing and design played a synergetic role in shaping the course of evolution of commerce. More probing resulted in more products, which could be just that much better to be used. In the initial stages when the pace of product offerings was slow, things appeared under control and none felt threatened.

The role-play is understood and like all good things, there are more claimants to all successful offerings and no blame takers for the disasters.

THE ARRIVAL OF HIGH TECHNOLOGY

The trends in technology forced a pace of adoption, which was confusing if not threatening. Mass manufacturing and marketing is only one of them.

What is more evident is that Designers have more control on the feelings evoked and perhaps needed that much more understanding of cause-effect relationships on product-user interactions. Users experience bewilderment in using some of the things, especially in home entertainment. One always wondered if it was meant for him or her at all. Nevertheless, marketers and designers have been using all possible techniques to adopt the latest technology to products. The understandings in gene research, immuno mechanisms and the ultimate possibility of genetic engineering seem to create more fear than awe (Barnett 2000). Somewhere along the line effect of genetically modified (GM) food seemed to evoke the primal emotion of a situation waiting to go out of control, leading to potential disasters. The original concept that designers will lead to a better world and have control on a different range of emotions seems less obvious. The situation is best described by the expression Designer Genes. For some time the arrival of this technology was welcome until perceptions started that somehow designers have moved over to the "other side", ignoring the user. The minor adjustments needed for nature to catch up did not seem to increase confidence. The simultaneous emergence of new strains of virus in nature somehow gave GM a grey shade. (Director, BACPE 2002).

Wherever visionaries invested and brought new technology seemed to be targets, best described by riots that had erupted in India against the seed maker Cargill and the gene research company Monsanto. It should be noted however that users seem to feel less threatened if they had an informed choice as reflected in the packaging and labelling legislation. The economy had to wait for just two plough-reap cycles to discover that all was well. At least in India, thankfully the gains of GM are slowly accepted (Monsanto 2002). Suddenly everyone wants to reap, the harvests of technology.

Branding Technology

This leads to the next stage; Branded Genes. While there is agreement on the investors right to return on investment (R.O.I.), there is haziness on patents off/to/on technology. Is it just a process or a product? What is clear is that brands can be created, fight for survival, and like all organic entities, die. But can be resuscitated at will. The incongruence of the two words is that they seem like different life forms depending on the perspective of the user. This issue seems to have caused only a small ripple and not an avalanche as the mysterious terminator gene. Which are associations with Freud's *Eros* and *Thanatos*. (White, 2002).

Technology in Home Products

Issues in electronics seem to have resolved themselves differently as in natural adoption is probably the World Wide Web. Zeltser (1995) mentions that CERN made the source code for its software publicly available. Academic interests have allowed technology to cross-pollinate, and come out with myriad opportunities. The costs of closed development are probably all the difference between Apple Mac and the IBM PC. Similarly, when it came to home entertainment, the VHS system somehow remained longer than Sony's betamax (Digital America 2002). In the very recent times, the rise and fall of free peer-share of MP3 music, involved the basic issue was of the stakeholders in the intellectual property rights issues of music sponsorship. The judgemental error, one cannot say was of the designers of the business model or that of the technologists, but everyone feels that music industry stopped a good thing, which was too good to last anyway (Rivera 2002). The latest artefact that has changed life is the mobile cell phone. But one notices because of the technical issues involved, all the offerings in the market place are nearly identical, except for what are perceived is small add-ons.

The DJINNS of Marketing

Marketing seems to have emerged independently with the proclaimed aim of satisfying needs. The miracle maker who could see that the channels of communication between the customer and the designer are open. The more notable of the models in the recent literature is the one from Dubberly (2002). He relates brand promise and perception by *design of experience*. Aaker (1991) deals exhaustively with brand equity. Janal (2002) focuses on the *power of positioning*. Few in this world have not been affected by Oglivies *imagery*. It seems that the magic of marketing or getting the Jeanie out of the bottle is getting clear. More new theories seem to baffle how any one made products at all! Is the future predictable? Noted among the big thinkers is Toffler (2002) when he says, "in the first phase, information technology revolutionizes biology. In the next phase, biology will revolutionize information technology. And that will totally, once again, revolutionize economies. Together these represent a turning point not just in economics, but in human history."

CONCLUSION

Human needs continue to grow and depend largely on the state of the society at any point of time. The most successful products continue to stay on and turn generic much quicker. Branding helps in keeping things focussed and can be best used to stretch our efforts in keeping up with the customer. More and more insights into minds of the users of artefacts, the context and situations, drive the trend of product development. What works appears always mystical. Keeping focussed on the customer will probably yield better pay-offs. The information superhighway has compressed distance and time. But regional influences will continue to shape products.

REFERENCES

Aaker, D, A.,1991, What is Brand Equity, in *Managing Brand equity*, (the free press), pp 15-16.
Barnett, A., 2002, GM gene threat to the third world. [Online] Retrieved on 20 June, 2002 from the web site http://www.guardian.co.uk/gmdebate/Story/0,2763,178404,00.html
Digital America., 2002, The Video Age, The VCR, [Online] Retrieved on 28 Feb., 2002, from the Web site of Digital America, available at http://www.ce.org/digitalamerica/history/history12.asp
Director, BACPE., 2002, Baba Amte Centre for People's Empowerment, and Secretary, Samaj Pragati Sahayog, [Online] Retrieved on 28 Feb., 2002, from the Web site of Organic Consumers Association available at http://www.organicconsumers.org/patent/cottonban091001.cfm
Dubberly, H, AIGA Journal of Design for the Network Economy [Online] Retrieved on 28 Feb., 2002, from the Web site of gain.aiga available at http://gain.aiga.org/pdf/Model_of_experience.pdf
Jana D.,2002, Unforgettable Internet Marketing, [Online] Retrieved on 28 Feb., 2002, from the Web site of gain.aiga available at http://www.danjanal.com/foolproof.html
Monsanto, (07-Dec-2001) Reuters, Author Unknown, [Online] Retrieved on 28 Feb., 2002, from the Website of biotech knowledge, available at http://www.organicconsumers.org/patent/cottonban091001.cfm
Rivera, E., 2002, TechTV Napster Deathwatch. [Online] Retrieved on 20 June, 2002, from the web site http://www.techtv.com/news/internet/story/0,24195,3381581,00.html
Sikes, 1983, Designer Jeans, [Online] Retrieved on 28 Feb., 2002, from the Web site of American Hemerocallis Society available at http://www.daylilies.org/ahs/designerj.html
Toffler, A.,2002, New Economy? You ain't seen nothin yet in *Toffler associates featured article* [Online] Retrieved on 28 Feb., 2002, from the Web site of Toffler Associates available at http://www.toffler.com/featured/fd_article_newec.shtml
Wheatley, A B., 1996, A Story of Silk, [Online] Retrieved on 28 Feb., 2002, from the Web site of Ann Braun Wheatley, available at http://www.globalpassage.com/storyboard/asia_connection/silk.html
White, T., 2002,Eros and Thanatos, [Online] Retrieved on 20 June, 2002, from the web site http://www.books-reborn.org/white/articles/2002_Eros.html
Zeltser, L., 1995, The World-Wide Web and beyond. [Online] Retrieved on 28 Feb, 2002, from the Web site of Lenny Zelster available at http://www.zeltser.com/WWW/index.html#Conclusion

DESIGN AND EMOTION: THEORETICAL AND ETHICAL ISSUES

Body and soul: the ethics of designing for embodied perception

Anthony Aldrich
University of Plymouth, UK

INTRODUCTION

The following paper makes an ethical argument for increasing the consideration of emotion in design - in particular, it addresses the question of how to improve the quality of our emotional experiences of design rather than just increase their quantity. It proposes that in order for this question to be addressed a reappraisal of our relation to design objects is required. Other writers have already intimated the need for such an approach. For instance, David Abram (1996) suggests that there is clearly a problem with the current tendency of design objects to rely too heavily on exploiting their capacity to symbolise and to evoke associative meanings. For example, the degree that design objects play upon our desired relationship with nature is now so ubiquitous that we are in danger of feeling alienated from the very empathy that they draw upon.

The paper addresses the above research question by undertaking a critique of how we consider design objects to be meaningful. It begins by investigating how the consideration of the design object as a 'text', which we then interpret, is actually a problematic approach for designers. From this critique, five inherent dilemmas within a 'hermeneutic' approach are articulated. Collectively these dilemmas clearly result in a narrowing down of our potential engagement with the design object and, ultimately, in the impoverishment of our relation with these objects, with each other, and with our world.

In the light of these findings the paper also discerns that a 'gap' or 'deficiency' exists in the consideration of our emotional relation with design objects. Furthermore, it is able to propose that this deficiency is in relation to our capacity to consider those meanings, which are derived from our 'embodied' experiences. The potential benefits of redressing this deficiency are explored in relation to a number of examples. As a result, the paper is able to highlight aspects, which characterise these examples and to articulate the wider positive implications of a fuller consideration of embodied perception in design.

THE BREADTH OF 'LIVED REALITY' AND THE PROBLEM OF CONTEMPORARY DESIGN

In an era when the validity and value of the meanings that we derive from our own experiences are continually being reduced through their being considered as merely relative and subjective opinions, it is reassuring to find affirmation of collectively held beliefs regarding the quality of our existence in Art, Poetry, Philosophy and

Religion. These disciplines continue to reveal our existence to be a rich, seamless, holistic, and emotionally charged condition that we experience with our whole self. We are reminded that our 'lived reality' has great breadth and that it is clearly something that we can only know fully through our body in conjunction with our mind. This concept of our existing primarily as a 'thinking body' is one that was developed by the French philosopher Maurice Merleau-Ponty (1964). His idea of the individual as being first and foremost a 'body-subject' is, among other things, a concept that aims to embrace the richness and breadth of how we experience our existence.

When one works with this broad and inclusive conception of 'lived reality' it soon becomes apparent how few designed products actually involve themselves with its breadth. In fact, all too often, they only address a very partial region of this realm - offering us shallow and insubstantial experiences. At present, one idea in particular holds sway; namely, that the meanings evoked by a designed object are located predominantly outside of itself. We tend to believe that design objects are meaningful only by reference away from themselves, towards other things. For example, the freedom that a new car may provide is typically signified in adverts by images of uninhabited panoramic landscapes. The car is not portrayed as an object in its own right, instead it is just a signpost to something else.

It is proposed here that this condition exists largely because of the way that we think we take meaning from things - in other words, it is due to our dominant epistemology, which at present could be described as being an hermeneutic one. Essentially, hermeneutics considers our reality as a text, which we are continually in the process of interpreting. However, as other authors have argued (Tarnas, 1991; Abram, 1996; Casey, 1993 and Johnson, 1987), a number of problems result from this position.

- The world of objects becomes separated from the world of people.
- All of our experience has to be bracketed as merely 'textual' interpretation
- All of our interpretations are regarded as being relative and subjective, with the consequence that the potential for collectively shared value judgements becomes obscured and diminished.
- Other modes of considering our lived reality become marginalised.
- Our conception of what constitutes our lived reality becomes diminished.

In a world where we are increasingly surrounded by designed objects it is disturbing to find that an hermeneutic epistemology prevents so many of them from being experienced as 'objects of quality in their own right'. This is not just a theoretical problem - its all too real consequences can be seen clearly in relation to our contemporary notion of art - and perhaps nowhere more representatively than with the Turner Prize. In 2001, the winning piece, by Martin Creed consisted of an empty gallery in which the lights were turned on and off at intervals. For those 'in the know' this was obviously a witty reference to the tradition of conceptual art that was articulated so clearly by Marcel Duchamp. Its value was seen to be in its capacity to refer to something outside of itself. Clearly, today, it is the 'signifying potential' of the work of art that counts above all else. However, some bizarre consequences follow from this situation. For example, the idea that there is potential for the art work to have other values that are a function of the 'object' or artwork itself are clearly regarded as unimportant or even rather quaint - the complaint that there is a lack of 'craft skill' in the making of an artwork is usually dismissed as an irrelevance by today's art critic. Nevertheless, the problem still remains that when we consider contemporary works of art and designed objects in terms of the breadth of potential experiential engagement that they might afford us, it is easy to see how the possibility of an aesthetic relation based upon bodily experience is denied. As a result the richness of our potential relation with the world is becoming impoverished.

Clearly, if we are to readmit the value of the immediate object then we need to find a different conception of how we take meaning from things - we need to find a different epistemology. As mentioned above, Merleau-Ponty's conception of the body-subject offers us such an epistemology. The question arises therefore - "what form will the design products of a body centred epistemology take?" In the following section I offer some examples.

DESIGNING FOR EMBODIED EXPERIENCE – SOME EXAMPLES

Perhaps 'Nature' is the resource par excellence for potential exemplars. Everything in nature appears to be 'true to itself' and exists very openly as an object in its own right. Typically we experience a directness, clarity and force in our bodily engagement with Natures' objects - with trees, pebbles, nests, seashells, waves, skies, etc. Fortunately, there are also many instances where our own productions call for the engagement of our bodies. For example, the iMac desktop computer by Apple. To the eye, the form and contours of its housing have an immediate perceptual logic and it even invites an empathetic caress. The tectonics of its panels and their junctions is similarly considered and affirms the gestalt integrity of the whole. Even its specific choice of translucent plastic, as opposed to clear or opaque, imparts to us the very nature of the material itself - thin, slightly flexible, potentially fragile, and with a capacity to return the warmth of our touch. Through our senses we understand it as an object in its own right and, moreover, as an object that appears coherent and even empathetic to our bodily relation with it.

In addition to visually perceived aesthetics, other design objects also exploit their actual physical 'fit' with our bodies. For example, the 'Dr Kiss' toothbrush or the 'Faitoo' range of kitchen utensils, both by Phillipe Starck, are so concerned with touch that they call out to be grasped. In architecture too, where the constraints of textual association are perhaps less pronounced, body perceptions can be explored at larger scales. For example, I recently witnessed a student of architecture apply this 'body perception' approach to the design of a pair of double doors within their project work. The two leaves of the doors were curved in plan - giving an effect rather like the meeting of two lock gates. As one approached the doors from one direction, with the curve facing towards you, it was clear that one had to pull in order to open them. As one approached the doors from the other direction, with the curve now facing away, it was clear that the doors had to be pushed. One's body would have immediately understood the forms - there was no need for signs telling us to push or pull. Quite clearly, our embodied perceptions do afford us a level of engagement with the world that is directly 'lived' rather than indirectly signified.

At present, in terms of mass-produced design items, the consideration of body perception appears to operate mostly at small scale. Even in larger products, such as cars, it is the smaller details that are attended too. For instance, there is the well-known case of the radii given to the folds in BMW car bodywork panels. It appears to have the capacity to persuade us as to the thickness, and therefore quality, of the metal being used. Similarly, other attributes, such as the sound made by closing its door is also considered. Unfortunately, the fact that the clarity and power of these meanings is derived through body perception is not so publicly articulated.

When one begins to consider valued aesthetic experiences in terms of body perception, the lists of items that result tend to read like a catalogue of 'design icons'. Stylistic differences become secondary, and one can establish common ground between objects as wildly separate as Arts and Crafts joinery and Sushi, or the counterpoint between a blank sheet of paper and the ink from a well weighted pen. Immediately that we begin to consider design products from the point of view of our intrinsically emotional embodied perceptions we readmit the richness, power and directness of our relation with things that we know to be the basis of our 'lived reality'.

TOWARDS EROTICS OF DESIGN

Ultimately therefore, if we are to reduce the ongoing impoverishment of our experiences and the negative commercial exploitation of the idea of emotion in design, it would appear that design needs to turn its attention away from obsessively manipulating the associative potential of names. It needs to give up on labels, expand its modes of communication beyond a concern with the merely visual, and break away from a narrow hermeneutic way of interpreting meaning.

Instead, we need to re-acknowledge that we have much to gain from paying attention to the meanings and experiences that are to be derived through our embodied perceptions. If we want to engage positively with the quality of emotion in design we should start attending to the way that our bodily perceptions already make sense of the world for us. We should grasp the full consequences of Merleau-Ponty's idea of the 'body-subject'.

Working in this way promises to readmit the emotional potential and ethical value of characteristics such as the inherent quality and meaning of materials, of the shape and fit of designed objects, of their weight, colour, texture, hardness, pattern, softness, reflectance, tectonics, sound, taste, movement, light, scale, aroma, to name but a few! Clearly, our bodily perceptions of these characteristics are the sorts of experience that involve us in our most direct and potentially richest engagement with our 'lived reality'.

However, this is more than a call for improvements in the 'science' of these matters, or of ergonomics, it is a call to designers to reacquaint us with the richness of our potential 'lived reality'. We need to ensure that design objects exploit their capacity to re-ignite our delight in the world and in each other. In returning to the question of how to improve the quality of or emotional experiences of design objects we would do well to heed Susan Sontag's words :

Our task is not to find the maximum amount of content in a work of art, much less to squeeze more content out of the work than is already there. Our task is to cut back the content so that we can see the thing at all. In place of hermeneutics we need an erotics of art.
Susan Sontag, 1964

REFERENCES

Abram, D., 1996, *The Spell of the Sensuous*, (New York: Pantheon).

Billington, R., 1990, *East of Existentialism. The Tao of the West*, (London: Unwin-Hyman).

Casey, E., 1993, *Getting Back into Place*, (Indiana: Indiana University Press).

Johnson, M., 1987, *The body in the mind*, (Chicago: Chicago University Press).

Merleau-Ponty, M., 1964, *The Phenomenology of Perception*, edited by C. Smith, (London: Routledge).

Pallasmaa, J., 1994, Six themes for the next millennium. *Architectural Review*, Vol 7, July, pp. 74 - 79.

Pallasmaa, J., 1996, *The Eyes of the Skin*, (London: Academy Editions).

Sontag, S., 1978, *Against Interpretation and other essays*, (New York: Octagon Books).

Tarnas, R., 1991, *The Passion of the Western Mind*, (London: Pimlico).

From Aristotle to Damásio: towards a rhetoric on interaction

João Branco, Nuno Dias
University of Aveiro, Portugal

Marco Ginoulhiac
University of Oporto, Portugal

Rosa Alice Branco
Matosinhos, Portugal

Vasco Branco
University of Aveiro, Portugal

INTRODUCTION

This project emerges from the University of Aveiro and congregates researchers of areas of knowledge such as philosophy, design, rhetoric and design management, isfocusing on two converging arguments. The first arises from the theories of António Damásio regarding the decisive role of emotion in decision taking. The second comes from the need to investigate further the potential of rhetoric in studying the theory and practice of design (specifically, interaction design), either as an architectonic art (as defined by Buchanan), or as an analytical tool.

Artefacts do not merely have an informative and communicative dimension, but above all, are appellative and persuasive requiring a rhetoric treatment of this interaction. Within this context, the focus of our research aims at Rhetoric on Interaction, seeing that interaction is the element which tends to qualify the surfaces of objects, either physical or virtual and a network of non-themed emotions.

We aim at outlining the shape of this project, its meaning, objectives and relevance, bearing in mind that both common sense and recent discoveries in neurology point to the absolute need to restore to emotion the role it was denied throughout many centuries.

RELATED WORK

Buchanan (Buchanan, 1989) states in the scope of design that "ironically, a unifying theory of rhetoric remains surprisingly unexplored". Furthermore, he distinguishes between three elements – logos, ethos and pathos ("the logos provides the backbone of a design argument, much as chain of formal or informal reasoning provide the core of comunication and persuasion in leanguage" – the ethos is the character of the products, "because in some way they reflect their makers (...) the problem is the way designers choose to represent themselves in products, not as they are, but as they wish to appear" – the pathos "is something regarded as the true province of design, giving it the status of a fine art (...) it collapses the distance between the object and the minds of the users") – in putting together an argument on design, adding that "they involve interrelated qualities of technological reasoning, character, and emotion, all of which provide the substance and form of design communication". Moreover, he infers that "The strongest designers, those who are most articulate if not always most persuasive, are concerned with discovering new aspects of the utility of emotional expression in practical life".

Studies on usability tend to format rules which essentially target logos. For example, Norman's characterisation (Norman, 1999) on a good interface for a tool – visibility, feedback, good mapping on functionality, a conceptual model on a tool easily built by the user – seems to denote the conviction that intelligibility is the only determinant of use.

If we cross Buchanan's opinion with the data supplied by Damásio, resulting from scientific experiments into neuroscience, one may simply conclude that an argument of design which devalues pathos will reduce the efficacy of logos. In other words, the design that refutes emotion from its argument by considering it superficial, tends to increase the cognitive charge on the product and hampers the set of decisions connected with their use. One of the challenges of this project shall be to prove this fact.

The significance of the emotional domain on the design of products has justified several studies. Desmet (2001) and McDonagh-Philp and Lebbon (2000), amongst others, attempted to capture and classify the emotional reactions of users. In these studies, models and methodologies are put forward, which, using the involvement of users as a point of departure, strive to inscribe on the process of design the gathering and analysis of emotional responses from the various protagonist typologies that are present. Methodological approaches are usually associated with the definition of useful instruments in the scope of design management, providing leads on tasks to be performed but do not intend to get involved in neither an analysis on the discourse of design, nor for that matter, in its construction. This shall be the aim of the project rhetoric.

Hummels and Overbeeke (Hummels 2000) selected five essential aspects in composing an aesthetic for interaction: functional possibilities and performance of the product; the user's desires, needs, interests and skills; general context; richness with respect to all senses; possibility to create one's own story and ritual. Furthermore, they propose methods to find "conditional laws with respect to the aesthetics of interaction" by means of "research through design".

One may state that from the point of view of a rhetorical approach towards interaction, the proposals described above are prescriptive as to the hypothesis on the inventio (discovery), and possibly, dispositio (arrangement) phases but seem absent from the elocutio phase (form of expressing). The project that we are designing begins with an analysis on the various scientific contributions that have been referred to and works towards a proposal on rhetoric of complete interaction where thought on the elocutio phase is inscribed.

ABOUT INTERACTIVE OBJECTS

A new generation of objects emerged, whose typology of interaction transcends the passiveness of use traditionally inscribed on shape and which lends itself to a colloquial relationship, mediated by an interface. These objects-quasi-individuals (Manzini 90) impart new challenges to the sphere of design, and furthermore, change the planning contexts of their gestation, mingling shape and behaviour as a presumption of desire and design. For the first time in the history of objects, they present us with a novel experience regarding the relationship with objects, configured as "hybrid entities half way between different polarities, between the material world of things and the immaterial world of informative fluxes" (Manzini 1990). The object/person dialogue, therefore, requires an interface between operation and use which normally matches simulations stripped of the familiarity that traditional everyday objects had gained. Through technological innovation, new products tend to become more specialised or witness successive increases in their uses. Nevertheless, they also become more incomprehensible. From this, one may conclude that technological innovation does not necessarily lead to new qualities. Consequently, the designer needs to find something that serves as a bridge between the machine world (sizes, velocity and operating criteria) and the person (senses, emotions, cognitive structures and cultural modules). For the culture of design whose main role has always been to provide quality to shapes in space, to be busy with interactivity means to venture into a new territory whose cultural references arise more from cinema, theatre or music. Which qualities must therefore be associated with interactive objects? May automation not become an alternative to decision? Over and above its functional performance, an interactive object shall reveal its quality as an interlocutor, that is, the degree of ease present in the relationship it encourages, the grounds where formal and behavioural aesthetic merge. In designing interfaces/interaction, one frequently resorts to metaphor. Nevertheless, rhetoric proposes a wider set of figures of style where the metaphor is merely an hypothesis. One of the lines of research of this project, therefore, handles the interpretation of those consolidated figures at a level of verbal discourse by striving to define (and implement) their equivalents within the scope of interaction.

FROM RHETORIC TO EMOTION

In Aristotle's Rhetoric, the argument was targeted at reason. Emotion is understood as the privileged element which the orator uses to steer the listeners away from rational deliberation. After the work by António Damásio (Damásio 1995), emotion surfaces as a biological function guiding rational behaviour in terms of survival and the way of living, namely, with respect to decision but with a wider reaching spectrum. It is necessary, therefore, to justify emotion as a substrate of reason and definitely dismiss the idea that emotion is the vicious foe of reason.

The existence of one's own body is the basis of behaviour that incorporates the art of living. The body determines by the regulating function of our organisms emotions, what is pleasant and unpleasant, a source of happiness and pain and what is decoded into actions requiring our mobility through either approximation or withdrawal with respect to the objects that surround us. Approximation reveals the appellative and persuasive capacity of objects, which must adjust the rationality of use to our emotions, so as not to disappoint our expectations as users after the purchasing act.

To mould the designer's emotions into the clients emotions reveals a dominant of appellation and persuasion. Nevertheless, persuasion in use introduces a temporal dimension where good performance becomes an empathic association with the product so that a positive emotion is sustained without yielding to a negative emotion. The designer, therefore, must be an anticipator and simulator of the other's emotions to whom he targets his project and this he can only achieve through dialogue between the first and second person, between a "you" and a "me" and never in the anonymity of the third person singular.

Our body is his design endowed with utmost mobility, especially the body's feeling-of-itself, which is the basis of our strategies for survival. The idea that we only have five senses is largely surpassed. The five senses are merely those which convey the exterior within us or in Whitehead's words (Whitehead, 1978), they are vectors which transport the "over there" to "over here". Nevertheless, the other senses called proprioceptives, provide us with indispensable information on ourselves with respect to the world and these are primitive in all species. Without them, we would never be able to feel that we are the ones who see, hear or feel.

Our work (Branco, 2002) lead us to conclude that in a great deal of voluntary behaviour there are physical laws which govern our gestures in an optimal manner. Furthermore, these preferences of the organism which are felt and act in accordance with the anticipation of emotions must be taken into consideration in our relationship with interactive objects. Using this base as a point of departure, this project will endeavour to respond to some issues which seem pertinent to us:

- May it be said that primary emotions are declinable for a whole series of sub-species resulting from the interactive relationship with objects?
- Which emotions may be controlled due to the maturity of individuals and sociability with external stimuli? What are the outcomes regarding the present saturation of the landscape of artefacts?
- If the artefact is not successful on first perusal (I like/dislike), is there any sense in speaking about emotion through use? Would this lead to loyalty to the brand name or to the typology of products or even to a purchase recommendation via "word of mouth" in the instance where the first is able to unleash the second?
- May emotion (especially when it crops up as an important agent over reason) be the mother of all explanatory variables of consumption regardless of whether they are intrinsic (personality, perception, memory, etc.) or extrinsic (fashion, social classes, reference groups, lifestyles, opinion leaders, etc.)?
- Is the symbolic and aesthetic configuration of a product and even the functional (practical) not the ideal field for transmitting emotions previously studied by the designer with the client or the user? What should be said about the contemplative function "discovered" by Alessandro Mendini for Alessi and about lines such as the "family follows fun or fiction" from the same company?

CONCLUSION

Our work, therefore, fits into a line of research which aims to detect and analyse the structuring elements of discourse on interactivity. By rejecting a merely technocratic point of view, a holistic vision is sought which incorporates logos and pathos in analysing the communicative process of interaction.

This research is not intended to replace the efforts that are being carried out in the field of usability, but indeed, complement them. The framework of reference, therefore, becomes wider and more complete by including components borrowed from other scientific areas seeking converging bridges which support the identification of basic communicative elements, transforming an interactive artefact into a cognitive and emotionally active agent, that is, into an intelligent rhetoric agent.

On the other hand, upholding a co-authorship for a process of design should imply interactive intervention by designers and organisations, the interpreters of the general and operational contexts of the project, as part of a shared overall strategic vision. From this presumption, it follows that the growing importance of the role of emotion in the process of design requires study and research in to effective communication between entities that are susceptible of feeling emotions which have sharing as their destiny. An unavoidable conclusion from the works of Hanna and Damásio, is that our humanity must also be the humility with which we accept that the body of reason does not dismiss the reason of emotion of a body.

REFERENCES

Aristóteles, *Rhetoric*, (Lisbon: Imprensa Nacional, 1999, *in Portuguese*)

Branco, R., 2002, *A Relação Causal na Percepção*, PhD Thesis, (Lisbon: Univ. Nova de Lisboa, *in Portuguese*)

Buchanan, R, 1989, Declaration by Design: Rhetoric, Argument and Demonstration in Design Practice. In *Design Discourse* edited by Victor Margolin (Chicago: The University of Chicago Press).

Damásio, A., 2001, *The Feeling of What Happens*. (Lisbon: Publicações Europa-América, 12th edition, translated into Portuguese)

Damásio, A., 1995, *Descartes' Error Emotion, Reason and the Human Brain.* (Lisbon: Publicações Europa-América, 11th ed., translated into Portuguese)

Desmet P., Overbeeke k., Tax S., 2001, Designing Products with added emotional value. *The Design Journal*, **4**, **1**.

Hummels, C., Overbeeke K., 2000, Actions speak louder than word: shifting from buttons and icons to aesthetics of interaction. In *Proceedings of the Design Plus Research Conference*, (Milan: Politecnico di Milano).

Manzini, E., 1990, *Artefatti. Verso una nuova ecologia dell' ambiente artificiale*, (Milan: Domus Academy).

McDonagh-Philp D., Lebbon C., 2000, The Emotional Domain in Product Design. *The Design Journal*, **3**, **1**.

Norman, D., 1999, *The Design of Everyday Things*. (Cambridge: The MIT Press, second printing).

Whitehead, A., 1978, *Process and Reality*. (New York: The free Press).

The total depravity of inanimate objects

Gordon Reavley
Nottingham Trent University, UK

INTRODUCTION

Gayle Hamilton's apposite description encapsulates the fatal allure of apparently neutral products and the effect that they can have on the user. The formula devised by frog design (Sweet 1999) that 'form follows emotion' acknowledges that product design should always include something extra: '….no matter how elegant and functional a design, it will not win a place in our lives unless it can appeal at a deeper level, to our emotions….the emotional element can be present in any number of ways: it may appeal to our desire for enhanced nostalgia….or it might be a tactile ergonomic experience….or it could be reinventing the familiar'. Dandavate *et al* (1996) observe that 'the emotional link might be as important or be even more important than usability because it creates satisfaction and awareness of the product and brand, and which prompts users to be loyal to that brand'. Investigating the connection between design and sensuality, Bayley (1986) observes that 'design is about closing the loop between desire and fulfilment – that's why it's like sex'. There are, perhaps, less tangible reasons why (particularly) Italian objects inspire such devotion in their *aficionados*. Holman (1986) argues that sometimes products can become objects of the emotions of consumers. This kind of situation is called the *aficionado* effect, observed among collectors….or those who become enchanted with technological aspects of products such as stereos, cars and computers. I want to explore the emotive qualities of objects in the interstices between design and the fine arts, using Italian design in the period after World War Two when designers such as Mollino, Pininfarina and Giugiaro brought design closer to sculpture than some mere prosaic sheet metal box or functional table ever could.

Farina liked to polish his mother's casseroles while the fundamental form of the *marmite* inspired Picasso and gave Farina a keen sense of sculptural form. Earlier in the 20th century, Picabia and the Futurist Marinetti understood the sexual attraction of cars, too. Sparke (1988) notes that after the Second War, designers who were disillusioned with pre-War Rationalism because of its associations with fascism took their cue from contemporary fine art, incorporating into their designs sensuous curves directly inspired by the abstract, organic sculpture of artists like Moore, Arp, Calder and Bill. The use of clay instead of wood encouraged the model makers to create flowing, complex curves that mirrored abstract sculpture.

More recently, the sculptured *nose* (the tendency towards anthropomorphism is always present) on an Alfa 147 or the sensuous curves of a Ferrari Testarossa (literally, red head) makes them closer to rolling sculpture than the globally-orientated, volume-produced cars and motorcycles that emerge from similar computer software programmes and the focus groups that manufacturers use as a safety net.

Clearly, people invest a great deal (emotionally as well as financially) in objects. Some drivers (and riders, too, because MV Agustas and Ducatis inspire the same emotions as Ferraris and Alfa Romeos) who buy Italian machinery cite just *looking* at them, as they would an artwork, as one of the principle pleasures of owning them. The architect Seth Stein designed a house that included a mechanical car lift so that the owners could bring their cars *into* the house, the better to appreciate their aesthetic qualities. (Arguably, though, aesthetic contemplation sometimes hinders practical usage; Croce (1967) notes that 'some new objects seem so well adapted to their purpose, and therefore so beautiful that people occasionally feel scruples in maltreating them by passing from their contemplation to their use'). Responses to surveys conducted by Csikszentmihalyi and Rochberg-Halton (1981) on how people attach meaning to objects indicated that stereotyped sexual roles influence the way we perceive and respond to objects in the environment. Interestingly, it was found that it was women who preferred objects of contemplation such as photographs or sculpture and tended to see objects as special because they were mementoes whereas men preferred things that could be interacted with such as television sets, stereos and cars. This would seem at odds with the notion that while traditionally, owners of fast and exotic machinery tend to be men, it is women who prefer the contemplation of objects. While colour, too, is important – Ferraris and Ducatis *have* to be red; MV Agustas *must* be the colour of hard metal and the sound of Italian machinery also contributes to the experience – it appears to be the close connections between the fine arts and Italian design that inspires such emotion.

Is it possible, then, to create an artefact, or artwork that embodies an emotionally rich experience? Moholy-Nagy (1947) believed that the goal of art is to form a 'unified manifestation....a balance of the social, intellectual, and emotional experience; a synthesis of attitudes and opinions, fears and hopes'. Where once the artist and designer were considered as a single profession, they became separated some time after the Renaissance, and Forty (1986) contends that the terms art and design have become confused and conflated in some people's minds and that there is 'a consequent idea that manufactured artefacts are works of art'. He goes on to argue that 'the crucial distinction is that, under present conditions, art objects are usually both conceived and made by (or under the direction of) one person, the artist, whereas this is not so with manufactured goods'. Performing both activities allowed the artist considerable autonomy, leading to the notion that one of the main functions of an artwork was 'to give free expression to creativity and expression'. It is evident, though, that objects conceived and made by an industrial or quasi-industrial process can be just as much a work of art and that many of the objects cited as examples here were conceived by one person as such.

However, there are no objective criteria when it comes to making art or designing an object - what people see and what they prefer is not determined by objective characteristics of visual stimuli. Croce (1967) argues that 'the beautiful is not a physical fact; it does not belong to things, but to the activity of man'....'natural beauty is simply a *stimulus* to aesthetic reproduction'. Here, following the model proposed by Alben (1996), aesthetic refers to whether an object is aesthetically pleasing and sensually satisfying....cohesively designed, and whether its spirit and style are in consistency. Croce also divides the beautiful in to *free* and *not free*; those objects which have to serve a double purpose, extra-aesthetic and aesthetic, the first setting limits and barriers in the way of the second so that the extra-aesthetic aspects of functional objects are impaired. This is not to imply that objects *per se* are imperfect since the mere presence of such works is enough to dispel such illusions. Neither does it mean that how a thing looks has no bearing on how it affects the viewer. Visual qualities obviously have a lot to do with how we react to an object or an environment. But our reactions are not direct 'natural' responses to configurations of colour and form. They are, as Csikszentmihalyi and Rochberg-Halton (1981) demonstrate, responses to meaning attached to those configurations.

It has become something of an orthodoxy that the fine arts embody emotive qualitities while design must be functional, rational or usable. More recently, though, the notion of 'form follows function' has been replaced by 'form follows emotion'. Venturi (1983) argues that one of the characteristics of postmodernism is an emphasis on richness of meaning rather than clarity of meaning. This leads directly to products that are emotionally rich and as a consequence, the pleasure derived from it will be satisfying. Nevertheless, the fine arts, sculpture and painting have long been thought to embody emotive qualities that designed objects do not. Gombrich (1963) (quoting Roger Fry who, using the wireless as an analogy, believed that the artist is the transmitter, the work of art the medium and the spectator the receiver) suggests that the artist broadcasts a message in the hope of reaching a mind that will vibrate in unison with his (sic) own, and that his medium (the work of art) is the only means to achieve this end. He continues 'the idea that art effects some kind of emotional contagion has been at the basis of all expressionist aesthetics ever since'.

The expressive artist, then, believes that there are such things as natural responses and points to the effects of forms and colours that suggest that, say, red (traditionally the colour of Italian sports cars and motorcycles) is exciting and that a certain shape (perhaps based on the contours of the human, female form) would initiate an emotional response. (Interestingly, though, in his theory of colour expression, Itten (1970) argues that 'the square corresponds to red, the colour of matter. The weight and opacity of red agrees with the static and grave shape of the square'). The square, though, is hardly a sensual shape and most Italian design such as the petrol tank on a Ducati or a Ferrari body panel are based on far more organic, sculptural forms. However, the equation of mood with colour becomes problematic when, say, an artist selects an equivalent shade to suit his/her mood. Logically, the recipient of the message should experience the identical emotion, since they would share the natural code of equivalence. It is exactly the superficial persuasiveness of this argument that has led to the abandonment of structure and to the increasingly 'natural' symptom in non-objective artworks and it appears to be these more abstract works that have influenced the formgivers. In some ways, this is understandable as 'representational' or figurative art always comes with its own baggage. A rather more persuasive argument is that an understanding of the language and context is essential before an affective empathy with the artwork or designed object is arrived at. For example, someone unfamiliar with more abstract work will not necessarily experience the same emotion(s) as someone fluent in this language.

Kandinsky, who began to paint in a non-objective way in 1908, believed that it was possible to represent internalised and spiritualised experience by means of shapes and colours and provides a useful model for this debate. Kandinsky contended that every colour has its proper expressional (i.e. emotional) value, and that it is therefore possible to create meaningful realities without representing objects. Ettlinger (cited in Gombrich, 1963) has shown convincingly how Kandinsky used unstructured colours and geometrical shapes in the belief that they were charged with emotive power.

It would appear, then, that if it has expressive qualities, and has not been made for purely venal reasons, most art is capable of inspiring an emotion in the viewer. Extrapolating from this, designed objects can thus fall into two categories: those that have been inspired by an artwork and those (necessarily more recent) examples, in which a designer has deliberately aimed for emotionally rich content. The first, some would argue more pure kind, can be found in the work of Carlo Mollino whose obsession with the erotic and the female body is a recurrent theme in his architecture, furniture and interiors. McDermott (1997) observes that the glass top of the Arabesque table (1949) is taken from a drawing by the Surrealist Léonor Fini, while the perforated frame has a similar organic quality to the reliefs of the sculptor Jean Arp. Brino (1987) notes that Mollino was an assiduous reader of Croce, particularly *The Aesthetic* which I have quoted from above, and that his work was informed by that of Matisse and that he admired Picasso. In his furniture, he used natural materials such as solid wood and marble but he also used wholly artificial materials like steel and glass, Plexiglass, and resins on woven cloth backings (Resinflex). By handling organic materials on a rational basis, he maintained his links with the neo-Futurists and the Constructivists who sought a conflation between art, design and society. A rather more base example is the Coca Cola bottle which the designer Raymond Loewy (1950) referred to as 'aggressively female....a quality that, in merchandise as in life, can transcend functionalism'.

Gaver writes: 'there is no such thing as neutral interface. Any design will elicit emotions from the user, or convey emotions from the designer, whether or not the designer intends this or is even conscious of it.' That there is no simplistic connection between the experience of an emotion and the conveyance of that same emotion to the user or spectator is not addressed, however. Gaver goes on to suggest five ways in which products could be designed to be more emotional. Firstly, by ensuring that an object's cultural values are consistent with some of the users' values; secondly, by designing objects to be used in a more emotional way; (an example might be handheld devices which allow for emotional exchanges between lovers); thirdly, by developing technology that could support emotional communication more effectively than traditional media; (interactive storytelling devices for use in hospitals, for example); fourthly, by embedding sensory devices that would recognise emotions into objects and lastly, by allowing products to have emotions themselves such as software that recognises and reacts to users' behaviour and emotions. Kotler, (1997) demonstrates that the focus of market segmentation in the 1990s has increasingly been on emotions and attitudes rather than on demographics or lifestyle factors. Manufacturers can now target consumers extremely accurately and anchor emotions to a specific product. However, it has taken them some time to catch up with what those designers who have taken their inspiration from the fine arts have been doing for some time.

REFERENCES

Alben, L., 1996, Quality of Experience. Defining the criteria for effective interaction design. *Interactions*, May-June, pp. 11-15.

Bayley, S., 1986, *Sex, Drink and Fast Cars: the creation and consumption of images*, (London: Faber and Faber).

Brino, G., 1987, *Carlo Mollino, Architecture as Autobiography*, (London: Thames and Hudson).

Croce, B., 1967, *The Aesthetic as Science of Expression and General Lin*guistic, (London, Peter Owen).

Csikszentmihalyi, M. and Rochberg-Halton, E., 1981, *The Meaning of Things: Domestic Symbols and the Self*, (New York: Cambridge University Press).

Dandavate U.. *et al*, 1996, Emotions matter: User empathy in the product society development process. In *Proceedings of the Human Factors and Ergonomics 40th Annual Meeting,* edited by C.J. Overbeeke, and P. Hekkert,, (London: Taylor and Francis), pp. 415-418.

Forty, A., 1986, *Objects of Desire, Design and Society 1750 – 1980*, (London: Thames and Hudson).

Gaver, B., 1999, Irrational aspects of technology: Anecdotal evidence. In *Proceedings of the 1st* Delft, *International Conference on Design and Emotion*, (Delft: Delft University of Technology), pp. 47-53.

Gombrich, E.H, 1963, *Meditations on a Hobby Horse and Other Essays on the Theory of Art*, (London: Phaidon Press).

Holman, R.H., 1986, Product as communication: A fresh appraisal of a venerable topic. In *Review of Marketing,* edited by B.M. Eris and K.J. Boering, (Chicago: American Marketing Association), pp. 106-119.

Itten, J., 1970, *The Elements of Colour*, (New York: John Wiley).

Kotler, P., 1997, *Marketing Management. Analysis, Planning, Implementation, and Control*. 9th int. ed, (Upper Saddle River, NJ: Prentice Hall International, Inc.).

Loewy, R., 1950, *Never Leave Well Enough Alone*, (Chicago: Simon and Schuster).

McDermott, C., 1997, *Twentieth Century Design*, (London: Carlton Books).

Moholy-Nagy, L., 1947, *Vision in Moti*on, (Chicago: Paul Theobald).

Sparke, P., 1988, *Italian Design, 1870 to the Present*, (London: Thames and Hudson).

Sweet, F., 1999, *Frog: Form Follows Emotion*, (London: Thames and Hudson)

Venturi, R., 1983, *Complexity and Contradiction in Modern Architecture,* (London: Architectural Press).

See me, feel me, touch me: emotion in radio design

Artemis Yagou
University of Thessaly, Greece

INTRODUCTION

According to the call for papers of the Third International Conference on Design and Emotion, "it is no longer sufficient to design good products or services; we all want to design experiences and generate pleasurable or exciting sensations". The call for papers also refers to emotion and experience as the "buzzwords of current design practice, research, and education", as well as to "this new design paradigm". However, recent research indicates that emotion and experience have already played a crucial role in design during the twentieth century in many product categories, but haven't received much attention from design historians and researchers. The radio set constitutes one of these product categories, where emotion can be identified as a crucial design factor.

The radio has been termed "a blind medium", as neither context nor message is visible (Crisell, 1994). It has also been claimed that in radio all the signs are auditory: they consist simply of noises and silence, and therefore use time, not space as their major structuring agent (Hawkes, 1977). However, the history of radio design reveals that the product developed to embody the radio medium, i.e., the radio set, has provided ample scope for the exploration and application of purely spatial, sensorial qualities. Although the usage of radio is primarily connected to sound and hearing, the design of radio sets clearly points to the fun of looking at, touching, and feeling the actual objects. This paper will present research on radio design and will discuss the relevant issues, by focusing on the role of sensorial and emotional aspects in the design of radio sets throughout the twentieth century.

TYPOLOGY OF RADIO DESIGN

Research by the author suggests that radio sets may be classified into five formal types. The first four of these types, which constitute the mainstream radio set production, follow a so-called functionalist approach. The majority of the very early radio sets have a quasi-technical appearance, directly influenced from scientific and technical equipment ("early domestic type"). Radios have later been housed into wooden cabinets to resemble furniture, which enter millions of homes and establish the typology of radio as a wooden box with dials ("classic domestic type"). At a later stage, the cabinets have been made of plastic, thus expressing modernity ("modern domestic type"). After the second world war, the emergence of transistor technology allows radios to become significantly smaller ("modern portable type"). These types present strong visual affinities to each other, as well as a sense of formal continuity (Yagou, 1999; 2002). The sets belonging to the last three types can be easily identified as radios and they represent a generic formal model for radio as a neat box with dials. Despite the functionalist approach, these radios bear strong stylistic references to various familiar iconographies, e.g., classicism or art deco. Such references may be interpreted as expressions of emotional qualities; they are however subsumed to the overall "functionalist" label.

The fifth group of radios has been termed "independent" by the author, in an attempt to define its essence by a single word. Another key term that has been employed to express these objects is "rupture". The main formal feature of "independent type" radios is precisely the absence of any specific standard, their limitless formal freedom. This group is extremely varied, eclectic, and in many senses marginal. It draws visual inspiration from nature, daily life, and popular culture. The respective radio sets take the form of animals, sunglasses, or cartoon figures, to give but a few examples. It is important to note that the "independent" type approach is not limited to radios, but may be observed throughout the twentieth century in most categories of technical consumer products, for example telephone sets in the form of animals or cartoon figures (Clark, 1997).

The origins of this type date back to the very beginnings of radio history, when this new product had not yet developed a specific typology. This period has produced some of the most representative radio designs of this approach. The Radio-Karte is a 1922 postcard crystal set, made by both German and English firms, which appears with a variety of picture fronts (Hawes, 1991). Felix, a British crystal set of c. 1923, is made after a popular character from early animated cartoon movies and it is tuned by moving the leg (Hawes, 1991). Later periods also provide interesting examples, indicating that this unconventional formal trend actually spans the twentieth century. Unless closely examined, the 1934 American Colonial New World globe in old ivory plastic is not easily identifiable as a radio (Hawes, 1991). The pop mentality of the sixties together with transistor technology triggers a whole new range of sensual and emotional designs. Typical examples are the transistor radio in the form of a beetle (Handy *et al*, 1993), and the transistor radio in the form of sunglasses (Handy *et al*, 1993). A similar case is the Sony radio of 1975/80 that looks like a microphone (Antique Radio Magazine, 1995). Defying formal and material conventions, such radios remain in the margins of mainstream culture. Rethinking of these sets, which resist classification, leads to the realisation that they actually possess significant sensorial and emotional features.

Furthermore, "independent type" designs and their sensorial qualities have permeated the more conventional, supposedly functionalist design typologies. A mains radio in bent plywood made by Emerson in the United States in 1938 might be mistaken at first glance for a typical wooden set. But its grotesque form with large conical loudspeakers is known to collectors as the Mae West (Hawes, 1991). It is a purely sensual design disguised as furniture. The American Emerson Snow White and the Seven Dwarfs mains radio, intended for the children's market, with imitation wood carving in a plastic material called "repwood", is another highly emotional design based on standard morphology (Hawes, 1991). The Radiorurale, a 1934 version of the Italian "People's Set", is an example of a functional form treated so as to exploit the nationalist feelings of farmers. It has a veneered cabinet with chromium fretwork featuring a sheaf of corn emblem and the fascism symbol

(Hawes, 1991). In the fifties, the British firm Roberts Radio produces small battery portables in real mink fur (Hawes, 1991). Another typical example of a plain sixties' set with an emotional touch is the Toshiba transistor radio decorated with lace (Handy *et al*, 1993). Finally, an archetypal example of conventional and emotional typologies merging into a single object is the portable receiver in the form of a pebble, manufactured by GE in 1997 (ID Magazine, 1997).

Therefore, it is obvious that "independent" type designs not only appear throughout the twentieth century, but they also influence mainstream designs. The rich sensorial properties of the unconventional radio designs lead us to the following question: Should we perhaps refer to an "emotional" type, which has no standard formal pattern but is primarily defined by the exploration of emotional and sensorial qualities?

INTERPRETING EMOTIONAL ASPECTS OF RADIO DESIGN

This paper has already mentioned the marginal character of "independent" or "emotional" type sets. The marginality of such sets is proved by the fact that design historians have usually ignored them. Sparke (1986), in her discussion of the development of radio design, writes first of radio as wooden furniture, then as a plastic cabinet representing modernity, and then as a miniature object, resulting from transistor technology. "Independent" type radios do not appear in her rather linear history of radio design. Likewise, Forty (1986) discusses radios as furniture and then as plastic and often portable objects evoking modernity. Heskett (1980) and Antique Radio Magazine (1995) follow a similar line of thought. All these historical accounts are primarily based on the mainstream production of well-known firms, such as Ekco, Murphy, Philips, Grundig, and Telefunken.

The author has been influenced to a great extent by such accounts and by the strictly modernist academic climate in which prior research was conducted. Due to such influences, "independent" type radios were discussed in that research from a highly critical point of view. Their unpredictable, fun, and often sarcastic forms seemed to represent a negation of the technology they carried. However, having identified the range and intensity of the "independent" forms phenomenon, the realisation came that it shouldn't be taken so lightly. This led to a substantial change of perception, and the concept of rupture, which was initially treated as negative, is now regarded as a deeper and more complex event.

The radios under discussion are often characterised as gadgets, where "a gadget is defined as an object that fulfils the search for physical and psychological comfort" (Larroche and Tucny, 1985). They might also be considered "kitsch", expressions of a reassuring, anti-technological language (Larroche and Tucny, 1985). But why does the "search for physical and psychological comfort" have negative connotations? Also, why does a "reassuring" design language have to be interpreted as "anti-technological", when it may simply being used as an aid to approach and understand new technology? An alternative, more positive interpretation might be developed.

Generally, different designs express various trade-offs between the beliefs that products should be "easily understood and used in an intuitive way" (Fiell and Fiell, 2001), and that products "must make pleasurable emotional connections with their end-users through the joy of their use and/or the beauty of their form" (Fiell and Fiell, 2001). Conventional radio typologies emphasise familiarisation and ease of use through a rational design approach, whereas "independent" type radios aim at familiarity through a radically different, sensual, and enjoyable approach. This is perhaps a response to the austerity of the Modern Movement and an attempt to undermine the supposed seriousness of mainstream radios. Furthermore, conventional types may be criticised for simply packaging technology. The "independent" designs under discussion may also be considered as packaging technology, though in a manner which is much more sensual and fun.

Both approaches attempt to generate familiar experiences and reassuring feelings towards new technology, which constitutes a potentially threatening reality. In order to facilitate the introduction of new technology (such as valve radio or transistor radio technology) to the wider public, "independent" or "emotional" radio designs emphasise sensorial and emotional aspects. In other words, they imply that a pleasurable object might be more appropriate to introduce a technological novelty to the public. Standardisation and ease of use become secondary, whereas the unpredictable and playful nature of objects arouses the interest of users and provokes more interaction between user and product. The technological object may be approached through humour, surprise, provocation, sensuality, and other properties not usually associated with technology.

The exploration of radio typologies brings to the fore another issue related to the reception and role of technology in daily life. The West prefers speed and effective use, whereas an element of play and the surprise of unexpected results are often experienced and appreciated in objects of the Japanese culture. According to the Japanese designer and thinker Kenzi Ekuan, a complex, multi-functional object of his culture "makes greater demands of its user, but is capable of an infinite extension of its possible functions according to the powers of the human imagination" (Dietz and Mönninger, 1994). This leads further to the possibility of designing a critical and even poetic dimension into technical products for everyday use, including contemporary electronic products. "By poeticising the distance between people and electronic objects, sensitive skepticism might be encouraged, rather than unthinking assimilation of the values and conceptual models embedded in electronic objects" (Dunne, 1999).

In this context, "independent" or "emotional" type radios are in a sense contradictory, because they are poetic in a rather mundane way! They certainly express substantial elements of feelings and imagination, but in most cases commercialisation overrides poetry. Oversimplifying or exaggerating the emotional approach carries the dangers of coming up with superficial styling solutions, of trivialising the technical object, and of eventually negating the poetry that potentially resides in technology.

CONCLUSION

This paper has discussed the design of unconventional radio sets throughout the twentieth century. These have been interpreted as expressions of design based primarily, if not exclusively, on emotion. Furthermore, it has been shown that the emotional approach has systematically influenced the design of mainstream radio production. The paper has also discussed how sensorial and emotional aspects of radio design have been used to mediate technology and technological change to the wider public. This mediation is regarded by the author as partially successful, because of the superficial or extreme nature of most design solutions. However, the emotional and sensorial approach to the design of technical products for daily use appears to be a very promising field for the exploration of poetic ways of living.

REFERENCES

Antique Radio Magazine, 1995, Cento anni di radio & design 1895-1995, (Maser: Mosè Edizioni).

Clark, P., 1997, *The phone – An appreciation*, (London: Aurum Press).

Crisell, A., 1994, *Understanding radio*, (London: Routledge).

Dietz, M. and Mönninger, M., 1994, *Japanese design*, (Cologne: Taschen).

Dunne, A., 1999, *Hertzian tales*, (London: RCA CRD Research Publications).

Fiell. C. and Fiell, P. (eds), 2001, *Designing the 21st century*, (Cologne: Taschen).

Forty, A., 1986, *Objects of desire – design and Society 1750-1980*, (London: Thames and Hudson).

Handy, R., Erbe, M., Antonier, A., and Blackham, H., 1993, *Made in Japan: Transistor radios in the 1950s and 1960s*, (San Francisco: Chronicle Books).

Hawes, R., 1991, *Radio art*, (London: Green Wood Publishing Company).

Hawkes, T., 1977, *Structuralism and semiotics*, (London: Methuen).

Heskett, J., 1980, *Industrial design*, (London: Thames and Hudson)

ID Magazine, July-August 1997, p. 199.

Larroche, H. and Tucny, Y., 1985, *L'objet industriel en question*, (Paris: Editions du Regard).

Sparke, P., 1986, *An introduction to design and culture in the twentieth century*, (London: Allen and Unwin).

Yagou, A., 1999, *The shape of technology – The case of radio set design* (in Greek), Unpublished PhD thesis, National Technical University of Athens, Department of Architecture, Athens.

Yagou, A., 2002, Shaping technology for everyday use: The case of radio set design. *The Design Journal*, in press.

Beyond emotions in designing and designs: epistemological and practical issues

Terence Love
Curtin University, Australia

INTRODUCTION

There is increasing focus on non-rational issues in designing, and in users' responses to designed products, systems and services. Many designers and design researchers have pointed to the importance of feelings in designing and the ways that humans perceive and interact with designed outcomes (see, for example, Cross, 1990; Davies and Talbot, 1987; Galle and Kovács, 1996; Lawson, 1994; Liu, 1996; Love, 2001, 1998; Tovey, 1997; Tovey, 1992). Research into the neurological basis of cognition by Bastick, Damasio and others have pointed to non-rational affective processes (feelings and emotions) being essential to, and driving, human cognition, behaviour, values, judgment and agency – key issues in design research (see, for example, Badgaiyan, 2000; Bastick, 1982; Damasio, 1994; Karr-Morse and Wiley, 1997; Knight, 1999; Shore, 1997).

The increased attention to non-rational processes has been dominated by a focus on 'emotions' (see, for example, ISRE, 2001; Massachusetts Institute of Technology -Affective Computing Research Area, 1999; Petrushin, 2000; Susac, 1998), and in design research this has been mainly from a rationalist, cognitive science perspective as evident in, e.g. "The Cognitive Structure of Emotions" by Ortony, Clore and Collins (1988).

This paper suggests design research is better served by focusing on understanding the key roles that emotions and feelings play in designing through understanding the physiological processes, because traditional "emotion-based" approaches are likely to restrict understanding and block the development of key areas involving close integration between humans' actions and designed products (e.g. ubiquitous computing systems). Human affective processes, based on physiological mechanisms offer an epistemologically useful cornerstone for design research and point to a research programme and theoretical foundation that addresses many fundamental, and often unacknowledged, epistemological contradictions in design theory.

EMOTION IN DESIGN RESEARCH

Traditionally, affective aspects of designing have been addressed in terms of emotions and feelings - often regarded as synonymous (Susac, 1998). In line with cognitive science, emotions (non-rational) have been regarded as separate from, and in most cases consequent to thinking (rational) and perceiving. Translated into design theories, this has frequently resulted in emotions being regarded as a direct result of the attributes of objects, situations or designs, or more unhelpfully, as *actual* attributes of objects, situations or designs.

Philosophical inspection indicates the relationship between thoughts and affects/emotions/feelings is not simple. Rationality and analysis require a prior process, sometimes called "intuition" or "insight", that depends upon processes in the realm of feelings/emotions/affects (Bastick, 1982; Damasio, 1994; Rosen, 1980). Recent research into human physiological and neurological processes, especially that undertaken through fMRI scanning of brain activity, has confirmed that this is in fact, how humans function (Damasio, 1994). The picture that is emerging from brain research is that the cognitive and affective mechanisms underlying human thought, judgment, creativity, feelings, actions, and motivations result from well defined neurological, hormonal and physiological processes operating in a structured architecture. These systems are not, however, simple or singular. A large number of parallel neurological, hormonal and physiological processes fulfill specific roles, and at the same time are reflexively connected with each other and their elements have a high level of reuse in other systems. That is, the cognitive, motor and affective systems are at the same time relatively separate, yet highly interlinked with feedback from different parts of each to different parts of themselves and to other systems (see, for example, Damasio, 1994; Macaluso, Frith, and Driver, 2002; Sloman 1998).

Prior to contemporary evidence from brain research tools, Bastick (1982) proposed a model of human thought intuition and action that consists of multiple reflexive relationships between feelings and thoughts. Bastick defined "feelings" as an individual's conscious and unconscious perceptions of subtle dynamic changes in their physiologically-based, somato-sensory and somato-motor processes and states (e.g. blood pressure, muscle tone, hormonal flux, breathing rate, blood vessel dilation, body skeletal kineasthetics, skin potential and heart rate). From this perspective, "emotions" are concepts, cognitive artefacts, referring to particular physiologically-based, and somato-sensory patterns. For example, the emotion "fear" consists of a particular concept referring to a particular pattern of high heart rate, high blood pressure, tense muscle tone, with specific facial and bodily kineasthetics. Other physiological patterns can be similarly identified for the remaining emotions. A key aspect of Bastick's theories is the way affects/feelings and thoughts have a reflexive physiological relationship: thoughts result in feelings, and feelings, as physiological somato-sensory and somato-motor states, cue particular thoughts.

More recently, on the basis of increasing knowledge about brain processes, Damasio (1994) has argued that thinking and feeling are different reflexive-linked aspects of a single complex process also involving humans' motor systems. He suggested that affective human systems are evolutionary building blocks, developed prior to and used as core aspects of, conscious reflective cognition. In other words, Damasio extended the multiple brain concept (that each person has and uses a brain that is a composite of brain elements from human, primate, reptile and other early stages of evolution) to include viscerally-based somato-sensory and somato-motor systems originating, in evolutionary terms, as the cognition and protection mechanisms (elementary brain) for earlier simpler lower-level organisms. Damasio postulated that physiologically, cognito-affective loops are doubled: sensation cues thoughts that in turn result in a secondary pattern of affects that in their turn cues other thoughts. For example, a pinprick results in conscious or unconscious thoughts that result in affects such as feeling faint (low blood pressure) that then cue other thoughts. Damasio's theories are confirmed by more recent brain research (e.g. Badgaiyan, 2000; Fabri, Polonara, Quattrini, and Salvolini, 2002; Kiehl, Liddle, and Hopfinger, 2000; Miller, 2000).

A DESIGN RESEARCH PROGRAMME BASED ON PHYSIOLOGICAL AFFECTS

There are many key issues that are epistemologically problematic in design theory: e.g. creativity, why styling works, how designers collaborate, why advertising works, cultural differences in interpretation, and acquisition of skills in designing. All of these are dependent on internal human processes that give rise to them. The previous lack of primary data about human internal processes has forced theories and research to be formulated using the only data available: second hand information from observation of humans' external behaviours. I suggest that this is has been a wrong-headed approach that is as practically and epistemologically problematic as trying to work out the program code for a word processor on the basis of documents produced by it. Inferring how humans undertake designing or interact with designed outcomes solely on the basis of observing their external activity (e.g. by protocol analysis) has been a major epistemological and ontological weakness of design research and design theories.

The new brain findings offer direct benefits. For example, the physiologically-based models of affective cognition of Damasio (1994) offer a direct understanding of the way that metaphor and allegory shape designing, design processes and aesthetics. Thus far, these issues have been relatively unsatisfactorily described in terms of the parallelism of metaphorical/allegorical elements (see, for example, Indurkhya, 1992), which does little to explain how they contribute to affective and cognitive changes in individual designers and to dispositional changes in particular designed outcomes. Understanding how metaphor and allegory 'work' provides the basis for new design theories and design methods to improve design practices and outcomes.

Understanding the physiology of affects/feelings in human cognition, action and agency offers a significant advance, and the basis for a major change to the theoretical foundations of design research that can potentially resolve many paradoxes, improve epistemological clarity and coherence, and offer integration with theories and research findings from related disciplines. In research terms, better theory about emotion/affects in designing and designs foreshadows improved clarity about other theory issues such as definitions of design (Love, 2001; O'Doherty, 1964). For example, it highlights the problems of using 'design' as if it has independent existence with intention and agency, e.g. "design has given the world a different..." Viewing designing in terms of the internal human processes also points up the weaknesses in using information theories and theories about problem and solution attributes as a basis for theories about designing (Love, 2001; Dilnot, 1982).

An alternative research and theory-building programme based on physiology must go beyond simply adding 'emotion' to otherwise rationalist research perspectives on designing and designs. In theory terms, such a research programme would involve three major changes. First, a theoretical framework that coherently includes relevant qualitative and quantitative biological, technical, social, environmental, ethical, aesthetical, cognitive and philosophical theories (Sloman, 1998). Second, definitions of key concepts (design, designing, design process, communication, collaboration etc) based on the physiological origins of human feelings, cognition, designing and object use (Love, 2001). Finally, but not least, a research approach in which theory making focuses on humans, rather than on external issues such as the attributes of objects, designs, problems, information or information flows (Love, 1998).

REFERENCES

Badgaiyan, R. D., 2000, Executive Control, Willed Actions, and Nonconscious Processing. *Human Brain Mapping,* **9**, pp. 38-41.

Bastick, T., 1982, *Intuition: How we think and act.* England: John Wiley and Sons.

Cross, N., 1990, The nature and nurture of design ability. *Design Studies,* **11**(3), pp. 127-140.

Damasio, A., 1994, *Descartes' Error: Emotion, Reason and the Human Brain.* New York: Grosset.

Davies, R., and Talbot, R. J., 1987, Experiencing ideas: identity insight and the imago. *Design Studies,* **8**(1), pp. 17-25.

Fabri, M., Polonara, G., Quattrini, A., and Salvolini, U. (2002). Mechanical Noxious Stimuli cause Bilateral Activitation of Parietal Operculum in Callostomized Subjects. *Cerebral Cortex,* **12**(4), pp. 446-451.

Galle, P., and Kovács, L. B., 1996, Replication protocol analysis: a method for the study of real world design thinking. *Design Studies,* **17**(2), pp. 181-200.

Indurkhya, B., 1992, *Metaphor and Cognition.* Dordrecht: Kluwer Academic Publishers.

ISRE. 2001, *International Society for Research on Emotions Recent Publications,* [html file]. International Society for Research on Emotions. Available: http://www.assumption.edu/HTML/Academic/users/tboone/isre/Rpub.html [2001, May].

Karr-Morse, R., and Wiley, M. S., 1997, *Ghosts from the Nursery: tracing the roots of violence.* New York: The Atlantic Monthly Press.

Kiehl, K. A., Liddle, P. F., and Hopfinger, J. B., 2000, Error processing and the rostral anterior cingulate: an event related fMRI study. *Psychophysiology,* **37**, pp. 216 - 233.

Knight, J., 1999, How stressful days steal your memories. *New Scientist* (26 June 1999), 6.

Lawson, B., 1994, *Design in Mind*. Oxford UK: Butterworth-Heinmann Ltd.

Liu, Y.-T., 1996, Is designing one search or two? A model of design thinking involving symbolism and connectionism. *Design Studies,* **17**(4), pp. 435-450.

Love, T., 2001, Concepts and Affects in Computational and Cognitive Models of Designing. In J. S. Gero and M. L. Maher (Eds.), *Computational and Cognitive Models of Creative Design*. Heron Island, Queensland.

Love, T., 1998, *Social, Environmental and Ethical Factors in Engineering Design Theory: a Post-positivist Approach*. Perth, Western Australia: Praxis Education.

Macaluso, E., Frith, C. D., and Driver, J., 2002, Directing Attention to Locations and to Sensory Modalities: Multiple levels of Selective Processing revealed with PET. *Cerebral Cortex,* **12**(4), pp. 357-368.

Massachusetts Institute of Technology - Affective Computing Research Area, 1999, *Affective Computing Research Areas,* [www]. Affective Computing Research Area, MIT. Available: http://www.media,mit.edu/affect/AC_research/ [1999, 1/6/99].

Miller, E. K., 2000, The prefrontal cortex and cognitive control. *Nature Reviews Neuroscience,* **1**(October), pp. 59-65.

O'Doherty, E. F., 1964, Psychological Aspects of the Creative Act. In J. C. Jones and D. G. Thornley (Eds.), *Conference on design methods* (pp. 197–204). New York: Macmillan.

Ortony, A., Clore, G. L., and Collins, A., 1988, *The Cognitive Structure of Emotions*. Cambridge: Cambridge University Press.

Petrushin, V. A., 2000, *Emotion in Speech: Recognition and Application to Call Centres*. Northbrook, Ill, USA: Andersen Consulting.

Rosen, S., 1980, *The Limits of Analysis*. New Haven: Yale University Press.

Shore, R., 1997, *Rethinking the brain: new insights into early development*. New York: Families and Work Institute.

Sloman, A., 1998, Damasio, Descartes, Alarms and Meta-management, *Proceedings Symposium on Cognitive Agents: Modelling Human Cognition, IEEE International Conference on Systems, Man, and Cybernetics* (pp. 2652-7). San Diego, CA: San Diego Cybernetics.Available: cogprints.ecs.soton.ac.uk/archive/00000717/00/Sloman_smc98.ps [28/10/2002].

Susac, D., 1998, *Emotional Machines: A review of Rosalind Picard's Affective Computing (1997, MIT Press)*, [www]. About.com Inc. Available: http://ai.miningco.com/library/weekly/aa070598.htm [1999, 2/6/99]

Tovey, M., 1997, Styling and design: intuition and analysis in industrial design. *Design Studies,* **18**(1), pp. 5-32.

Tovey, M., 1992, Intuitive and objective processes in automotive design. *Design Studies,* **13**(1), pp. 23-41.

Towards a better world

R R Gheerawo, C S Lebbon
Royal College of Art

INTRODUCTION

This paper focuses on applied research projects run by the Helen Hamlyn Research Centre (HHRC) at the Royal College of Art (RCA) where an empathic approach is taken to the design process. The paper is a description of the practical manner in which this user-centred design research method can result in creative design thinking that addresses user needs and aspirations whilst increasing business opportunity.

THE USER

Respect for the user's feelings and their needs and aspirations are not usually given much priority in a design brief. Designers generally design for themselves, to their own aesthetic values and to their own likes and dislikes, and this often leads to design exclusion. Moggridge (2001), a founder of the successful design and development firm IDEO, said that 'we all find it much easier to develop products and services for ourselves than for other people. Entrepreneurs, marketing people, designers and engineers are often selfish in this way.'

Design Exclusion

There are many groups whose physical and emotional needs are not addressed by mainstream design. Obvious groups include older people, disabled people or those recovering from an illness or an accident. However, anyone can suffer from the consequences of user-unaware design that does not cater for the emotional state of the user or the surroundings.

The push-button London Transport ticket machine provides a good example. The machine features a plethora of buttons but no obvious place to start the process of ticket buying, immediately putting emotional pressure on the user in the crowded, frantic station environment. As users become more frustrated with the machine, they become less able to decipher the ticket-buying process, or even cope with the whole situation. In an observation study by Gheerawo, nearly 70% of people who approached the machine, regardless of age or ability, found it difficult, and in some cases impossible, to make a transaction. Most were pressured into giving up and seeking help from the station staff.

In a further study carried out by Gheerawo in conjuction with London Transport, a user group including older people and visually impaired people explored the ticket buying process with special emphasis on the emotional mindset of the traveller. The results from this research informed the subsequent design brief. Since most of the users stated that they approached the station with their destination foremost in their thoughts, the process started with a 'choose destination' option. The resultant prototype machine enabled 90% of users to buy a ticket unaided, and, by designing the whole experience around the intuitive expectation of the user, effectively removed the worry of failure or of getting lost in the process.

By including these 'non-physical' needs of users, especially those outside mainstream design, the overall user experience will be more positive. This also has added business benefits as customers will view an emotionally satisfying product or service as desirable, and previously untapped markets will be catered for. 'Inclusive design', a nominal framework for this process is now advocated by the UK government, as a process whereby, 'designers ensure that their products and services address the needs of the widest possible audience.' (DTI Foresight, 2000)

Ergonomics has played an important part in helping the designer understand and respond to user characteristics, but there is a danger of seeing ergonomic data as sufficient, when in fact human beings have aspirations, spiritual dimensions and cultural characteristics that design should address.

So how do we look past physical dimensions and tap into the feelings and aspirations of users? A large number of methods exist, from observing people in situ, to focus groups and long-term prototype testing. The key is in using the right method at the right time in the design process, an approach in which quantitative data and qualitative user insights can be combined to meet user need, whether physical or emotional. Work at the Helen Hamlyn Research Centre (HHRC) is exploring practical ways in which this can be realised.

THE HELEN HAMLYN RESEARCH CENTRE

At the HHRC, design is carried out with small groups of critical users, In this context, critical or key users are those who sit outside mainstream consumer groups, often with capabilities towards the extreme of the bell curve. Recent work has explored alternative ways to create products, services and environments that respect user opinion and aspiration. One important vehicle for achieving this has been the Research Associates Programme where new graduates of the RCA collaborate with industry and voluntary sector partners on year-long design research projects centred around the user and with vast potential commercial influence within the partner organisation.

The projects employ a range of design research methodologies tailored to identify and include the needs and requirements of diverse and often marginalised groups of users in the design process. These methods include questionnaires, expert consultation, user diaries, interview, observation 'in situ', testing with prototypes, and research 'kits' requiring a range of responses from photographic to emotive.

By reflecting on some of the outputs from the Research Associates Programme it is possible to see the way in which centring the design process around the user can help identify new business opportunities and seed product innovation whilst enhancing the user's lifestyle and meeting some of their emotional needs. The following case studies illustrate this process.

Case Study 1 – Kinder: Quality Time At The Workstation

A design project with research partner Kinnarps (Scandinavia's largest office furniture manufacturer) looking at the physical and psychological implications of long hours of sedentary office work. Pascal Anson, the designer, adopted a human-centred approach from the start. General observation of behaviours and attitudes in London offices lead to the development of a photo diary where people were asked to record instances in the office to do with tiredness, bad posture, food, drink, storage and greenery over a period of fourteen days. This more human and personal approach viewed the workers as people, not commodities, and Anson, therefore, aimed to tackle their negative feelings and frustrations along with recognised design issues.

He discovered that the office was a place of imposed anxiety and stress, a lack of time to eat, and despondency with the artificial surroundings. On a more fundamentally emotional level was the need for friendship, value, comfort and security that many people felt were lacking. Further observation and investigation revealed that people tried to do much more in the office besides work, but the environments were too sterile to support these efforts. The result was a hoard of negative feelings when in the workplace. Anson rationalised these concerns and distilled them into five questions – can I work and eat properly at the same time, is my 'stuff' safe, where is the 'nature' in all this plastic, should I be sitting like this, and do I take a break or carry on tired? These were then written into the project brief and triggered five design proposals for furniture and desktop products to fit into most existing workstations and office layouts.

'Mini Desk' addressed posture by allowing users to work whilst sitting or standing, thereby encouraging sedentary workers to move between positions with resultant physical and mental health benefits (see Figure 1), 'Mini Store' was a redesign of the traditional pedestal incorporating storage and transport of personal items in and around the office, thereby keeping 'stuff' safe (see Figure 1). 'Mini Bed' was designed for short rest periods to be taken at the desk without making the user feel self–conscious (see Figure 1), 'Mini Café' encouraged people to take a break over lunch, interact with colleagues and not spill food and drink on their laptops (see Figure 1). 'Mini Garden' divided two adjacent desks, acting as a metaphorical garden hedge over which 'neighbourly' conversations could take place whilst giving co-workers some control of the office environment through the planting, growing and maintenance of the garden (see Figure 1).

Anson (2001) states that '(User feedback) was the most intense period of learning', and the five design proposals were further tested with real users to ensure that they not only met the physical design requirements such as space and safety, but also addressed user concerns about health and emotional well-being.

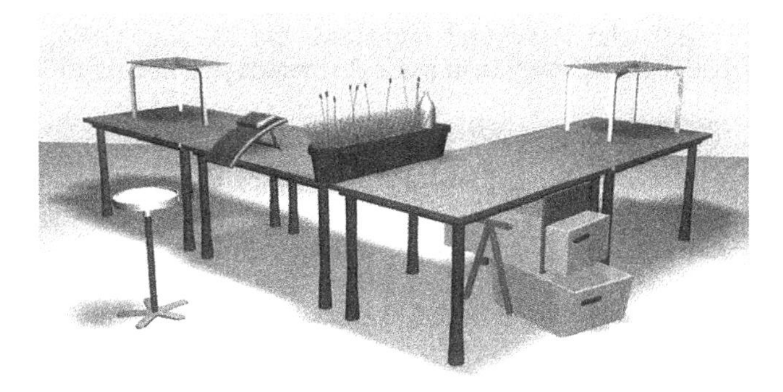

Figure 1 Mini Café, Mini Desk, Mini Bed, Mini Garden, Mini Store

Case Study 2 – Off The Wall: Organising The Transferable Workplace

A design innovation project, partnered by Esselte, the world's leading manufacturer of office supplies, that created four storage and transport systems to allow the transfer of work between the office and home.

The designer, Arash Kaynama, picked a key user group of 7 people already familiar with transferring work between home and their primary place of work – teachers and schoolchildren. Central to the project was the matching of the product design solution to user characteristics, personality and emotional need. Kaynama began by observing the selected range of users and documenting the different ways in which they solved the common problem of transporting work. Interviews followed to flesh out the observations, and began to reveal the personal approach and emotional attitudes each person carried when organising their work.

What became clear was that each approach was highly individual and could not be addressed by a generic solution. Kaynama therefore rationalised the users into four distinct personality 'types' and developed the design concepts to suit each one. At this critical stage, he expanded the user group to include an architect, a freelance designer, art dealer and office administrator, so the sample would not be limited to one area of work. Kaynama (2001) documented user comments that inspired the personality types; Mr Organised – 'I am expected to have a section for everything', Ms Individual - 'I don't have an official filing system at home or at work, but I know where all my things are', Ms Flexible – 'I work on my bed', and Mr Particular – 'Dad customised the computer table so that it had a shelf underneath for storing paper'.

The final proposals closely reflected the user mindset and personality traits. 'Mr Organised', a compulsive labeller with compartments for everything was given the Shaker Office where files are hung from the wall (see Figure 2). 'Ms Individual' a messy storer of eclectic objects with an instinctive knowledge of where everything is was given the Tetris Office, a highly customisable storage system of interlocking shapes (see Figure 2). The Magnetic Office (see Figure 2) was designed for 'Ms Flexible' who will work anywhere including the kitchen table or floor. The work tools here glowed in the dark and were magnetic so they could be attached to fridges or filing cabinets. For 'Mr Particular', a creator of custom-made order who is beyond organised, the Missing Office, a pegboard system was created where every item has a designated position (see Figure 2).

Figure 2 From left to right: Shaker Office, Tetris Office, Magnetic Office, Missing Office

CONCLUSION

These case studies demonstrate that developing products and services with critical user groups can increase design innovation, especially if the designer can see past the physical and cognitive and into the cultural and personal. By writing these emotional and aspirational user needs into the design brief, good experiences, especially good consumer experience, can then result, as the product begins to work for the user at both functional and emotive levels. his approach also brings important business advantages. Because there are obvious benefits to the user, customer satisfaction and positive perception of the company will increase. Broadening the range, and therefore number of users of a product and service means that revenues will also increase.

Importantly, within the Research Associates Programme, companies have been encouraged to adopt an inclusive approach to key areas of their business as a result of involvement with the programme. Raymond Turner, Group Design Director at Heathrow Airport remarked after two Research Associates proposed solutions to counter wayfinding problems in airport terminals that 'we're involved in the Research Associates Programme not because it is the politically correct thing to do but because it is the economically sensible thing to do.'

REFERENCES

Anson, P., 2001, *Kinder: Quality Time at the Workstation – Final Report,* (London: The Helen Hamlyn Research Centre), p. 15.

DTI Foresight., 2000, *Making the Future Work for You,* (London: Department of Trade and Industry).

Kaynama, A., 2001, *Off the Wall: Organising the Transferable Workplace – Final Report,* (London: The Helen Hamlyn Research Centre), pp. 15–16.

Myerson J (ed.) (2001) *The Helen Hamlyn Research Associates Programme Show and Symposium Catalogue 2001.* (London: The Helen Hamlyn Research Centre, London, p 63

The "in between" factors needed for tomorrow's product creation

G J van der Veen, M E Illman
University of Northumbria, UK

BACKGROUND AND SUBJECT MATTER

Over the last few decades, the rapid progress of technological possibility has led both industry and academia to realise that the notion of what determines a products functionality needs to be re-examined. At this moment, much of a product's functionality has been reduced to purely materialistic concerns and therefore offers little in terms of engagement and interaction. Where a focus on hard (technologically based) product functionality was once enough to differentiate one company's (or industries) offering from another, the new ubiquity of affordable technology has meant that matching or surpassing the functional (quantitative) standards of a competitors products will do little to grab the attention of a more knowledgeable, mature and demanding consumer led society.

Pine and Gilmore (1999) have surmised worldwide industrial developments in a number of business sectors to show a seismic shift away from the age of the 'Service Economy' towards a bold and rapidly emergent 'Experience Economy', However, by word of warning, they comment;

> "This transition from selling services to selling experiences will be no easier for established companies to undertake and weather than the last great shift from the industrial age…the question, then, isn't whether, but when – and how – to enter the next stage of economic value".

THE NEW ECONOMY: INTRODUCTION

Numerous eminent publications have focused debate on the buying public's readiness for change and highlighted the growing need for industry to adapt to a new customer focused imperative which exceeds and challenges notions of current product and brand values (Marzono, 1998). Collectively, these studies have begun to re-address and re-examine, not just industry, but also the role of respecting society and the way through which people interact with, enjoy and value the products they purchase. Coupled with this, there is also clear evidence supporting the fact that consumers are rapidly becoming more knowledgeable and demanding in their purchasing habits (Markham, 1998). This is now being seen through the established routes of commerce and is putting greater pressure on both consumers and manufacturers. On the one hand, consumers are consistently being forced to re-align their expectations of acceptable quality compared to financial outlay, while company success in numerous sectors now depends more than ever on attracting attention in increasingly price sensitive and competitive markets.

In order to understand how to satisfy these new demands, the authors carried out extensive research into two key-factors that lie behind any given design. These being the Contextual (Emotion-based) and Physical (Direct interaction with one or more of the senses) characteristics (or attributes). The importance to this paper is the clear evidence that highlighted a need for the design profession to focus on the *translation* between physical interaction and the context of realising the experience through product use as this was seen to be a major avenue through which eventual consumer satisfaction would finally be derived.

DESIGNING FOR EMOTION: THE EMOTIONAL PARADIGM

According to the psychologist Oatley, traditionally 'emotions' had been categorised as 'extra's' in Psychology on top of other serious cognitive functions like learning, thinking, language etc. After in depth research into emotions, these conclusions changed to show that emotions are important processes that *link* things and happenings that are important for us to the world.

Two main families of theories on Emotions are however still present. One is the theory of 'Basic Emotions' (Jordan, 2000; Lazarus, 1991); covering species-specific packages of genetic programs, appraisal by goal-relevance and coherence with physiological expressions. By going through this three layered structure of emotions, it becomes relevant to explore the link towards physical perception at each level. The second is the 'Componential Theory of Emotions', stating that emotions are; less structured, acting on a social dynamic system, culturally variable and that appraisal is based on features where there is a low physical correlation with other aspects of emotion.

While it is not our intention to validate the true nature of each theory, our research interests lie in the development of professional design methodologies which will indicate the relevance of particular emotional attributes for future manufacturing. It is expected that their relationship towards everyday products will ultimately generate a more sustainable satisfaction in consumer society; fulfilling cultural and personal needs in balance with succeeding generations of consumer expectations.

The authors suggest that product interaction has the possibility to be built up from more than one specific emotion. In accordance with other situations in life, it is suggested that product creation should *not* only be focused along the lines of today's 'Pleasure Principle' (Jordan, 2000). This states that pleasure is *the* basic need which drives consumers consumption. The development of our research hypothesis was therefore; that future product satisfaction could and should exist through evoking positive and negative emotional cues and levels of experience within the same product. These will either support, or suppress the final experience of the product (e.g. aggressively pronounced shields could emphasise the preciousness of it's contents).

The authors' research into product creation follows the decision tree of primary appraisals based on the three feature levels of goal relevance, goal congruence and ego involvement (See figure 1). Due to the fact that human brain activity, thinking and memories, are processed in a multi layered structure, making associations by executing parallel activities, each product should contain and reflect, multiple, positive and negative emotional attributes whereby eventual product satisfaction will be derived from designed (or constructed) 'evil' as well as 'goodness' attributes in the very same product.

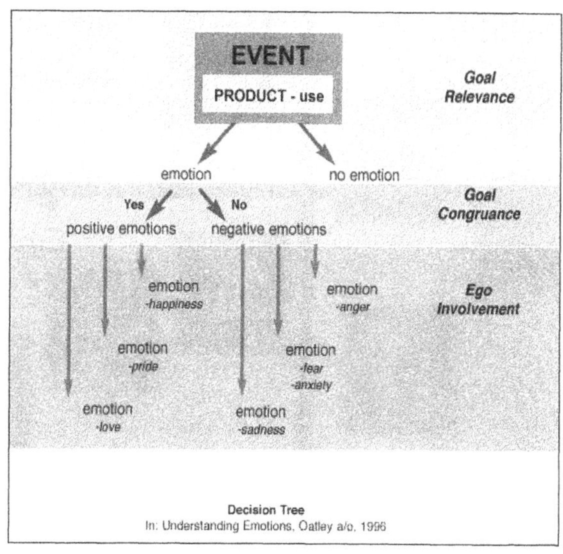

Figure 1: Decision tree model

DESIGN MANAGEMENT: STEERING THE DEVELOPMENT OF TOMORROWS PCP

Through the nature of the profession, the designer is already known to act as a 'linguist' between other disciplines and this will place him/her well in respect to communicating and discussing the widened functionality of ALL product attributes through tomorrows Product Creation Process. The authors have already described how they foresee future product functionality as the *created* interplay between Physical (hereafter referred to as Sensorial) *and* Contextual (hereafter referred to as Emotional) product attributes. This will help manufacturers develop consistency in respect to interrelation between attribute functionality to ensure they become a clear key criteria for the design, purchase or use of consumer goods. The basis of this methodology consists of transferring the following expression into a tool.

$$P = (E,S)$$

(Where P = perception, E = emotional attributes, S = sensorial attributes)

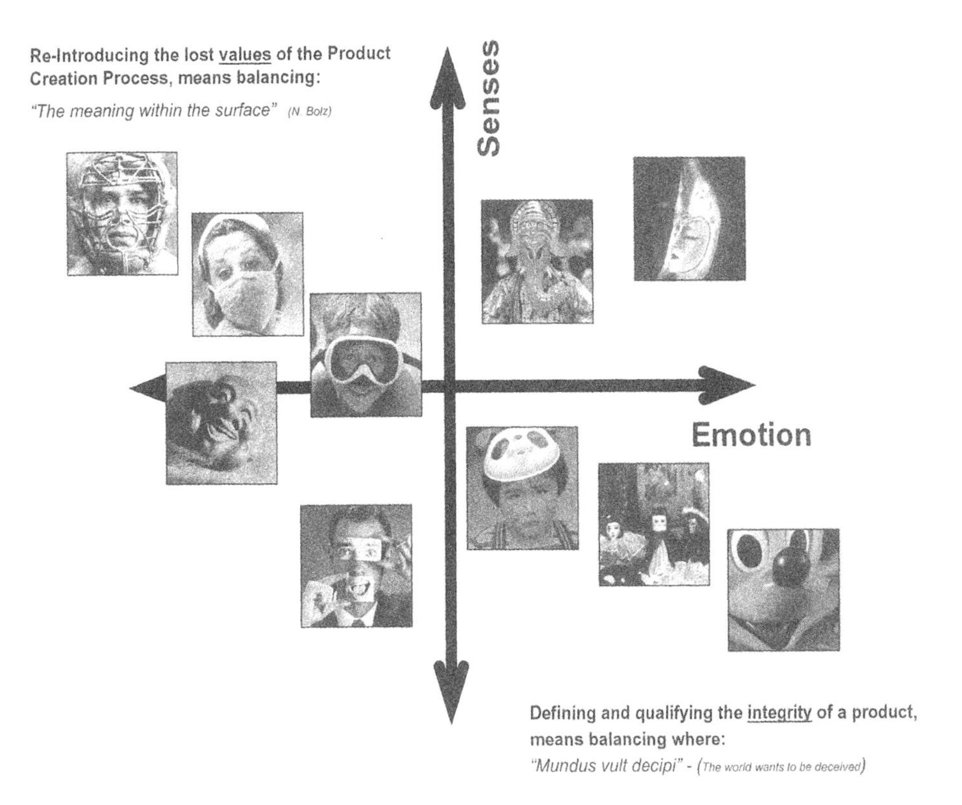

Re-Introducing the lost values of the Product Creation Process, means balancing:
"The meaning within the surface" (N. Bolz)
Senses
Emotion
Defining and qualifying the integrity of a product, means balancing where:
"Mundus vult decipi" - (The world wants to be deceived)

Figure 2: Sensory & Emotional Strategic Product Concept Map

By successfully integrating, playing with or balancing both types of attribute, the authors are now using the tool to determine the intensity (or emphasis) of each different attribute and consequently intensify the eventual satisfaction of a product. Figure 2 highlights how future 'Masks' (i.e. the (second) skin of a product) could be strategically developed and managed within an industrial setting, depending on the required emphasis of physical and emotional attributes.

EARLY RESULTS: STUDENT CONCEPTUAL PRODUCT CREATION PROJECT

To explore and develop the tool further a student design programme was constructed to monitor how many entirely new product ideas could be created for a little considered room, the bathroom.

The group, consisting of thirty five students, was randomly split into eight groups, with each group being assigned one of the eight known forms of basic emotions (Oatley and Jenkins, 1996). After special 'Attribute Creation' sessions, all students went through a similar, pre-defined, product creation process. Following these sessions the students' developed Design Briefs and individual studies into the creation of a product based on a designated emotion which had to be represented primarily through one key form of sensory interaction. Other limiting factors were that the product had to be a consumer product capable of communicating or reflecting the different consumer 'affects' centred around the emotional cue. Influences of other senses were admitted (as long as they were perceived to be of a secondary nature) but no cross-over developments between other emotions were allowed.

It is unfortunately beyond the scope of this paper to fully describe all our findings, but some of the preliminary results are shown in figure three below. These include a drinking ritual focused on anticipating the melting process of ice, through to a 'disgusting' toothbrush holder that cleverly allows a whole family to share just one 'ever clean' brush.

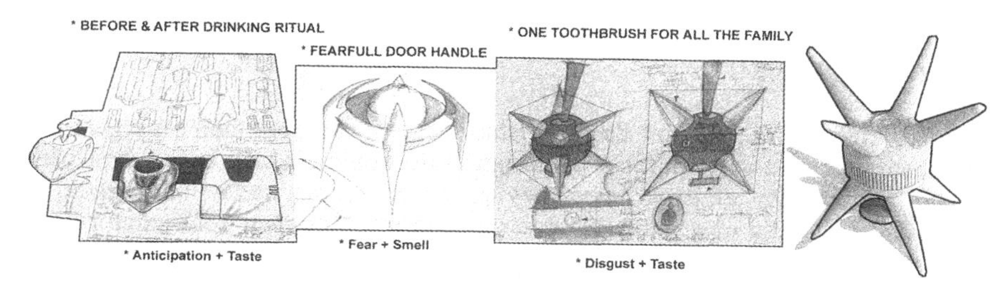

Figure 3: Constructing Products through Emotional Product Creation Process

Interestingly a key finding has been that groups focused towards negative emotional conditions (e.g. anger and fear) showed more difficulty in starting up their design process, than those focused on positive emotions. Students found it harder to be affiliated towards a negative emotion. Literature supports a similar attitude by describing that Western Society is very oriented towards the creation of happiness and beauty as opposed to exploring or uplifting negative situations.

As phrased in the beginning half of the paper, this research began as a hypothesis that while the Pleasure Principle is a good starting point for looking towards 'new' product functionality, it is too narrow a position to create sustainable consumer satisfaction in the longer term. These first findings have indicated that it is possible to construct entirely new products on the basis of using different as well as multiple Emotional Attributes as the starting point for new product creation.

REFERENCES

Pine, J.P., Gilmore, J.H., 1999, *The Experience Economy*. USA: Harvard Business School Press.

Marzano, S., 1998, *Thoughts*. Philips Design, V&K Publication.

Markham J.E. (1998) *The Future of Shopping: Traditional Patterns and Net Effect*. London: Macmillan.

Jordan P, (2000) *Designing Pleasurable Products*. London: Taylor and Francis.

Lazarus R.S, (1991) *Emotion and Adaption*. New York: Oxford University Press.

Oatley K, Jenkins J.M, (1996) *Understanding Emotions*. London: Blackwell.

Questioning the validity of emotion in design: a critical examination of the multi-faceted conditions of its historical emergence

Deniz Patlar, Aren Kurtgözü
METU, Turkey

INTRODUCTION

Even a brief examination of design history reveals the continual emergence and disappearance of a large number of design philosophies, which based their claims on the authenticity of experience they promised to the users. Throughout this "incessant recycling of arguments" one can observe that even contradictory terms were proposed as recipes for achieving authentic experience in the user-product relationship (Martinidis, 1991). For example, 'ornamentation in design' was declared either as harmful or indispensable to the authenticity of the experience of users. While the proponents of functionalist aesthetics denounced ornament as crime, as an accidental rather than essential quality of designed products, others were arguing for ornament as a basic emotional need of human beings. For example, William Dyce wrote in the *Journal of Design* in the late nineteenth century that,

> Ornamental art is an ingredient necessary to the completeness of the results of mechanical skill. I say necessary, because we all feel it to be so. The love of ornament is a tendency of our being. We all are sensible, and we cannot help being so, that mechanical contrivances are like skeletons without skin, like birds without feathers – pieces of organization, in short, without the ingredient which renders natural productions objects of pleasure to the senses (Heskett, 1987, pp. 21-23).

Note, however, that there were strong social ideals behind either of these design philosophies because neither of them initially presented themselves as yet another style. On the other hand, many critics believe that adopting the same term in a contradictory manner as the solution to the same problem relegates these design philosophies to the status of style. For example, when retrospectively evaluating the modernist design principles, Dormer (1991) declares the maxim 'form follows function', one of the most powerful and influential design arguments fashioned in the most truthful manner, as having been merely a style among others. We suggest that there is strong evidence for believing that 'design and emotion' is destined to become a style among others within the contemporary consumer society. This is because, (1) 'design and emotion' shares the same structural mechanism of discursive formation with design arguments that arose in the past and (2) suffers from a neglect of the dynamics of the prevailing ideology of consumption.

In recent years, we have started to witness an unprecedented emphasis placed on the concept of emotion in design. The need for an emotional appeal in designed products has been unanimously voiced and designers have been called for creating products that address human emotions. These accounts generally suggest that people suffer from a neglect of their emotional needs in our contemporary consumer culture. That is, too much attention directed to material achievement in industrial world has given rise to a spiritual impoverishment of modern individuals. For example, Walker (2000) blames the excessive emphasis on "utilitarianism, economic efficiency, competition, and progress in today's societies and today's world of design" for the elimination of a concern for our "inner, spiritual side". Many design critics and scholars, in turn, have started to develop generative design frameworks that take possible emotional responses of users to products into account. While, for example, Djajadiningrat *et al.* (2000) proposes the technique of "interaction relabelling" in which "the exaggerated emotional attitudes" of imagined users are taken "as the basis of design", the leading human factors specialist Jordan (1998) has recently turned towards "pleasure with products" as a contribution to a holistic understanding of human needs.

What is common to all these diverse and usually divergent undertakings is the frequently unfounded assumption that the truth of our beings lies in our emotional experiences. More generally, however, these accounts of 'design and emotion' seem to be motivated by a need to address the dissatisfactions that the modern consumer society has brought about. For this reason, we prefer discussing 'design and emotion' not in terms of formalist design criteria but from the perspective of the social role it assumes. In other words, 'design and emotion' is to be judged on its present and future capacity for providing solutions to the social and cultural problems it addresses rather than on the formal integrity of its products.

For this aim, we should first put 'design and emotion' in its proper context of consumer culture, in which it operates and against which it raises its arguments.

Modern consumer culture is characterised by a crisis of identity for which the endless and purposeless circulation of commodities is responsible. It is argued that the goods manufactured and marketed within such a culture can not satisfy the users' deeply embedded human needs because these goods are "manufactured and calculated in relation to profit rather than arising organically from authentic individual and communal life" (Slater, 1997).

Certainly, the individual became free since s/he can exercise freedom of choice in a market that offers a vast array of articles, promising unlimited means of satisfaction. Nevertheless, in a culture where there is a lack of binding, authentic values, the human search for satisfactions is doomed to failure, resulting either in 'dissatisfaction' or 'false satisfaction'. The result is the disappearance of spiritual and humanistic values in a daily life motivated solely by the compulsion to consume. In this context, 'design and emotion' emerges as one of the proposed solutions of the design discipline to the above-mentioned problems of modern consumer culture.

We should, therefore, ask:
1) How can the authentic human experiences be recreated?
2) Where shall this critique of modern consumer society look in order to find its real sources of value?

To find its real sources of value, this critique should evoke both spatial and temporal "elsewheres", such realms where 'human essence' and 'human nature' are believed to exist in their purest forms (Lury, 1996). These 'elsewheres' include,

a) The past which is lost with the advent of modernity (tradition, craft civilization, pre-industrial values, pre-technological culture, eastern cultures, etc.) The exploration of this source demands what Slater (1997) calls as "a nostalgic science of society," which is "forced to identify with the past as a source of values for the critique of the present". However, such revivalisms can frequently border on the invention of tradition rather than discovery of the past.

b) Nostalgia for some current practices where the authentic experiences are believed to still exist. This includes such pseudo-traditional cultural practices as cooking, explorations of nature, the obsession with the housekeeping, etc. These sources are explored in order to design and produce goods of authentic value.

c) Finally, such conceptions as 'human nature' or 'human essence' are explored as sources of authentic values. They are believed to exist deeply in our souls despite constant attacks of the shallowness of modern consumer culture. Indeed, what is essential to our nature is also subject to invention.

We observe that these three 'ideological' sources are being indiscriminately explored by 'design and emotion' in order to bestow emotional values upon products. However, the problem with these 'authentic' sources of the self is that they are also the feeding ground for the interests of the system of consumption. The modern individual is continually called on to pursue the traces of h/er authentic self in these sources by the ideology of consumerism.

At first sight, there is nothing wrong in this pursuit of authentic self, because, after all, if the needs are not based on an authentic ground, there can be no end either to the needs or to the frustrations they give rise. As is well known, consumerism thrives on the constant renewal of needs. Its motto is to 'keep up with the Joneses', based on the continual search for the satisfaction of ever newer needs. The dissatisfactions of modern consumption can only be remedied through the discovery of an authentic ground on which to base humanistic needs. In this regard, many proposals ranging from nourishment and shelter to the need to be esteemed or emotional satisfaction have been made in order to ground our needs in human nature. Design has been involved in this search from its very inception. From the perspective of designers, only the needs arising from the authenticity of human values are real and can be legitimately pursued.

Yet, we suggest that this search for one's inner self in order to reveal one's authentic, true identity is also the central tenet of the ideology of consumption. Among other things, the ideology of consumption benefits from the idea that consumption alienates human beings to their true nature. Indeed, it encourages individuals to repeatedly look at the void in themselves and somehow to discover an essence that belongs to themselves. Yet, there is nothing waiting passively to be discovered in the inner self of the modern individual. It is rather that one should ceaselessly rediscover one's essential self.

It should be remembered, however, that the market offers a great variety of products, which correspond to and address these invented 'authentic' selves. In this respect, Foucault was right to characterize the exercise of power in modern societies as dependent upon the production of truth at the level of both individuals and institutions. According to Foucault (1981), the modern individual is forced to produce the truth, even the truth about himself. In other words, modernity is a way of life in which individuals are continually incited to search for and discover their authentic selves, to catch their true selves in the act (was not psychoanalysis precisely this?), to produce true discourses on their selves and to raise demands in the name of these authentic selves. This obsession with the authentic self is the common feature of any modern discipline involved in the exercise of power through the production of truth. Certainly, design is one of these modern disciplines. The natural outcome of this ideology of the authentic self is an individual who is continually encouraged – indeed, forced – to make choices. As Slater (1997) also notes, "liberal society appears as a social space which requires and produces a 'choosing self', rather than one in which a naturally choosing self is liberated".

Furthermore, the critique of alienation does not run counter to the interests of consumption ideology. For inasmuch as there is an anxiety about alienation existing in the lived culture, the individuals will all the more voluntarily engage in the process of self-discovery. In this regard, the binary opposition of 'authenticity-alienation' is the motive power of consumption, which thrives on and provokes the anxiety about identity. This mechanism runs perfectly because the ideology of consumption is based on the popular acceptance that every arising need can be satisfied via commodities offered in the market. According to this ideology, for virtually every need, there exists a corresponding product in the market (Bauman, 1990). This is because modernity both epistemologically and practically technicizes daily life. Thanks to the correspondence established between need and the commodity form, even the needs and experiences that one believes to be authentic are thought to be satisfied by finding a corresponding object in the market. This, in fact, is nothing but what Marx called as the fetishism of commodities.

'Design and emotion' is no exception. Emotion is among values that the ideology of consumption exploits. Thanks to what Baudrillard (1999) calls as the "fun morality" and the rise of consumer hedonism, pleasure ceased to be about the satisfaction of needs and became an ideal experience to be pursued for its own sake. Consumerism responds to this transformation through "the intensification of emotional experiences, which are understood as located in the internal world of the self" (Slater, 1997). Therefore, rather than engaging the users in a spiritual and prolonged interaction with products, 'design and emotion' runs the risk of becoming a fashionable style, a catchword employed by advertising for the marketing of luxury products to an elite culture.

In a certain sense, this is the faith that awaits virtually any design argument in consumer society. If the quest for an authentic self and experience is a market for consumption, in order for consumption to continue, it is not enough to find and decide on a single type of authentic experience. These experiences should be constantly rediscovered, invented ever anew, and the vicious circle of authenticity and alienation should continually be supplied with new materials. This task is entrusted to a set of cultural intermediaries who are entrenched in strategic positions in the system of consumption. In other words, the system of consumption demands that new and authentic 'lifestyles' should be created and put into market circulation within the logic of obsolescence by certain experts. That is to say, there should be some cultural actors and agencies who create new identities, bestow authenticity upon them (authenticate them) and then call them into crisis again. Design, as one of these cultural intermediaries, is given this mission in the contemporary system of consumption.

The search for authentic experience in design has been voiced differently by a number of mottoes appeared in design history. Therefore, design is one part of what Clifford calls as the "machine for making authenticity" which are situated in the modern "art-culture system" (Lury, 1996).

Of course, there are other cultural intermediaries such as advertising, trend-setting magazines, the entertainment industry, alongside design. They are too involved in the authentication of the concept of emotion in different ways. While designers try to create products with built-in emotions, market psychologists and trend setting magazines continue the project of authenticating the self in terms of emotions by highlighting EQ (emotional quotient) tests. IQ (intelligence quotient) is outmoded because the identity of the individual is decided to be situated in its soul instead of its mind. In conclusion, we suggest that not only products are designed to demonstrate greater emotional appeal but also the modern consumer is designed as an emotional being.

REFERENCES

Baudrillard, J., 1999, *The consumer society: Myths and structures*, (London: Sage).

Bauman, Z., 1990, *Thinking sociologically*, (Oxford: Blackwell).

Djajadiningrat, J.P., Gaver, W.W. and Frens, J.W., 2000, Interaction relabelling and extreme characters: Methods for exploring aesthetic interactions. In *Proceedings DIS2000.*

Dormer, P., 1991, *The meanings of modern design: Towards the twenty-first century*, (London: Thames and Hudson).

Foucault, M., 1981, *Power/knowledge: Selected interviews & other writings 1972-1977*, (NewYork: Pantheon Books), pp. 94.

Heskett, J., 1987, *Industrial design*, (London: Thames and Hudson).

Jordan, P.W., 1998, Human factors for pleasure in product use. *Applied Ergonomics*, **29** (1), pp. 25-33.

Lury, C., 1996, *Consumer culture*, (Cambridge: Polity Press).

Martinidis, Petros, 1991, The Incessant Recycling of Arguments: Structural Affinities between Scientific and Architectural Discourses, (Unpublished Paper).

Slater, Don, 1997, *Consumer culture and modernity*, (Cambridge: Polity Press).

Walker, S., 2000, How the other half lives: Product design, sustainability, and the human spirit. *Design Issues*, **16** (1), pp. 52-59.

EMOTION IN DESIGN

When you feel, the brain blinks: an analysis of brain waves generated by various behaviours and creation/imagination

Seung-Hee Lee
University of Tsukuba, Japan

INTRODUCTION

This study aims to provide an understanding of the *Kansei* evaluation model in human sciences. *Kansei* means subjective criteria work on the emotional aspects of events (Lee, 2000) and the high order function of the brain to impart sources of inspiration, intuition, pleasure/displeasure, taste, curiosity, aesthetics and creation. In a previous study (Lee, 2001), we found several patterns for evaluation when people appreciated paintings or products.

This paper describes how *Kansei* works on design approaches and the relationship it has with personality through brain waves. It introduces experiments on brain waves when people engage in various behaviours. This idea originated in the human sciences but has recently been extended to interdisciplinary fields including the medical and neural sciences.

We discuss some results that we obtained for concentration using brain waves and how the personality influences creation and various behaviours and creation/imagination. From the correlation between creation/imagination and various behaviours through brain waves we show the possibility of developing a support tool for creativity.

WHAT IS *KANSEI*?

The word "*Kansei*" can be interpreted in various ways and has been used in many research-related areas concerned with not only design but other fields as well. It is a word that inclusively encompasses the meaning of words such as sensitivity, sense, sensibility, feeling, aesthetics, emotion, and affection. It also has a high order function in the brain providing sources of inspiration, intuition, pleasure/displeasure, taste, curiosity, aesthetics and creation. Figure 1 shows the etymology of *Kansei* and *Chisei* interpreted from Chinese characters, both of which are processed in the human minds when it receives information from the external world. As you see in Figure 1, *Chisei* works to increase knowledge or understanding that matures through verbal descriptions of logical facts. *Kansei* also works to increase creativity through images that evoke feelings or emotions.

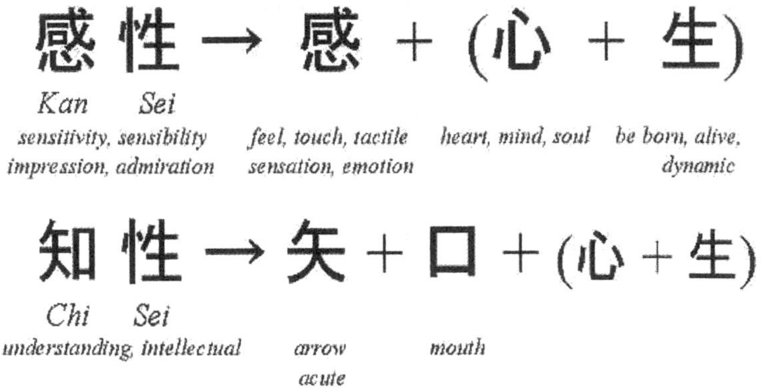

感性 → 感 + (心 + 生)

Kan Sei

sensitivity, sensibility impression, admiration — *feel, touch, tactile sensation, emotion* — *heart, mind, soul* — *be born, alive, dynamic*

知性 → 矢 + 口 + (心 + 生)

Chi Sei

understanding, intellectual — *arrow acute* — *mouth*

Figure 1 The Etymology of *Kansei*

Personality

When a design approach uses *Kansei* engineering, it focuses attention on the behaviours of people when they perceive images or objects, including products, and it reveals how their personal preferences or cultural bases affect their feelings. To date, we have conducted several different experiments on measuring people's *Kansei* when they have appreciated objects or products (2D and 3D) and our findings have clarified our evaluation model.

Personality has been used as one of the main factors in finding evaluation patterns. It expresses the unchanged character of preference when people perceive objects. In this study, a personality test was done before the main two sessions. The personality test was the 'Maudsley Personality Inventory' developed by H. J. Eysenk. It has two main evaluation dimensions, neuroticism-stability (N) on the Y axis and extraversion-introversion (E) on the X axis. Elements of the test were critically, statistically, and theoretically data.

Various behaviours and brainwaves

The experiment in the first session was based on a comparison of brain waves that were derived from different behaviours, such as reading sentences or comic, or calculating spatial forms, which people experience in their daily lives. There is the possibility of developing a tool to support creativity using behavioural stimuli obtains through brain waves for each form of behaviour.

Sketching,sculpting, creating a new form and drawing an image

The subjects were divided into two groups for the experiment for the second session. First, both groups were assigned a task to express their impressions after appreciating a product. The manner of appreciation was evaluated under three different conditions, 1) the product was monitored in QTVR on computer, 2) it was looked at but not touched, and 3) it was looked at and touched. In the final task, the design-experienced group created a new design for the product (called Creation), and the design-inexperienced group drew a sketch of the main character of the comic seen in the first session (called Imagination).

EXPERIMENT

How the brain works, when you see, when you write, and when you think
In the first session, the subjects engaged in various ordinary daily behaviours, such as calculating a spatial mathematical form, understanding sentences, reading a Manga (Japanese *Doraemon* comic), watching an animated movie of the comic, solving puzzles (2D and 3D), and touching a doll. Fig.2 shows a subject wearing a brain-wave set up and the tasks for the first session.

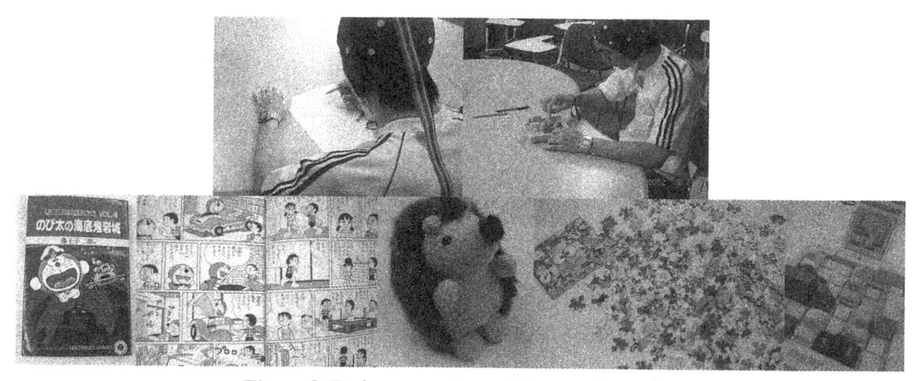

Figure 2 Brain-wave setup and first-session tasks

The brain-wave device used in this experiment (ESA-16) had 14 channels over-all (Fig. 3) and detected three main waves, Alpha, Beta and Theta. When a wave was detected, its frequency was indicated by colors, gradated from red to blue. Red indicates the highest frequency up to 100. Generally, Alpha waves are generated when people feel relaxed. Beta waves are generated when people feel stressed or tense. Not much is known about Theta waves, but recently these were found to exist when people concentrated on tasks.

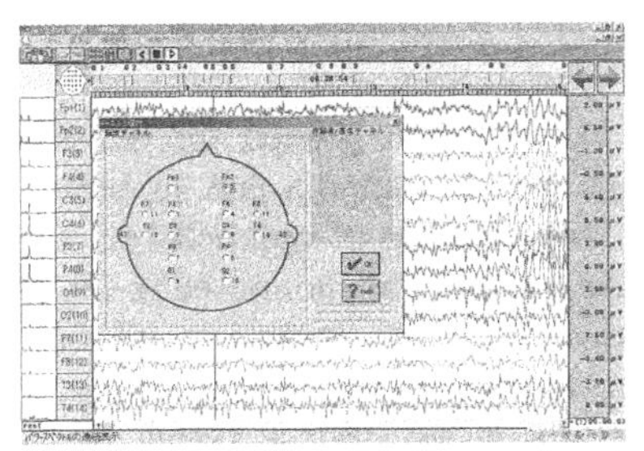

Figure 3 Device interface for experiment

How the brain works, when you feel and express your impressions by sketching and

Sculpting

In the second session, an Air Mac Station was used as the object and there were three sets of observation conditions. The first group observed the object in QTVR, the second group only looked at it, and the third group looked at and touched the object (Fig. 4). The session attempted to reproduce the different situations when people encounter various products in their day-to-day lives. Their task was to draw and sculpt their impressions of the object (Fig. 5).

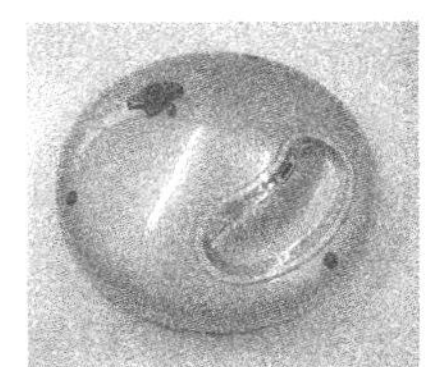

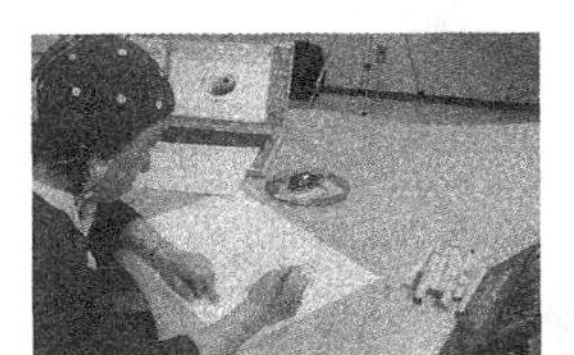

Figure 4 Air Mac Station

Figure 5 Drawing and sculpting impression of objects

The drawings and sculpted models were evaluated by personality as Fig. 6. Shows. Extraversion is on the X axis (E-, E0, E+) and Neuroticism is on the Y axis (N-, N0, N+).

In our previous experiment (Lee 2000), E+N- type personalities were able to make more creative impressions of objects. In Fig. 6.1, the sketches and sculpted models in the circle at bottom right have more unrealistic shapes than in the other personality area. Personality seems related to impressions in this experiment.

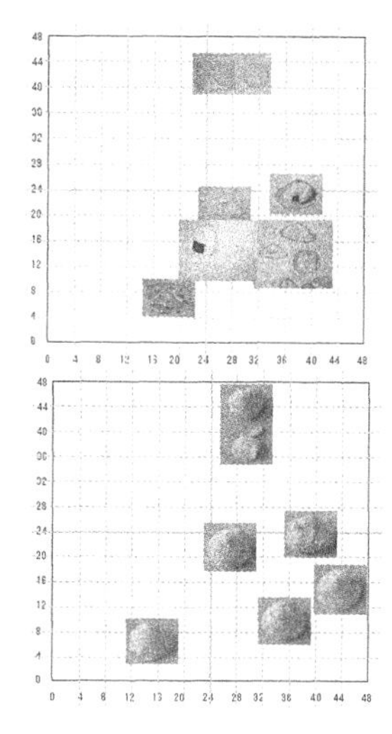

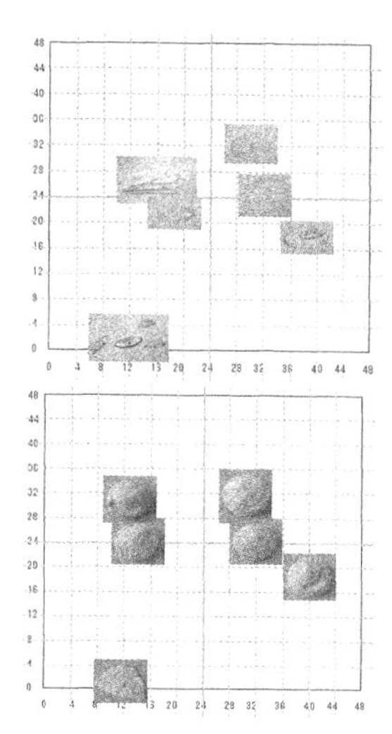

Figure 6.1 Drawings and sculpted models by female subjects **Figure 6.2** By male subjects

The brain waves for creation and imagination in the second session will be discussed and compared with the data for various behaviours in the first session.

DISCUSSION

We obtained the most common results from the brain waves when the subjects were doing 3D puzzles and when they were doing creation /imagination from the same subjects. In Fig.7 both tasks needed concentration (Theta at the left) but they also showed relaxation (Alpha in the middle) and tension (Beta at the right) at the same time. Some objects reproduced by the design-experienced group created a high frequency for all three waves while the subjects were making sketches and sculpting models.

High frequency up to 100

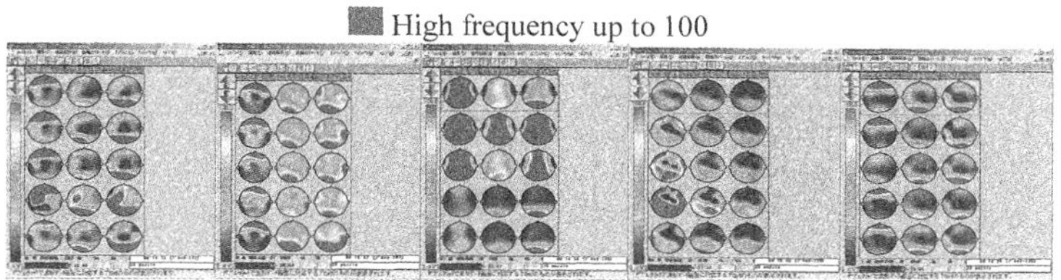

Figure 7.1 Brain waves when doing 3D puzzles by five subjects

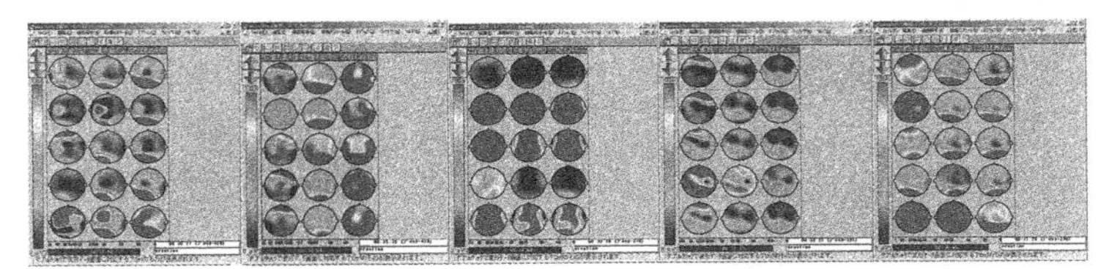

Figure 7.2 Brain waves when doing creation/imagination by the same subjects on Fig.7.1

In this experiment, the data were analysed by taking into account each subject's personality and high frequency brain waves. Subjects who were in the N-(stable) area, tended to relax easily in new circumstances and could easily concentrate on the task. The results indicate that, in supporting creativity, tactile experience is useful in increasing neural reactions. Creation may arise from a situation involving concentration with a little tension within a relaxed setting. Future studies on the correlation between creativity and personality monitored through brain waves, needs detailed tasks with professional, design-experienced subjects.

REFERENCES

Lee S.H., Stappers P.J., Harada A., 2000, *Kansei* Appreciation of Observing 3D Objects, Proceedings of XVI Congress of the International Association of Empirical Aesthetics 2000, pp.83-84

Lee S.H., Okazaki A., Hirotomi A., Harada A., 2001, Modeling *Kansei*: An Analysis of how People Appreciate Art through a Remote Controlled Robot, The 5th Asian Design Conference International Symposium on Design Science (CD-ROM)

Tachibana T., 2001, Brain Research, *Asahi News Paper*, pp.153-168

MPI Research Lab., 1996, New Personality Test, Seishin Shobo, pp.7-15

Introducing the student designer to the role of emotion in design

Howard G Denton, Deana McDonagh, Sheila Baker, Paul Wormald
Loughborough University, UK

INTRODUCTION

> *No matter how elegant and functional a design is it will not win a place in our lives unless it can appeal at a deeper level, to our emotions.* (Sweet, 1999)

Products satisfy needs beyond the functional. These needs may include aspirations, emotions, cultural and social needs. It is crucial that designers are equipped to deal with the emotional domain of products and users. Carlson (1997) considered emotion to be short-term waves of feeling arising without conscious effort, whereas mood, is defined as a longer termed state with less intensity. The authors' working definition of the 'emotional domain' is the range of states of mind (which may influence the body) and which are influenced by internal and external stimuli. These effects tend to be transient although an individual will have a tendency towards a particular state, such as melancholy, happiness, calmness and so on. But even here various stimuli can cause a significant swing in state of mind. But this is a highly individual effect for example, one person may find the new iMac computer delightful and cheerful, where as others, may find it irritating and repulsive.

The authors recognise that this concept of the emotional domain may be a mind shift for many students. Anybody who has engaged in and completed a significant design project, such as that usually undertaken at A level in the UK will have had a strong emotional involvement with the project, whether it be aesthetically or technologically based design work. However, the authors experience of working with new undergraduates indicated that post-A level students did not seem to recognise the full implications and potential for emotion in design work. Teachers of Design and Technology in UK schools appear not to explore this domain to any extent, possibly due to the intangible nature of the concept. Teaching students to recognise the value of an emotional input to design work and to employ such approaches successfully is not straightforward. Therefore, in planning, teaching and learning staff need to approach the exploration of such concepts over an extended period of time and using a range of appropriate techniques, some of which are discussed in this paper. It should certainly be introduced early in an industrial design degree programme and be centred on activities, which generate discussion and enable them to explore the concepts within the designing of products. The aim of this paper is to describe and analyse the way in which students are introduced to the emotional domain in relation to product design at Loughborough through the first semesters design practice experience. The module is described and staff thinking and objectives are outlined. The student background is described. The method employed in analysis is described. Results are given and discussed.

BACKGROUND

The Department of Design and Technology at Loughborough University, takes approximately 130 undergraduates on to its programmes of industrial design and technology each year. The practice of design is developed through two first year modules, the first of which is the subject of this paper and has been the subject of several papers (Baker *et al* 2000: McDonagh-Philp and Lebbon 2000).

The first exercise, the 'Outwoods' project has multiple aims including introducing students to each other and to the staff team. The project employs a fantasy context: that the students are small, nomadic groups of five, travelling open countryside and using portable shelters. These are designed and constructed in a nearby wooded area. Students are encouraged to use their imagination and emotional responses in relation to this context, to relax, be comfortable and unconstrained. The shelters are not simply functional, they must also reflect the chosen 'character' (e.g. priests, herbalists, warriors) and stated emotions of each group in their form. This means they also have a sculptural, expressive form.

Teamwork is used to encourage an openness of ideas and communication. This results in students sharing and developing their responses to the characterisation and emotional aspects of the design task. Students are encouraged to explore the aesthetic possibilities in both the overall form and the detail of the structures. For example, bindings should be aesthetically pleasing; materials should be used and combined in harmonious and pleasing manners; these are emotional aspects of design. Immediate post-exercise evaluation (Denton 2002) revealed whilst most students find the fantasy context difficult to grasp it is, interestingly, males who are more accepting. Subsequent projects expand on this approach in different contexts to help develop the sensitivity to design and emotion.

The second exercise, the hand project, firstly requires the students to design and make an articulated model of a human hand, using only white card. Secondly, they use found materials to make a hand, which should express an aspect of personality or character. The first section is an exercise demanding precision, observation and modelling skills. Students are encouraged to work in pairs (again encouraging discussion to assist in externalising/expressing their emotional responses to the design tasks) and explore the manipulation of card in an expressive way as well as to open up their preconceived ideas of simulating movement. This often results in sculptural and innovative responses.

The third exercise requires the production of a 3D-card model to hold a pen, which will sit on a desk. They have to design the artefact so that it reflects the 'essence' of themselves (e.g. personality, influences, style, preferences and energy). This two-day project further refines ideas presented in the previous projects. It has two aspects. The first is to produce a 'crisp' desktop penholder. This requires cleanliness and precision. Students explore the emotional adjectives that are used to convey 'being held' and to communicate this expressively by the use of form, texture, scale, colour and their innovative manipulation of card.

A further exercise focuses on the use of colour to effect the subliminal message of a product. Students work in groups (discussion again) to produce a mood board, working in a given colour. This is supported by lectures on colour symbolism. Individually, the students use this mood board to re-style a product.

METHOD

In order to access students' reactions to emotional aspects of design, it was decided that a discussion based methodology was most appropriate as by then the group had become familiar with each other and open communication. A focus group was conducted with a sample group (8 females and 13 males), all in the 18-19 age range. The focus group was guided by a set of questions whilst allowing fluid discussion to encourage students to raise issues important to them. Notes were made by two independent scribes and immediately transcribed for analysis. A summary of the findings are presented below and discussed in context.

RESULTS/DISCUSSION

What do you think is the relationship between design and emotion?

Students recognised that users have emotional responses to products beyond the functional. They reported that design can create an image, a style and 'spark a feeling' in the user. Colour and texture were identified as important elements. They stated that both the designer and user want to use a product to express their own personality; to project an image of self or one they aspire to (the power of branding). The students also recognised that people react individually and that a designer needs to be able to 'cater for a range of emotions in one design'.

These findings indicate that the student group were, by then, ready to discuss such concepts and appeared to have grasped a *basic* understanding of the relationship of design and emotion.

What role do you think emotion has in design?

Two main issues emerged. Firstly, emotion was perceived as the 'fuel for design'. This student statement was supported within the group and illustrates a positive role for emotion in terms of driving concept ideation. Secondly, students perceived emotion as linking the 'character' of a product with marketing as an important selling point. This point was followed up by some students who noted that a designer/marketer 'play on emotions to seduce the buyer'; a negative and rather cynical view for an undergraduate designer, but reflects a positive moral stance nonetheless. They emphasised emotion in design enhancing the quality of life for the user, which was revealing.

The group had, therefore, recognised roles for emotion in design from a number of perspectives, and yet, had a healthy moral regard for the dangers of 'seducing' the buyer by manipulating their emotions.

Has your view changed because of the module?

They felt that introducing and integrating the emotional domain enhanced their design solutions while helping them to overcome 'superficiality', meaning they now perceive industrial/product design to be more than just as a mere styling activity. Some reported that they had previously at a school level tended to design a product from the 'inside out' – 'I got it working and then covered it'. Now they appreciated the value of a more holistic approach. Students considered that the rapid sketching and exploration pushed by module staff, whilst at first difficult, is now helping them explore ideas further and actually promotes the integration of emotion. Slow, deliberate drawing appears to limit the expression and integration of emotion at the concept generation stage. Many of the students reported that their work at school level was not helping them to communicate design ideas and they tended to be extremely reliant on the use of words and text rather than sketching and modelling. They felt their ideas were now more realistic and that they were more in tune with real world constraints. Colour was specifically highlighted as a tool a designer can use in relation to emotion, though their appreciation of this was at a basic stage. There had been a number of elements of exercises, which involved the exploration of colour and its emotional impact.

Brainstorm which activities have involved emotion

The group listed all the activities conducted; students recognised that all had an emotional content but they particularly singled out the mood board exercise, the hand project and the nomadic brief. They felt that individual student input by its nature is emotional. They recognised that staff encouraged this.

What particular aspects of the module work has had the most impact in relation to design and emotion?

The module had led to them considering the personality of the designer being of paramount importance (sympathetic, proactive, imaginative, creative, team member, expressive, worldly, wide knowledge). They expressed the need to 'live and breath design'. Emotion was perceived as essential to 'good design' as well as the designer being emotionally involved within design, they identified the need for the designer to be aware of users emotions. This raises issues on a designer's ability to design for others outside their own experiences.

Has anything had an adverse effect on design and emotion in this module?

The main area of concern for the students was the way in which they found themselves immediately immersed within the module. They expressed the desire for more time to develop their drawing skills, gain more confidence and have more time to bond with fellow students. Staff do not necessarily agree that a 'gentler' approach would be appropriate. One intention, was to punch through the usual 'noise' of the first month or so of undergraduate life to establish the key points of design practice, including the value of the emotional domain.

Describe the perfect designer

The students identified that being sympathetic as a designer is a crucial attribute, alongside being able to work concurrently on a number of tasks, and thinking on a number of levels at any one time. Emotion was perceived as being an important input to design, which differentiated designers from engineers. They felt that it is advantageous to be emotional for a designer but that this may not help the engineer.

'Designers become dream catchers'
'Emotion fuels design'

CONCLUSIONS

There were strong indications from the group that the focus on emotional dimensions within design work at the University were new to them and a mind shift. At school level (pre-degree) students, on the whole, were not explicitly encouraged to respond emotionally to project work or to have discussed this dimension with teachers or amongst themselves. This, of course, would require further investigation before any firm conclusions could be drawn. Certainly the findings indicated that students recognised the emotional domain as an overt objective of the staff within the module described. Whilst not all found it easy to adjust to this approach, the post-module focus group brought a unanimous agreement that the objective was sound and related to 'good' design practice and was appropriate to undergraduate training. The student cohort responded well to the introduction and integration of this paradigm and recognised the importance of the emotions within design.

The one negative area uncovered was that of students questioning the moral value of what some perceived as a 'seduction' by manipulating emotions. This is an interesting point and certainly such moral questions must be a part of any under graduate design training. The question of where seduction in a negative sense starts and a positive process of giving the purchaser or buyer a product that they can use to enhance their lives and project their personality is an important one.

Students in the early stages of a degree programme may feel exposed and vulnerable. Nevertheless, staff perceive this ability to communicate and share emotional reactions as being central to the development of sensitivity in design. Therefore, it is essential that this importance is signalled overtly in the early stages of design training, and the process is continued throughout design training.

REFERENCES

Baker, S., Wormald, P., McDonagh-Philp, D. and Denton, H., 2000, Sthira, sukha: Introducing industrial design undergraduates to softer design issues. In *International Conference on Design Education,* (Curtin University of Technology, Australia), edited by C. Swann and E. Young, pp. 94-102.

Carlson, R., 1997, *Experienced Cognition,* (NJ, USA: Lawrence Erlbaum).

Denton, H. G., 2002, Starting a design course with a bang: warming up a new group and ensuring key principles are internalized at the start of an undergraduate programme. *The Design Journal,* **4** (3) pp. 41-49.

McDonagh-Philp, D. and Lebbon, C., 2000, The emotional domain in product design, *The Design Journal* **3** (1) pp. 31-43.

Sweet, F., 1999, *Frog: Form Follows Emotion,* (London: Thames and Hudson).

The role of emotion in design reflection

Dieter K Hammer
Eindhoven University of Technology, The Netherlands

Isabelle M M J Reymen
University of Twente, The Netherlands

INTRODUCTION

Reflection on design processes performed by designers is called design reflection. In our view, this kind of reflection aims at answering essential questions like "Is my design answering the stakeholder concerns?", "Am I solving the essential problems or am I wasting time on irrelevant aspects?", "Does the result feel satisfactory or are further iterations necessary?", "Does my design obey the rules of conceptual integrity and aesthetics?", and "Is my design process appropriate for the problem?". Design reflection is important since it can improve the design process and the product being designed (Reymen, 2001). It can also help the designers to learn from their experiences, i.e. their thoughts and feelings, and to improve their professional capabilities. Recent design research recognised the need for stimulating reflection, including the development of supporting methods (Badke-Schaub et al., 1999; Reymen, 2001; Schön, 1983; and Valkenburg, 2000).

Reflection is, however, often interpreted as evaluating the design rationally, giving no explicit place for emotions. For answering the questions mentioned above, we state that both feelings and thoughts are important. We advocate a balanced approach in which both *rationality* and *emotions* play a role. The underlying idea is that we hope that balanced answers to essential questions lead to balanced design decisions and to a balanced design process. The goal of this paper is to explore the possibilities of letting emotions play a role in design related reflection processes. The exploration is partially based on our experiences with a method that supports reflection on design processes; a description and discussion of the method can be found in (Reymen, 2001).

This paper introduces the concepts emotion, reflection, and design reflection and explores their relations. Based on these insights, the paper continues with describing a prescriptive model of a reflection process in which emotions of designers and stakeholders play an important role.

DESIGN, REFLECTION, AND EMOTION

From a social-psychological point of view, reflection and emotion are related to each other. Rosenberg (1990) illustrates that reflexivity (the process of an entity to act back upon itself) is a central feature of determining the nature of our emotions (emotional identification), of attempting to regulate their display (emotional display), and of seeking to control the experiences of these emotions by producing effects on our minds and on our bodies (emotional experience). Mills and Kleinman (1988) describe a variety of ways in which people experience their thoughts and feelings. Their typology demonstrates four ways in which an individual may respond to a situation: reflexive and emotional, unreflexive and emotional, reflexive without feeling, and neither reflexive nor emotional. The typology is based on individuals experiencing themselves in various situations. The authors state that circumstances shape peoples emotional/cognitive style. These circumstances can be situations of uncertainty, group membership, or historical period. People usually develop their own emotional/cognitive style and can thus respond differently to emotion-provoking situations.

These studies evoke many questions when we apply them to the field of designing: "Are the four ways in which an individual may respond to a situation also manifest in design processes?", "Are there important differences in the emotional/cognitive styles of designers and do they relate to differences in the personality of designers?", and "Which circumstances (design situations) give rise to which type of emotional response?". Mills and Kleinman (1988) state that situations of uncertainty are more likely than stable situations to evoke emotional responses. The fact that designers often have to deal with uncertainty is an indication that emotions are an important aspect of design processes. If we want to let both rationality and emotions play a role in design reflection, then, we must give designers the opportunity to express not only their thoughts but also their feelings. In the remainder of this section, we explore how we can relate emotion and design reflection. The questions mentioned above should be taken into account for further research.

In line with Reymen (2001), we consider reflection on a design process as an introspective contemplation on a designers perception of the design situation and on the remembered design activities. A design situation is defined as the combination of the state of the design process, the state of the product being designed, and the state of the design context at that moment. The goal of a reflection process is to plan appropriate future design activities. Design activities are appropriate if they are performed effectively and efficiently with respect to the design goal. The goal of reflection can thus be considered as investigating the future, based on the past, the present, and the current design goal. Looking back can help to analyse what went good and wrong and why this happened. Looking forward means thinking about further developments of the product being designed and about the activities that are necessary for this purpose.

To perform balanced design reflection, in our eyes, a holistic view on the design process is required. We believe that such a holistic view must be based on the viewpoints of the stakeholders of the design. The notion of stakeholders, stakeholder concerns, and related views is borrowed from software architecting and, for example, described in (IEEE, 2000). We consider the designers themselves also as important stakeholders; for, as professionals, they are very much related to their own designs. Typical stakeholder concerns are functionality, quality, cost, time-to-market, organisational issues, and the touch-and-feel of the design. In the remainder of this paper, we call the views of the designers and stakeholders *stakeholder views*.

Stakeholder views should incorporate both rationality and emotions and should be based on the involvement of the whole personality of the respective person. *Rational aspects* of a view are usually expressed in a formal way, for example, by models and formulas. Engineers are in general less experienced in expressing the *emotional aspects* of a view, for example, by expressing their feelings in words or artistic utterances. We consider here emotions that are related to the product being designed, the design process, the design team, and the design context. Important emotions are for instance related to the touch and feel of the object of reflection, to its aesthetic or artistic value, to the feelings about its usability, and to the feelings about its sensibility in a given context. Depending on the situation, these emotions can be positive or negative. Our approach differs from the usual way of describing views only in terms of abstract models. The problems that raise when designers are concentrating on abstractions are especially visible in software engineering. This mentality is one of the reasons why computer systems are notoriously failing to fulfil the expectations of their users, who are much more thinking in terms of their daily business. Of course, abstractions are important in engineering design. They are, however, only related to our thinking and not to our feeling.

To develop a holistic view on the design process, many stakeholder views need to be combined into *a lively image* of a design or a design process. We consider a lively image to be an inner image that is build from a number of stakeholder views that is sufficient to allow the designer to have not only an intellectual but also an emotional relation to the object of reflection. This means that not only technological views are incorporated, but also views related to the expected future use of the artefact and to its environmental, social, ethical, and psychological impact. The lively image should be the basis for planning future design activities. To do so, it is important that the designer gets a deeper meaning of the lively image, both rationally and emotionally. We call the creative act of apprehending the deeper meaning of a situation *intuition*. If the creative act needs to be more than a spontaneous free association, it must be carefully prepared. This means that the design and design process are first carefully evaluated from all relevant points of view by building a lively image; a lively image is thus a prerequisite for intuition. Next, the designer must be abandoned for some time in order to create room for deeper insights.

Summarising, emotions should play an explicit role in design reflection. Contrary to the more usual reductionistic ways to deal with design, our approach is holistic in several ways. First, the reflection is based on the input (views) of all stakeholders. Second, rationality and feelings are both taken into account. Finally, the designer is involved as human being and not only as intellectual being.

A REFLECTION PROCESS

In order to include emotions explicitly in a design reflection process, we divide it into the following five steps: defining the questions to be answered by the reflection process, collecting the relevant stakeholder views, building of a lively image, investigating the deeper meaning of the lively image and answering the initial questions, and drawing of conclusions. We group the first three steps into a preparation phase and the last two steps into a conclusion phase. Between the two phases, a break should take place. Such a break simulates 'natural' reflection processes in which some incubation period is necessary before conclusions can be drawn. An explanation of each of the five steps is given below.

The *preparation phase* starts with defining a number of essential *questions*, as indicated in the introduction. These questions can concern the current and the desired state of the product being designed, the design process, and the design context. In the second step, the relevant stakeholders, their concerns, and the related *stakeholder views* must be defined, depending on the questions defined in the first step. The thoughts and feelings that are relevant for the different views can, for example, be worked out by means of checklists and models. The third step concerns imagination. Here, the designer tries to integrate the various stakeholder views into a *lively image*. To do so, they have to lean back for a while and make a synthesis of the views in order to get an as complete as possible image.

Break between preparation and conclusion phase: during the break, things that are not directly related to reflecting or designing should be performed. Designers can also communicate with other designers or stakeholders for completing or checking their views on the design situation and design activities. We assume that during this break, the reflection process continues, but in an unconscious way.

The *conclusion phase* starts when the lively image reveals its essence by means of intuition. Grasping the essence of a design or design process usually takes some time and probably several reconsiderations of the image and its views. The designer is now ready to *answer* the essential questions formulated in step 1. Also here, the designers may use their own feelings to validate the answers. The reflection process ends with drawing *conclusions* and defining the further actions to be taken.

CONCLUSIONS

Summarising, we introduced a prescriptive model of a reflection process that incorporates emotion by giving attention to the feelings of designers and stakeholders. The result should contribute to a balanced approach to design reflection in which both rationality and emotions play a role. To really support well-balanced design reflection in concrete design processes, aids for each of the five steps of the reflection process must be developed and the reflection process must be integrated into the design process. The latter can be obtained by performing the reflection process at the beginning and end of design sessions, as proposed in Reymen (2001). Special attention must be given to the attitudes needed for well-balanced reflection and to personality traits related to expressing emotions. As already mentioned, further research should therefore also concentrate on studying the usefulness and consequences of the typology described in Mills and Kleinman (1988) in the field of designing, and more specific, for design reflection.

A main limitation of our approach is that it only works when the designers are well trained with respect to emotions and intuition. This means that they must (a) take these phenomena seriously, (b) be able to observe them carefully, and (c) differentiate between different types of emotions and their meaning. Unfortunately, most engineering programmes do their best to achieve the opposite. Another problem is that feelings can be based on prejudices, the personal situation of the designer, and the context. A designer can thus make mistakes when they base decisions on feelings. However, the danger of making mistakes may be reduced when the designers are well prepared, i.e., if they follow the process described in the paper, which includes taking into account the viewpoints of several stakeholders.

REFERENCES

Badke-Schaub, P., Wallmeier, S., Dörner, D., 1999, Training for designers: A way to reflect design processes and cope with critical situations in order to increase efficiency. In *Proceedings of the 12th International Conference on Engineering Design*, München, Vol.1, edited by U. Lindeman et al., (München: Technische Universität München), pp. 205-210.

IEEE Standard 1471-2000, 2000, *Recommended Practice for Architectural Description*, (New York: IEEE Press).

Mills, T., Kleinman, S., 1988, Emotions, reflexivity, and action: An interactionist analysis. *Social Forces*, **66** (4), Chapel Hill: University of North Carolina Press, pp. 1009-1027.

Reymen, I., 2001, *Improving Design Processes through Structured Reflection: A Domain-independent Approach*, Ph.D. Thesis, (Eindhoven: Technische Universiteit Eindhoven).

Rosenberg, M., 1990, Reflexivity and emotions. *Social Psychology Quarterly*, 53(1), Washington: American Sociological Association, pp. 3-12.

Schön, D., 1983, *The Reflective Practitioner: How Professionals Think in Action*, (New York: Basic Books).

Valkenburg, R., 2000, *The Reflective Practice in Product Design Teams*, Ph.D. Thesis, (Delft: Delft University of Technology).

The development of empirical techniques for the investigation of design perception

S E W Crothers, R B Clarke, J A I Montgomery
University of Ulster, Northen Ireland

INTRODUCTION

This paper describes the development of an empirical method for measuring design perception. It is argued that by adopting an experimental approach for assessing perceived character it is possible to test specific hypothesis with respect to design perception. This study aims to test the fundamental assumption that design changes affect product perception. Perceived character is assessed by asking participants to evaluate kettle silhouettes on the basis of given keywords. This provides experimental evidence that changes in form affect product perception. Beyond this, the study is exploratory in nature, it is intended to establish an experimental basis for further study and guide future research efforts.

Despite the increasing utilisation of consumer research and market testing, design still relies heavily on the intuition of the designer. It has been suggested (Montgomery, 1999) that, "The desired objective of the designer is the accurate mapping of design intention onto consumer perception". Achieving this objective requires understanding of consumer interpretation. Furthermore if, as Krippendorff argues, design is the creation of meaning as well as the creation of a physical object (Krippendorff, 1989) then designers must look beyond the visual and investigate how their creations are interpreted by the intended audience. This would suggest the need to develop an empirical method for assessing perceptual response to design – a sort of 'perceptual ergonomics' to complement intuitive knowledge. This is not to say that intuition has not been hugely successful or that scientific enquiry is inherently superior, just that intuitive ability varies from person-to-person and as such can be unreliable. Participant based perception experiments can be used to lend credibility to intuitive understanding of design by allowing formal hypothesis to be formulated and tested under scientific conditions.

This study seeks to test the fundamental assumption that design changes affect product perception. It describes an empirical method for the measurement of design perception and provides experimental evidence that changes in form affect product perception. Beyond this the study is exploratory in nature, it is intended to establish an experimental basis for future study and guide future research efforts.

Experimental Method

This study is based on a convenience sample of 80 undergraduate students drawn from a range of courses in the Faculties of Engineering and Arts at the University of Ulster. Participants were exposed to a set of 12 kettles[1] based on 3 'primary silhouettes' designated 'Alpha', 'Beta' and 'Gamma'. Modifications were then made to the spout, handle and base of each 'primary' type in order to generate the set. Recent evidence suggests that prototypicality may have a positive affect on the formulation of preference (Veryzer and Hutchinson, 1988). It is possible then that prototypical preconceptions may also affect perceived character. If the participant could readily identify the prototype[2] of each 'primary' type, it is possible that the perception of the product would be affected. So to prevent the artificial formulation of prototypical preconceptions and preferences no two variations of each 'primary' kettle type were positioned adjacent to each other.

To reduce the risk of perceptual response being affected by anything other than changes in form all other extraneous visual information was suppressed. In this experiment the test set was presented in silhouette (as shown in Table 1).

Table 1: The set of kettles used as the experimental stimulus

A	B	C	D
E	F	G	H
I	J	K	L

[1] The kettle was chosen as the stimulus because it is a familiar small domestic product, which it is assumed, all the participants will be familiar.

[2] The concept of prototypicality originated in computational cognitive psychology. Essentially prototypes are mental constructs that facilitate object recognition and can best be described as an, "idealised internalised representation." It should be noted that a prototype does not necessarily correspond to any physical object, it is an abstraction, a composite of many examples of a given object.

Consumer interpretation of product character was assessed using adjectival keywords. The use of keywords enabled the structured assessment of perceived character. Exposure time was not considered to be of critical importance. So no limit was imposed on response time. Keywords were presented, using the same size and type of font (Arial). Each participant received a response sheet with written instructions, which were read aloud to the group. Participants were instructed, "For each keyword your task is to determine which design is *most* representative of that keyword, *moderately* representative of that keyword and *least* representative of that keyword. The keywords were not printed on the response sheet so that evaluation would reflect the perceived character of the product rather than previous responses. It is conceivable that in assessing femininity a participant may reverse the responses already given for masculinity. For the same reason, keywords were presented strategically to so not to 'lead' the evaluation i.e. terms with obvious associations such as, 'masculine' and 'feminine' were not presented sequentially.

Comments on Methodology and Anecdotal Evidence

Participants appeared to have no difficulty identifying the *'most'* and *'least'* extreme but many stated that they found it difficult to classify *'moderate'* examples. The results support this observation in that the peak frequency for the *'most'* and *'least* rankings was 22 (28%) and 23 (29%) respectively, while the peak frequency for the *'moderate'* ranking was significantly lower at 15 (19%). This lower frequency supports the anecdotal evidence that participants found it difficult to classify *'moderate'* examples and is significant because the greater the consensus the more compelling the argument. This may suggest the need to eliminate the moderate, asking only for extremes, or exaggerate differences between the images.

Comments by the subjects provide useful insight into the evaluative process. These comments suggest that, while general agreement with respect to perceived character does exist, it is impossible insulate evaluation of design from subjective influences. For instance, kettle E (see figure 1) was considered to be feminine by some because it looked 'pregnant' and masculine by others because it looked as if it had a 'beer belly'. Furthermore, this kind of comment illustrates that, for non-functional criteria at least, evaluation may make reference to the social and cultural context. One mechanism being the utilisation of analogy.

Figure 1 Kettle E was compared to a pregnant woman and a darts player

Hypothesis

It has been noted (Veryzer and Hutchinson, 1998) that, "surprisingly little experimental research testing specific hypotheses with regard to aesthetic response to product design." Using accepted statistical reasoning (McClave and Sincich, 2000) this empirical method for the measurement of design perception can be used to test such formal hypothesis. The objective of this first experiment is to provide evidence that changes in form influence the perception of design. This intention is captured in these formal hypotheses.

H_o: Changes in form *do not* alter the perceived character

If this null hypothesis is correct and changes in form *do not* alter perceived character then there is an equal probability ($P_A = P_B = P_C \ldots\ldots = P_L$) of each kettle being seen as representative of a given keyword:

$P_A, P_B, P_C \ldots\ldots P_L = 1/12$

That is, each kettle has an equal chance of being considered representative of a given keyword. So the expected value or mean number of participants responding to each keyword is given by:

$$E(n_A) = n_A \times P_A = 80 \times 1/12 = 6.7 \text{ (1 d.p.)}$$

Similarly, $E(n_B) = E(n_C) = 6.7$

Where $E(n)$ = Number of participants expected in a random (equal) distribution
And n = Total number of participants in the sample group

H_1: Changes in form alter the perceived character of a product.

The Chi Squared statistic (written as χ^2) measures the degree of disagreement or mismatch between the data and the null hypothesis. The further the observed responses are from the expected value (6.7) greater the mismatch and the larger χ^2 will become. Therefore, large values of χ^2 imply that H_o is false and that changes in form *do* alter perceived character. It should be noted that when dealing with sample groups of this size statistical methods/reasoning should be applied and interpreted with caution. However, confidence in the conclusions and in the experimental method itself will be borne out through repeated application with sample groups of varying size and composition.

Beyond establishing that changes in form affects design perception this paper is exploratory in nature. It is intended to establish an experimental basis for future study and guide future research efforts.

CONCLUSIONS

The development of this empirical method assessing perceived character provides a solid foundation for the investigation of factors that influence evaluative response. It has been shown that, using this method, it is possible to test formal hypothesis under scientific conditions. Specifically, experimental evidence is presented which supports the intuitive supposition that changes in form affect the perceived character of a product.

Perhaps more importantly this experiment demonstrates how participant-based experiments can be used to complement and support intuitive understanding of how design is interpreted by the consumer. It is hoped that such experiments could be used to, not only support, but also build on our intuitive knowing and ensure the effective mapping of design intent onto consumer perception.

REFERENCES

Krippendorff, K., 1989, Product Semantics; A Triangulation and Four Design Theories. In *Product Semantics*, (Helsinki: University of Helsinki)

McClave, J., T., Sincich, T. 2000, *Statistics,* (London: Prentice-Hall).

Montgomery, J., A., I., 1999, *Do real relationships exist between product design and typography?,* (Belfast: University of Ulster Thesis).

Veryzer, R., W., Hutchinson, J., W., 1998, The influence of unity and prototypicality on aesthetic responses to new product designs. *Journal of Consumer Research*, Vol.24, March

DESIGN AND EMOTION WORKSHOPS

Sleeping policemen: a workshop in cathexically affective design

Cristiaan de Groot, Graham Powell, Ben Hughes
Nowhere Foundation, UK

WORKSHOP PHILOSOPHY

Feelings of unrest are emerging from the peaceful land of warm, fuzzy and transparent interfaces? Questions are being asked about approaches to the affective design of technologies that pervade our lives. Should one think twice before hopping on the bandwagon of 'in-your-face' interface? Is 'always on' always right? Are Nielson and Norman not normal?

The development of our relationship with technology has been complicated and arduous. Industrial designers tend to be first in the queue to adopt innovation and exploit potential for a notional 'user' or 'consumer.' The consequent triumphs in the development of user-interface appear to promote an unquestioning acceptance of a future where computers look like animals and animals look nervous. Whereas this tends towards a cathartic, passive and ultimately unsatisfying relationship with technology, there are some that think we should not take it lying down.

Sleeping Policemen explores objects that demand more from their audience than a spiral of desire and disappointment: cathexic and emotionally engaging objects, which might take you by surprise, engage your attention to the task at hand, increase your focal awareness of the interface and the media behind it. The *Sleeping Policemen* concept is defined in relation to these, together with the accompanying distinction of an emotional barometer which has catharsis and cathexis at either end of the scale.

Figure 1: Sleeping Policemen have visited your drill...

The results of this explorative process aim to act as a catalyst enriching and extending design discourse, and in time consciousness. Redirecting the gaze from product to process and championing the new business of design as a central and essential contributor to social learning and cultural vitality.

WORKSHOP NARRATIVE

A preliminary presentation of the *Sleeping Policemen* was delivered to bring the workshop participants into a first stage understanding. Objects were presented during the introduction to further demonstrate the concepts of catharsis and cathexis both in use and in creative design practice.

Teams were formed to examine a number of existing designs, initially from drawings and then moving onto real artefacts. This was done by labelling the physical attributes that could be described as *Sleeping Policemen* with the use of special adhesive markers. After presenting their analytical results the teams moved on to the design phase whereupon they were given a cardboard box and a set of arbitrary and exotic materials – and a design brief. The brief was to design a television set that would enable the user enact a 'mindful' relationship with the medium of television.

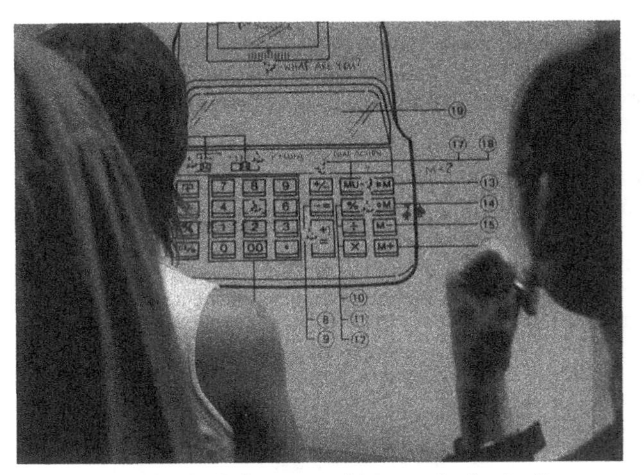

Figure 2: Sleeping Policemen participants at the first hurdle...

After an energetic period of designing, scenario-building, and making, the teams then reconvened to present back their design concept. By acting out a scene from a prospective interaction between a user and a *Sleeping Policemen* inspired television the groups were able to bring their understanding and learning 'to life', for themselves and their bewildered audience.

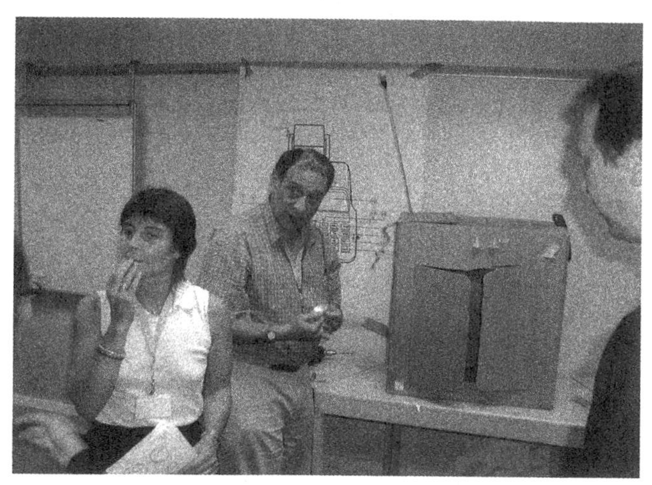

Figure 3: Minor adjustments to the 'Cube d'Cupchik'

All of the concepts presented were at odds with accepted design expectations in terms of the embodiment/reproduction/synthesis of emotion in design – they were *Sleeping Policemen.*

The Sleeping Policemen Team:

Cristiaan de Groot is a founder and the Design Director of the nowherefoundation (www.nowherefoundation.org). He is also a visiting fellow at Goldsmith's College, London. Cris is a founder and the secretary of the Design Transformation Group (www.designquest.org).

Graham Powell is an experienced designer and manager who runs his own design company 'Guinea Pig Design' (www.guineapigdesign.com) in Leamington Spa, UK. He also tutors (P/T) at the Royal College of Art and Aston University. Longest serving DesignQuest junky.

Ben Hughes is the course director of MA Industrial Design at Central St.Martin's, London. He dreams regularly of being the author of the best selling 'Useability, Hughesability' - a manual for the integration of mainframe and brainframe.

Close encounters of the first kind: meet the material world

Marieke Sonneveld
Delft University of Technology, The Netherlands

INTRODUCTION

A human view on objects:
Objects are made to be used, to serve us, and to do that in the most appropriate and pleasant way. If there is a hierarchy, and sure there is, humans rule the material world. This is what humans like to believe.

An object's view on humans:

Humans are made to use objects, to give meaning to their existence. Unfortunately, they often do so in the most unpleasant, inappropriate and disrespectful way possible. It is only because objects have infinitely more wisdom than humans, that they support this situation in silence and resignation. Because of this wisdom to accept life the way it is, objects are at the highest level in the hierarchy of importance.
And this is what humans do not want to know.

The legend of 'The objects in power' (Belcampo, 1955) confronts us with the old knowledge of the secret life of objects, and with the danger of ignoring it. When they revolt the dependant and defenceless position of humans becomes evident and painful. On the other hand, awareness of this powerful position of objects makes us sensitive to their affective attitude towards mankind, and increases the intensity of the relationship between mankind and his surrounding material world.

Children know. They talk to their toys and cuddles, and are fully aware of the power objects have to love and support them or to destroy them. Some musicians know too. When making music, they sense the instrument performing together with them, in their own manner, never completely under control, and never completely on their own.

How come we lost our sensitivity to this secret life of objects, to their affective behaviour? Moreover, can we get it back? The good news is yes, we can, and your sense of touch is the key to this material world of experience.

THE WORKSHOP

To touch is to be touched. Physical interactions with products are truly experienced as a two-way interaction, where both user and product express their character and feelings.

In the workshop, the tactile aspects of this encounter were explored: How do we perceive and express personality in interaction? To get more insight into the tactile experience, such experiences were first generated and analysed in inter-human communication. These experiences 'opened' our senses, as to be ready for the next phase.

Next, tactile aspects of interaction with products were explored. The participants went out and met the different doors controlling the ins and outs of the conference building. The encounters were divers and intriguing. Arrogant doors that would kick you out when you were already leaving, sneaky doors, doors with humour and stupid doors, they were all present. Perhaps the door that would shiver when you passed it was one of the biggest surprises. Was it thrilled or disgusted?

CONCLUSION

The workshop concluded with the design of new doors that would have their own appropriate affective behaviour, given the context of use. For example doors of old libraries that would welcome you as an old friend or doors of meditation rooms that would let you calm down and come to yourself before letting you in.

Marieke Sonneveld is researching the tactile experience of objects at the Delft University of Technology, department of Industrial Design. Her work is oriented towards the Design Practice. Starting point for the design of sensitive and sensual objects is the development of the sensitivity of the designer himself.

Figure 1: The workshop participants went out to meet the doors in the conference building

The brand is the product: the product is the brand

Matthijs van Dijk,
KvD, The Netherlands

Jeroen van Erp
Fabrique, The Netherlands

Additional facilitators: Talea Bohlander & Pieter Desmet

INTRODUCTION

Branding, the buzz word within the global design society at this moment, as it was within the world of advertising over the past 25 years. In recent years a very fragile relationship has developed between design and branding. We do believe that it is very interesting getting to know each other.

Every respectable company has a mission statement from which the holy brand values are developed. These values should express everything the company stands for or wants you to believe they stand for. Often they are developed only to act as guidelines for advertising or graphic design briefings in order to evoke the right emotion in all the printed matter. In most cases, advertising is used to create an 'add-on' emotion to the product. It is strange that brand values are rarely used to steer a design process of the product itself. We believe that it is interesting to use them as a starting point in order to create a product that evokes emotions that match with the company goals. Heineken advertisements show you a world, which is young, dynamic and gives you a lot of fun. The beer bottle itself does the opposite. Volvo cars of the '80's, with their 'chunky' bumpers express safety, which was the keyword in their mission statement.

WORKSHOP APPROACH

In this workshop 6 strong brands were chosen, each represented by a product; Apple with the i-Pod MP3 player, Freitag with a football made of used canvas from trucks, the Freelander from Landrover, the Walkman from Sony, a copy of the National Geographic and a box with Lego construction toys. The aim of the workshop was to design a product, or at least a concept of it, which evokes the emotion or experience that fits the brand.

Six teams were formed and in the first part the participants were asked to define the key values of the brand, derived from the intrinsic quality of the given products. These values should be emotion or experience related. One key value had to be picked out.

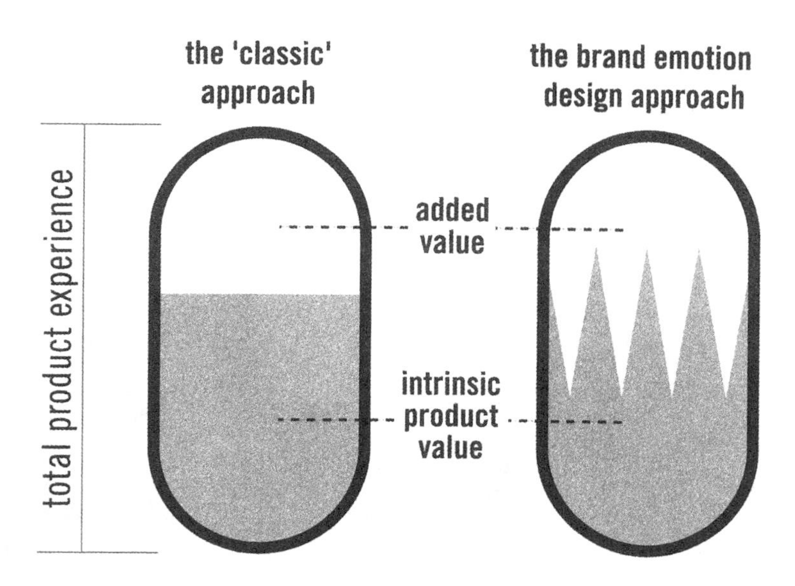

Figure 1: the Brand Emotion Design Approach

In the second part the customers were asked to think about the concern of people that buy the products of this specific brand.

In the final part of the workshop the groups were asked to design a 'dish cleaning tool' this elicits the original Brand emotions in its use.

Although time was short, the results of the workshop showed that the Brand Emotion Design Approach could be very effective. Brand experience and product experience can be one of a kind!

Matthijs van Dijk combines his partnership at Reframing and Design Agency KVD (founded in 1993) with a academic position at the faculty of Industrial Design Engineering. The different roles provide tools developed in the academic world to be applied in real life, thereby providing each other continuous feedback and further development.

Jeroen van Erp graduated at Delft University, faculty of Industrial Design in '88. In 1992, he co-founded Fabrique, a multidisciplinary design agency in which the design disciplines (graphic, industrial and new media) are closely interwoven.

Figure 2: presentation of the results

AUTHOR INDEX

SUBJECT INDEX